MANAGING INNOVATION AND ENTREPRENEURSHIP IN TECHNOLOGY-BASED FIRMS

WILEY SERIES IN ENGINEERING & TECHNOLOGY MANAGEMENT

Series Editor: Dundar F. Kocaoglu, Portland State University

MANAGING INNOVATION AND ENTREPRENEURSHIP IN TECHNOLOGY-BASED FIRMS

MICHAEL J.C. MARTIN
Dalhousie University

A Wiley-Interscience Publication

JOHN WILEY & SONS, INC.

NEW YORK   CHICHESTER   BRISBANE   TORONTO   SINGAPORE

HD
45
M35
1994

Library of Congress Cataloging in Publication Data
Martin, Michael J.C.
    Managing innovation and entrepreneurship in technology-based firms
  / Michael J.C. Martin
      p.   cm.
    Rev. ed. of: Managing technological innovation and
entrepreneurship. c1984.
    ISBN 0-471-57219-5 (acid-free paper)
    1. Technological innovations—Management.   2. High technology
industries—Management.   3. Industrial management.
4. Entrepreneurship.   I. Martin, Michael J. C. Managing
technological innovation and entrepreneurship.   II. Title.
HD45.M35   1994
658.5'14—dc20   93-36588

Printed in the United States of America

10 9 8 7 6 5 4 3 2 1

To my family
and
Past, present, and future students and readers

PREFACE

Man is limited not so much by his tools as by his vision. Historians
tell us that the notion of the earth as round had been discussed for
500 years before Columbus' time. What Columbus did was to translate
an abstract concept into its practical implications.

RICHARD TANNER PASCALE AND ANTHONY G. ATHOS,
THE ART OF JAPANESE MANAGEMENT

This book is an update of one published in 1984.[1] Since that time, a significant
quantity of new material has appeared in the field rendering an update essential. Its
original *raison d'etre* remains, however. The declining economic growth rates
coupled with high unemployment and (until recently) high inflation of the last two
decades, as well as the global impacts of impressive Japanese new management
approaches, have aroused both government and business concern for technological
innovation. These concerns were reinforced by the ending of the Cold War. Although this event was mainly determined by the emergence of more enlightened
Russian leadership in the Kremlin, it was also influenced by what could be described (after Robert Ardray) as the *technological imperative*. That is, recognition
that military strength is nowadays based more on better technology, rather than
bigger territory. Therefore, effective technology management and, more specifically, the effective management of the new technology invention and innovation
process is now recognized as crucially important to government, business, as well
as enlightened management and engineering schools.

As a result of much research performed since the 1950s, the complex technological and socioeconomic process whereby a technological invention is converted into
a socially useful and commercially successful new product is now less obscure than
it was a generation ago. This improved understanding is of little practical value
unless it is disseminated to present and future innovation managers through the
educational system. This requirement raises another issue. Both managers and
management teachers have displayed a historic *technology-aversion*,[2] while many

---

[1] Michael J. C. Martin, *Managing Technological Innovation and Entrepreneurship.* Reston, VA: Reston
Publishing Company Inc., 1984.
[2] Wickham Skinner, "Technology and the Manager," in *Manufacturing in the Corporate Strategy.* New
York: John Wiley & Sons Inc., 1978.

academic engineers and scientists display a complementary *management-aversion*, subscribing to the belief that management is unworthy of serious scholarly attention. This means that the management of technology has been a neglected subject in both the management and engineering schools of universities, so that their two educational systems have evolved with largely separate languages and cultures. The issues and problems associated with this linguistic and cultural gap were discussed in two reports in the United States.[3,4] More recently, initiatives have also been taken by the American Assembly of Collegiate Schools of Business (AASCB), the National Consortium for Technology in Business (NCTB), and the American Society of Engineering Education (ASEE) to help bridge the gap, and the AASCB has also set up a task force to study the matter.

If technological innovation is to sustain future economic growth and employment in competitive international markets, the gap between these two cultures must be bridged by individuals capable of effectively managing the process. It is sometimes argued that individuals without prior engineering or science backgrounds cannot become effective technological managers. However, as Skinner pointed out, individuals without such technological backgrounds can manage technology, provided they:

> can learn to understand and deal effectively with the technology of their industry . . . What is usually sufficient is a framework consisting of a few basic concepts and a set of questions that lead to the acquisition of that knowledge and those insights needed by a manager.

The purpose of this book is to help bridge this gap by providing a framework to future and present technological innovation managers with or without engineering/science backgrounds. It has grown out of my experiences developing and teaching courses on Managing Technogical Innovation and Entrepreneurship to mixed classes of business (mainly MBA), engineering, and science students from the senior undergraduate to the postdoctoral level. These experiences are written about elsewhere.[5] This book can therefore be used as a required text for similar or related one-semester courses in management/business, engineering, and science schools of universities and colleges. I have also used it for executive education courses for both managers and R&D staff. Alternative course frameworks are discussed in teaching notes provided by the publisher. The book is also written for professional readers, general and functional managers and entrepreneurs in high-technology firms, R&D professionals and managers in both firms and government laboratories, as well as administrators in government agencies responsible for stimulating and regulating

---

[3]Dwight M Baumann, Ed., *National Conference for Deans of Engineering and Business.* Washington, DC: National Science Foundation, Purchase Order Number ISP-802087 A01, 1981.

[4]National Research Council, *"Research on the Management of Technology: Unleashing the Hidden Competitive Advantage,"* Manufacturing Studies Board, U.S. National Research Council, 2101 Constitution Avenue, Washington, DC 20418, 1990.

[5]Michael Martin, "Teaching Business Students Technological Innovation Management," *Business Graduate,* **13**(2) (May 1983).

high technology. Technological innovation is typically enacted in a global context, so I have tried to avoid an unduly ethnocentric, exclusively North American perspective. Being an expatriate "Brit," and a resident of Eastern Canada, where fishing still remains a major industry, is possibly advantageous in this respect. The source materials are largely "trawled" from both sides of the North Atlantic "pond" and elsewhere, and so are relevant to the needs of readers outside of North America.

Chapter 1 begins with four vignettes describing the innovation–entrepreneurship process which commercializes new technology to illustrate the well-founded contention that it is much more than invention. After defining it, Chapter 2 suggests a conceptual foundation for viewing the process and its science–technology–commercialization linkages congruent with this contention. Such linkages are illustrated in the impressive achievements of the semiconductor industry, which is grounded on the physics and chemistry of the solid state, over the past 45 years. In this chapter, similar science–technology–commercialization linkages are reviewed at a conceptual level, tracing a thread from Popper's evolutionary methodology of science (first postulated in the 1930s), through to Moore's ecological treatment of business competition, a conceptualization which appeared just before this text went into production.[6]

Chapters 3 and 4 view the innovation–entrepreneurship process in technology-based firms from the corporate–general management perspective. Traditionally, this technological base has been viewed rather narrowly as engineering or R&D versus manufacturing and sales. In Chapter 3 a description of the technological base of a firm is provided which is broad enough to encompass the requirements of the process to provide a factual background to later chapters. Chapter 4 begins by arguing that corporate management's responsibility to stimulate and manage innovation creates the need for an innovation management function and for technological plans and strategies which permeate the organization. It discusses the planning process and alternative technological strategies which may be pursued by a firm. Technological and social forecasting techniques are now used to aid the long-term planning of innovation. Also, most people now accept that innovations must be thoroughly and exhaustively evaluated to identify potentially harmful impacts on individuals and the ecological and social environment. The two Appendices to Chapter 4 describe the approaches and some of the techniques of technology forecasting and assessment, respectively.

Chapters 5 through 8 address the R&D setting of the process. Chapter 5 briefly discusses the overall organization, budgeting, and planning of the R&D activities in larger firms. R&D management has both technicoeconomic and behavioral aspects, which are dealt with in turn. Chapter 6 discusses approaches to the project evaluation process, especially emphasizing the importance of the marketing function's role in the process. Its Appendix reviews project selection techniques. Chapter 7 focuses on some of the human resource aspects of R&D and project management. It outlines the alternative career development paths available to R&D scientists and engineers,

---

[6]James F. Moore, "Predators and Prey: A New Ecology of Competition," *Harvard Business Review*, **71**(3), 75–86 (May–June 1993).

so it may be of specific interest to readers who are contemplating a career change from R&D to management. Chapter 7 also discusses other behaviorial and management requirements of a project as it progresses through the innovation process. Since scientific invention is the basis of the innovation–entrepreneurship process, its Appendix discusses creativity and how it may be nurtured and stimulated in organizations. Chapter 8 deals with nonbehavioral aspects of project management and approaches to accelerating the innovation process. Chapter 9 moves to the operations setting and discusses the problems of transferring technological know-how from an R&D to a production environment.

Chapters 10 through 12 focus on the entrepreneurial aspects of the process. Chapter 10 discusses the important role that successful, growing high-technology firms play in generating economic wealth and jobs and discusses the spin-off phenomenon, notably associated with Silicon Valley and Route 128. The success of a spin-off is markedly dependent upon the entrepreneurial aptitudes and skills of its founders. Entrepreneurship may be viewed as the commercial expression of the creative process described in the Appendix to Chapter 7, so Chapter 10 also examines the personality and biographical characteristics of technological entrepreneurs. Many scientists, engineers, and technological managers employed in larger firms, government, or universities aspire to set up their own high-technology businesses, possibly to exploit their own inventive ideas commercially. Chapter 11 examines, in some detail, the problems, pitfalls, and rewards of setting up a new high-technology venture, partially based upon personal observations in some 50 such firms. Clearly, larger high-technology firms do not wish to lose all the valued scientists, engineers, and technology managers to spin-off ventures. Furthermore, innovation requires what is now called *intrapreneurship* to succeed in large as well as small firms. Established high-technology firms must sustain a climate conducive to intrapreneurship if they are to sustain innovation and retain staff. The approaches that established firms have taken to do so are discussed in Chapter 12.

The book concludes with a review of some of the strategic aspects of the technological innovation–entrepreneurship process. The growing costs, coupled with reduced product development and life cycles, have dictated that firms in some industries must enter into R&D collaborations through consortia. Moreover, notably in the biotechnology industry, two firms must often pool their complementary assets to enact the innovation process. Therefore, Chapter 13 deals with the topic of technology acquisition and strategic alliances. Finally, Chapter 14 provides a summary of the requirements for sustaining innovation and entrepreneurship in technology-based firms, based upon the concepts of Chapter 2 and recent findings on the subject.

I have included fairly extensive literature citations to support the observations in most chapters for readers who wish to explore individual topics in more depth. I suggest that they be ignored in a first reading so that the gists of the discussions can be followed, then, if required, be consulted later according to the reader's special interests. I also apologize to those writers whose works have not been cited. The volume of literature in the field precludes the citation of all the important contributions. This volume also precludes dealing with topics which are relevant, but not central, to the theme of the book, which essentially concentrates on innovation and entrepreneurship at the individual firm level. Discussions of the public policy aspect of the subject and the relationship between technological innovation and long waves in world economies have been excluded. These important (but sometimes contentious) topics are best discussed by researchers associated with specialized agencies in the field, such as the Science Policy Research Unit (SPRU) at Sussex University in England.[7] Also, because I lack a legal background, legally specialized topics such as patents and licensing, as well as environmental, health, and safety regulation, have not been examined in any detail.

No book is written without help from students and colleagues. I express my thanks to the 500-plus bright, mature, and industrious students who have made teaching my course such a pleasure (for me, anyway!). I am sure some of their ideas have "spun off" here! These thanks are unquestionably due to four of these former students who are now experienced MBA graduates: Dr. David Othen (now also a research colleague) who provided many helpful comments on the 1984 text, and Gail Henderson Angel who generously and graciously corrected many of the grammatical and stylistic errors in it; together with Michael Yanai and Betsy O'Neil whose terms papers contributed to the third and fourth vignettes in Chapter 1 of this book. I also thank Tom Clarke of Stargate Consultants, Ottawa for stimulating my interest in the field, as well as providing an excellent bibliography, which is a highly recommended seminal source[8] and Dundar Kocaoglu of Portland State University for his support. I thank Danielle Foley, James Mann, Theresa Walsh and others at Dalhousie University for help in preparing the manuscript, as well as Frank Cerra, Lisa Van Horn and John Wiley & Sons for converting this manuscript into a finished product. Needless to say, I am to blame for all its faults. Finally, I thank my wife for her forbearance.

Alvin Toffer has described education as a vision of the future. It is often claimed that the age of the heroic inventor–entrepreneur is past, but the evidence of the recent explosive developments in the solid-state electronics and computer industries suggests otherwise. If this book helps just one individual to become a successful technological entrepreneur who creates satisfying employment and products for others, it will have been worth it!

MICHAEL J. C. MARTIN

Halifax, Nova Scotia, Canada

[7]Christopher Freeman, John Clark, and Luc Soete, *Unemployment and Technological Innovation: A Study of Long Waves and Economic Developoment.* Westport, CT: Greenwood Press, 1982.
[8]T. E. Clarke and Jean Reavley, *Science and Technology Management Bibliography 1993.* Ottawa, Ont.: Stargate Consultants Limited, 1993.

CONTENTS

PART I
THE TECHNOLOGICAL INNOVATION PROCESS

Part I describes the technological innovation process, beginning with some illustrative examples. It then examines Bright's conceptualization of the process. Kuhn's and Popper's analyses of the evolution of scientific knowledge are discussed and linked to more recent descriptions of technological evolution by Abernathy–Utterback and Sahal. These treatments together provide a framework for viewing the process in the corporate setting.

──1
PROLOGUE: THE TECHNOLOGICAL INNOVATION PROCESS: FOUR VIGNETTES

It is a little like playing poker when it is the only game you want to play and all you know about the game is that the deck is stacked. You don't know how it is stacked but it is stacked and you must play.
MONTE C. THRODAHL, SENIOR VICE PRESIDENT, MONSANTO CO.
UNTITLED PRESENTATION ON THE INNOVATION PROCESS
IN DWIGHT M. BAUMANN (ED.), *NATIONAL CONFERENCE FOR DEANS OF BUSINESS AND ENGINEERING*

I do not believe that any amount of market research could have told us that the Sony Walkman would be successful, not to say a sensational hit that would spawn many imitators. And yet this small item has literally changed the music-listening habits of millions of people all around the world. Many of my friends in the music world, such as conductors Herbert von Karajan, Zubin Mehta, Lorin Maazel, and virtuosos like Isaac Stern, have contacted me for more and more Walkmans, a very rewarding confirmation of the excellence of the idea and the product itself.
AKIO MORITA WITH EDWIN M. REINGOLD
AND MITSUKO SHIMOMURA, *MADE IN JAPAN*

1.1 INVENTION AND INNOVATION

In the Preface, it was implied that general managers of high-technology businesses have a responsibility to nurture a continuous succession of new products, based upon a continued succession of technological innovations, and therefore need to be familiar with the innovation process. It was also suggested that students of science, engineering, and management who plan (or have already started) to pursue careers in high-technology industries should recognize the main features of this process, particularly if they aspire to reach senior/general management positions or set up their own independent high-technology business later. Therefore, to begin, we examine the process in a little detail to identify the combination of elements required to generate a commercially successful technological product.

Most high-technology companies are "science based" in that their innovations develop out of new ideas and inventions generated by scientific activity, either within their own R&D function or elsewhere. However, at the outset, it is important to recognize the distinction between scientific *invention* and technological *innovation*. A scientific invention may be viewed as a new idea or concept generated by R&D, but this invention only becomes an innovation when it is transformed into a socially usable product. Lay persons, probably because of the mystique which surrounds science, generally view invention as a relatively rare event, and assume that once it has occurred, the process of innovation can be completed in a straight-forward manner. In actuality, the converse situation obtains. All who have worked in R&D will agree that it is intellectually and emotionally demanding and frustrating, but even so, the R&D community is quite prolific in generating inventions, and companies can rarely afford to fund all promising R&D projects (see Chapter 2, Section 2.3). It is the subsequent path to technological innovation which is typically fraught with numerous obstacles to be overcome, if the R&D invention is to be commercially successful.

Managing the innovation process may perhaps be compared with breeding and training a horse to finish in the first three places in the English Grand National or other steeplechases. Although many horses may be bred and trained to come under "starter's orders," in any year only three horses perform well enough over the fences to win the first three places in the race, and win prize money for their owners and bets for their backers. Managing the innovation process in larger companies essentially requires breeding a stable of promising R&D inventions, some of which are developed into product innovations in the exception that they will prove to be commercially successful "winners."

To illustrate some of the features and obstacles of this innovation process, five examples will be given. There are numerous other cases and research studies cited in the literature, two of the most useful of which are Layton *et al.*[1] and Langrish *et al.*[2] Pilkington[3] describes his personal experiences in pioneering the development of the float-glass process to commercial success, and so presents innovation through the eyes of the innovator. Several examples are also described in a special issue of *Research Management.*[4]

1.2 MARCONI AND THE DEVELOPMENT OF RADIO TELEGRAPHY[5,6]

Radio grew out of what is now called experimental and theoretical physics. Its origins can be traced back to the "science" of classical Greece as well as China, but the foundation was laid by the physician to Elizabeth I, Dr. Humphrey Gilbert, who really began the systemic scientific study of the phenomena of electricity and magnetism. Human knowledge of the properties of these phenomena grew steadily between the sixteenth and nineteenth centuries, but it was Michael Faraday who first discovered an interrelationship between the two of them. Like many brilliant experimental physicists, Faraday was no mathematician, and it was James Clerk Maxwell who provided the mathematical formulation of this interrelationship with his electromagnetic field theory, including the proof of Maxwell's equations and the existence of electromagnetic waves. Heinrich Hertz provided the experimental verification of the existence of these waves, Sir William Crookes suggested their potential utility as communication tools, and Sir Oliver Lodge provided a practical demonstration of their use in a lecture at the Royal Institution in 1894. By this time, numerous inventive scientists and engineers were exploring and experimenting with the properties of this promising new tool.

Technological innovation typically requires the collaborative effects of numerous workers, but often its ultimate success can be attributed to the sustained efforts of one or more inventor–entrepreneurs who pioneer the development of a new high-technology industry. The inventor–entrepreneur is twice-blessed by Providence with both technological and commercial creative skills and has the vision, capability, and energy to improve upon and commercially exploit an invention. Examples of such inventor–entrepreneurs are Alfred Nobel (dynamite), W. H. Perkin (aniline dyes), Herbert Dow (industrial electrolysis), Alexander Graham Bell (the telephone), and Thomas Edison (electric light and power supply). Such a role can be credited to Marconi in the development of radio-telegraphy.

Guglielmo Marconi was born in 1874 in Bologna, Italy of Irish and Italian parents. His mother was Irish, born Annie Jameson and a member of the famous Irish whiskey family, and his father was a well-to-do Italian landowner and businessman. It seems reasonable to conjecture that Guglielmo's family business background influenced his entrepreneurial outlook (see Chapter 10 for a discussion of this factor). He was educated in England and Italy, studied physics at university, and conducted experimental work under Professor Righi of the University of Bologna. He displayed inventive skills from an early age and, having read the work of Hertz, produced a crude form of wireless communication when he was only 22 years old. He quickly recognized the potential utility of wireless telegraphy in ship-to-ship or ship-to-shore communication where wired telegraphy was obviously impossible. He therefore offered to demonstrate his invention to the Italian Government, but it was not interested. Acting upon the advice of the Jameson family, he decided to pursue his ideas in Britain—the home of the world's largest mercantile fleet and Queen Victoria's Royal Navy, and where his mother's family connections might prove useful.

Once residing in England, he made contact with other workers in the field and began a program of practical demonstrations of wireless communication to various British Government departments in 1896. In 1897 he filed the first patents for wireless telegraphy which were held by "The Wireless Telegraph and Signal Co. Ltd." which he founded and capitalized in the City of London in that year. It had a nominal capital of 100,000 £1 shares of which Marconi held 60,000. The remaining 40,000 were sold in a public issue yielding £40,000 which, after £15,000 in start-up expenses, provided the first £25,000 of working capital. Thus began a decade and a half of unremitting and frustrating efforts to establish the embryonic innovation. Marconi quickly allied himself with other talented scientists, engineers, and businessmen, and filed or purchased a succession of patents which appeared crucial to the technological development of radio telegraphy, including Dr. (later Sir) J. A. Fleming's development of the thermionic diode, based upon earlier work by Thomas Edison.

Throughout this period, Marconi was punctilious in maintaining the technological integrity of his company, never publicizing new inventions until they had been fully proven, or making promises which he could not "deliver." In his choice of public demonstrations of wireless communication, he did display a considerable flair for publicity, however. By 1898, he had established wireless intercommunications between Bournemouth (in southern England) and the Isle of Wight, 14.5 miles away. In the winter of that year, the eminent Victorian statesman William Ewart Gladstone lay dying in Bournemouth surrounded by the world's press corps. This event could be compared to the week prior to the death of Sir Winston Churchill 67 years later. Unfortunately, a winter snowstorm hit southern England, severing wired telegraphic and telephonic communication between Bournemouth and London, so the reporters were unable to file their latest reports of the stateman's sinking health to their newspaper offices in London. Marconi quickly determined that wired telegraphic or telephone services (via undersea cable) were still maintained between the Isle of Wight and London. He therefore allowed the press corps to file their reports by *wireless* communication from Bournemouth to the Isle of Wight and thence to London. This action doubtlessly won him many friends amongst the media people of the day.

The Isle of Wight was to provide him with the site for invaluable publicity again in the summer of that year. The Prince of Wales was convalescing from an injury aboard a yacht moored a few miles off that island. Queen Victoria, who was resident at Osborne House on the island at the time, naturally wished to be kept informed of her heir's progress. Intervening hills made visual signal communication impossible, so crude wireless stations were quickly erected on both the yacht and at Osborne House to report the Prince's progress, and over 16 days, 150 messages were exchanged. Doubtlessly, on this occasion, Queen Victoria was more than "amused."

Although these two incidents could be labeled as fortuitous, but useful, publicity stunts, they helped to establish the usefulness of wireless telegraphy. By the early 1900s, it had proved to be of limited marine use and Marconi was engaged in a succession of demonstration projects of increasing import. By 1901, Marconi was ready to demonstrate transatlantic radio communication. Despite the fact that on September 17, 1901 a severe gale destroyed the antennae masts (costing £50,000) at the British transmission station at Poldhu, Cornwall, a temporary replacement was quickly installed, and on December 12 of that year, a signal was successfully transmitted from that station to a receiving antenna on Signal Hill, St. John's, Newfoundland. The first transatlantic wireless communication had been achieved.

Unfortunately, any euphoria arising from this success was short-lived. The first reaction to the successful transmission was that solicitors for the Anglo-American Telegraph Co. claimed that Marconi had infringed that company's monopoly over communications in Newfoundland. Although this claim was, to say the least, disputable, Marconi decided to accept the offer of Alexander Graham Bell (the inventor of the telephone) of facilities in Cape Breton, Nova Scotia. Both the Canadian federal government and the government of the Province of Nova Scotia supported his work, and on December 5, 1902 signals were successfully transmitted across the Atlantic in an easterly direction from Glace Bay, Nova Scotia to Cornwall. On December 16th the following messages were transmitted:

> His Majesty the King. May I be permitted by means of first wireless message to congratulate Your Majesty on success of Marconi's great invention connecting Canada and England.
> —Minto [Lord Minto, Governor-General of Canada]

> Lord Knollys, Buckingham Palace, London. Upon occasion of first wireless telegraphic communication across Atlantic Ocean, may I be permitted to present by means of this wireless telegram transmitted from Canada to England my respectful homage to His Majesty the King.
> —G. Marconi, Glace Bay

At the same time, Marconi was establishing another station at Cape Cod, Massachusetts with the support of the U.S. government. On January 18, 1903 the following message was transmitted from Cape Cod to Cornwall:

> His Majesty King Edward the Seventh (by Marconi's trans-Atlantic wireless telegraph) in taking advantage of the wonderful triumph of scientific research and ingenuity which has been achieved in perfecting a system of wireless telegraphy I send on behalf of the American people most cordial greetings and good wishes to you and all the people of the British Empire.
> —Theodore Roosevelt, White House, Washington

The age of transatlantic wireless communication had arrived!

It would be gratifying to report that the fortunes of Guglielmo Marconi and the company (now called the Marconi-Wireless Telegraph Co. Ltd.) were now made, but this was not so. The path of technological innovation, like that of true love, seldom runs smoothly. Wireless telegraphy was certainly attracting the support of governments and the business community. Several governments were by now providing limited support for demonstrations of its marine and military utility, and Marconi was able to launch a U.S. company which later became the Radio Corporation of America (RCA). The path to ultimate technological and commercial success was still strewn with many obstacles, however. First, Marconi and his colleagues recognized that many technological problems had yet to be solved, so most of the earnings of his companies were channeled back into what is now called R&D. Second, as in Newfoundland, the established wired telegraphy services were reluctant to surrender their monopolies and markets to the fledgling innovation, so continued efforts were required to change national legislations to allow competition from wireless communication. Third, Marconi faced competition from other wireless telegraphy companies (notably Telefunken in Germany) who were pursuing their own technological and commercial developments. In 1906, Dr. Lee de Forest in the United States added a third electrode to the diode tube to produce the triode tube—a device capable of signal amplification as well as rectification. Dr. de Forest, who was not affiliated with the Marconi companies, patented his invention and Marconi recognized that it could be of crucial importance in the development of radio telegraphy. He therefore sued de Forest, claiming that the triode did not incorporate a new inventive principle, but was a variation of the existing diode tube and therefore an infringement of Fleming's patent. There followed several years of protracted litigation which imposed a further drain on Marconi's time, energy, and financial resources. In 1908 the ordinary shares of the company, for which investors paid up to four pounds, dropped to six shillings and three pence (about a dollar and a quarter), so it was becoming increasingly difficult to raise further capital from the stock market. Fortunately, in 1910 and 1912 the affairs of the company were transformed by two chance blessings that came in tragic guises.

In 1910 the wife murderer Dr. Albert Crippen evaded arrest by the English police by boarding the SS Montrose bound for Canada, traveling under a false name and with his mistress disguised as a boy. The murder inquiry and subsequent search for Dr. Crippen, after the police dug up the dismembered remains of Mrs. Crippen in the couple's London home, received widespread publicity. While at sea headed for Canada, the skipper of the Montrose recognized his infamous passenger and radioed this information back to London. Scotland Yard sent a detective on a faster ship to arrest Dr. Crippen on his arrival in North America. The details of this gory murder, compounded with marital infidelity, the dramatic circumstances of Crippen's arrest, and his subsequent trial and execution whetted public interest in radio-telegraphy. This interest was intensified with the occurrence of a major maritime tragedy a short time later.

In April 1912, the "unsinkable" RMS Titanic set sail on its maiden voyage from Southampton to New York. On Sunday night, April 14, the ship struck an iceberg and sank with the loss of 1517 lives—constituting the greatest peacetime maritime disaster in history. Ironically, the SS Californian was close by, but because the radio operator had gone off duty after working 16 continuous hours, the ship did not pick up the Titanic's radioed distress signals. Had the Californian had a radio operator "on watch," most of the lives lost would have been saved. Prior to that tragedy, wireless telegraphy was being adopted only cautiously by society in general and maritime interests in particular. Because of the publicity surrounding the Titanic's maiden voyage and social prestige of some of the passengers drowned, political and social pressures forced the shipping industry to install wireless telegraphy and require a 24-hour radio watch on all ocean-going vessels. From then on, despite further setbacks (including the "Marconi scandal" of 1912–13, surrounding the terms of a proposed British government contract to be awarded to the company), the company's fortunes steadily prospered. World War I witnessed the widespread adoption of radio communication in many applications and, shortly after that War, the birth of the radio broadcast industry. The Marconi companies played important roles in most of these developments, and grew to be a leading and respected presence in the radio, and later, the electronic industries.

1.3 **THE EMI CAT SCANNER**[7,8]

By the early 1970s EMI or Electrical and Musical Industries Ltd. was a British-based multinational corporation with sales in excess of one billion dollars and diverse product bases. These ranged from the Golden Egg Group Ltd. (a leading British restaurant and hotel chain) through serious and popular phonograph records (including music by the Beatles), movies (including "Murder on the Orient Express") to high-technology electronics products. EMI had initiated half a century of stereophonic records going back to the 1920s. In 1936 EMI developed the first high-definition television system upon which current television technology is based. EMI also pioneered airborne radar during World War II and developed the first all solid-state computer in the United Kingdom in 1952.

Ninety percent of EMI's R&D expenditure was marked for specific projects, but the balance was used to fund promising "ideas" generated by R&D scientists and engineers. In 1967, Godfrey Houndsfield, one of EMI's senior research engineers who had played a key role in developing the solid-state computer mentioned above, was working on the problem of programming the computer to recognize a written word and display it on a screen. During his pattern recognition research, he discovered how inefficient the current methods were for collecting and storing information. He realized, for example, that only 1% of the information carried on an X-ray beam transmitted through a patient was put to clinical use, the "shadow" picture produced by the differential opacity of X-rays to body tissues being extremely inefficient. He surmised that there should be a way of utilizing this "wasted" data by exploiting computer technology, and therefore obtained financial support from the "ideas" fund to explore a different approach. Dr. (now Sir) Godfrey Houndsfield subsequently shared a Nobel prize in medicine for this work.

Houndsfield passed a narrow pencil beam of rays through a pig's head (the "patient") which detected the radiation transmitted (and thereby absorbed) and stored the information in computer memory. The radiation source and detectors were mounted opposite each other on a common gantry, and all three moved in translation. The pencil beam of rays passed through a thin "slice" or plane of the subject, and a total scan of the whole slice was completed by the repetition of the irradiation-detection process 160 times as the gantry linearly traversed the subject (see Figure 1.1). The gantry was then successively rotated one degree at a time and the scan repeated until a half-rotation had been completed, by which time 160 by 180 or 28,000 observations had been made. Houndsfield developed computer software to process these data and display an integrated "picture" of the cross-sectional slice of the pig's brain from these 180 multi-angled scans. In this first laboratory "lash-up," it took nine days of irradiation and two-and-a-half hours of computer time to produce one picture. Pictures of further slices could be obtained by changing the traversing plane.

The results obtained were sufficiently promising for the British Ministry of Health to support further work with a small development grant, and in October 1971, a program of clinical trials was launched at an internationally recognized center in neurosurgery and medicine, the Atkinson Morley Hospital in Wimbledon.

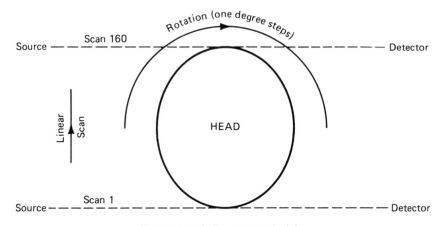

Figure 1.1. CAT scanner principle.

Work was confined to brain scans because, in those early days of development, it was the only bodily organ that could be kept still long enough for the scanning procedure to be completed. Within six months, Dr. James Ambrose, the consultant radiologist who supervised these trials, was absolutely convinced that the technique of computerized axial tomography (CAT) constituted the greatest advance in radiology since the discovery of X-rays in 1895.

The advantages of the technique over the existing methods were the following:

1. The CAT scans provided diagnostic detail and accuracy far superior to that offered by conventional techniques, and so offered the potential for diagnostic refinement and improved therapy.
2. The technique was noninvasive (it did not require foreign substances to be injected into the brain to produce a diagnostic picture), was safe, and not unpleasant for the patient.
3. The technique was rapid (by then, scan times had been reduced to a few minutes) and straightforward and could be frequently performed as an out-patient procedure. Thus, it was likely to be cost-effective.

By April 1972, EMI recognized that it had developed an "invention" of major humanitarian, medical, and commercial promise, but also recognized that the continuance of the innovation process presented them with a dilemma. Clearly, the inventive promise shown warranted continuation on humanitarian grounds, but was fraught with commercial risks. EMI estimated that a further £6 million of investment would be required to create a viable business operation to manufacture and market CAT scanners. The domestic British market centrally directed through the National Health Service would be too small to justify this investment, so that major international markets (notably the United States) would have to be created. Even then, it was expected that it would take some time to recover the initial investment.

Moreover, the marketing of CAT scanners internationally created its own problems. Despite being a large corporation in the electronics industry, EMI lacked a major market presence and track record in medical electronics in the big markets of the United States as well as in West Germany and Japan, where EMI would face some of its major traditional competitors such as Siemens, Philips, CGR, and General Electric. Given the complexity of the scanner technology, EMI would have to establish a sales and servicing network to persuade users (hospitals) to purchase an item of equipment at a price of about one quarter million dollars which could be expected to have a radical impact upon radiological practice when, at this point in time, its efficiency had been demonstrated only to one doctor who had performed a clinical trial on 70 patients! Discussions with the medical profession in Britain and Europe indicated that the North American market would be the critical one, so further extensive investigations were conducted there—following the sound military (and commercial) maxim that time and effort spent on reconnaissance are seldom wasted. Given the novelty of the technology, it was virtually impossible to make an informed estimate of potential sales, and the estimate made after these discussions was for sales of 6000 over a decade beginning in 1973.

Although by 1972 the corporation had built up a substantial patent application portfolio, they had yet to be granted one patent, and so had no real idea of the strength and depth of their patent protection. Once the impact of the innovation was generally recognized, EMI could expect strong competition from a number of potential sources—the traditional competitors already in medical electronics cited earlier, plus small entrepreneurially oriented companies with appropriate skills and possibly major corporations in the pharmaceutical industry. The latter were used to incurring high R&D expenditures to support pharmaceutical innovations, but were now cutting back on their R&D investment in the United States because of changes in government regulations for the testing of new drugs, so they too might seek to invest "spare" R&D dollars in promoting nonpharmaceutical medical innovation. EMI did consider licensing the invention to one of the established medical electronic suppliers (and was "courted" to do so by most of them), but decided to "go it alone." EMI reasoned that it had established a three year lead in the technology and that this time could be exploited to establish its credibility and market presence. EMI therefore established a manufacturing capability in the London area and concentrated on their major marketing effort in the United States. They identified the top 30 institutions in the radiological field there, and concentrated on installing EMI scanners in half of these. They reasoned that the first-hand experience the doctors and students would gain from the successful operation of their scanners in these institutions would generate a supportive momentum throughout the whole country.

An important step in the implementation of this market strategy was taken in November 1972. The corporation set up an exhibit at the Radiological Society of America Congress in Chicago, and Dr. Ambrose also presented a paper on the clinical results he had obtained. The positive impact on the radiological profession was immediate and demand for CAT scanners "took off." By 1975, the scanner market was estimated at £40 million per year and EMI had an order backlog of £55 million. It was anticipated that the corporation would secure a worldwide market share of at least £100 million per year and possibly several times this figure. During 1972–1976, EMI's turnover in electronics was quadrupled to £207 million and pretax profits increased from £1 million to £26 million. The London stock market began to mutter about "another Xerox." Initially, because of lengthy scan times, only brain scanners could be manufactured and marketed, but in the mid-1970s, body scanners were produced. By 1978, the scan time had been reduced to 3 seconds.

During the mid-1970s, as had been anticipated, other companies entered the scanner market. Ohio Nuclear was a small entrepreneurial company which quickly launched machines, and GE established itself in the market by developing its own capabilities. Thus, EMI faced toughening competition. More important, however, was the changing attitude of the U.S. government. The enthusiastic acceptance of CAT scanning by radiologists coupled with the competitive characteristics of the U.S. health care market meant that many hospitals sought to purchase their own scanners rather than share facilities with a competitor, regardless of the underlying health care needs. For instance, two hospitals in the same U.S. city *both* stated that they would buy an EMI scanner provided they were the first hospital in the city to receive one. EMI displayed the wisdom of Solomon and sent two teams to that particular city to install a scanner simultaneously in each hospital.

Faced with a continued inflation in health care costs, the U.S. government felt that there was an unnecessary proliferation of these expensive units, and in 1976, the newly elected President Jimmy Carter, introduced a *certificate of need* requirement before a hospital could purchase a scanner. This effectively reduced the U.S. scanner market by a half, just at the time that EMI was expanding its production to maintain its strong market penetration. During the late 1970s, the corporation ran into other problems too. Their American phonograph record sales fell off because it failed to maintain a stable of talented young recording artists and was reluctant to record and market the profanities of the *punk rock* fashion. EMI was also losing money in its Australian TV set manufacturing operation. In December 1977 EMI's ordinary share dropped by 17% in one week, and this slide continued over the next year or so as the financial community speculated about the corporation's future. Given the underlying strength and resources of the corporation, its declining financial status made it a potentially attractive take-over prospect. Speculation ended in late 1979 when it was announced that the Thorn Group Ltd. had merged with EMI to form Thorn-EMI Ltd. The new corporation immediately began to rationalize its operations and indicated its willingness to withdraw from the CAT scanner business. Both GE and Toshiba were keen to purchase EMI's scanner operations, and in April 1980, the sale of EMI Medical to GE was announced. In the press release announcing its withdrawal from medical electronics, Mr. Peter Laister, Managing Director of Thorn-EMI, said:

> It is obviously a great disappointment to all of those who have been associated with this business to have to withdraw from this field. EMI's highly innovative work— especially in CT scanning—has helped to transform important areas of medical diagnosis and treatment and has contributed greatly to improved health care around the world. Although the net cost of this withdrawal will be substantial, it should be borne

in mind that in the two years up to 30 June 1979 losses on these activities had totalled 26 million since which time lossess continue to be incurred.[9]

1.4 THE SONY WALKMAN*

The Sony Corporation is a phoenix that arose from the ashes of Japan's defeat in World War II. Its cofounder, Akio Morita, was born in 1921, the first son of Kyuzaemon Morita and fifteenth generation heir to one of Japan's finest and oldest sake-brewing family businesses.[10] As a boy, he shared his mother's love of Western classical music, and he would often listen to 78 rpm records of such works on first a mechanical, and later an electric, phonograph. His favorite hobby was building electronic equipment. Given these twin interests, he was fascinated by improvements in recording equipment and, as a schoolboy, successfully built his own electric phonograph and radio receiver. He failed to build a magnetic wire recorder because he was then unaware of its underlying principles. Needless to say, his best subjects at school were mathematics, physics, and chemistry so he chose to study physics at Osaka Imperial University, whose faculty included Hidetsugu Yagi, the inventor of the Yagi antenna and K. Okabe, the inventor of the magnetron.

By then, World War II was raging and, although the Moritas had opposed the military faction that plunged Japan into war with the United States, they loyally supported their Emperor and country. The young Akio managed to continue as a student by combining his studies with a research project to aid Japan's war effort. After graduation, he joined the Imperial Japanese Navy as a lieutenant and met Masara Ibuka, the future cofounder of Sony. The latter was 13 years Akio's senior and an excellent electronics engineer with his own company. Ibuka had developed instrumentation to detect submerged submarines from low flying aircraft, and this instrumentation was a precursor to the present day magnetic anomaly detectors (MADs) used in such work. Together, Ibuka and Morita researched the design of heat-seeking devices, and this research was again a precursor to the U.S. Sidewinder missile. On August 8, 1945, they heard the news of the new devastating bomb that had been exploded over Hiroshima. Morita immediately realized it was a nuclear bomb, and was amazed at America's technological prowess since he had believed that it would take another 20 years to develop such a weapon. He recognized that Japan had been defeated by superior American technology and also implicitly realized that, as a small, resource-poor island nation, his country's future lay in developing its own technological prowess.

Tokyo Telecommunications Research Laboratories

Life was tough in the aftermath of World War II, and Morita was fortunate in securing a junior faculty appointment in the Physics Department of Tokyo Institute of Technology. More importantly, he renewed his professional association with

---

*This vignette is largely based upon the Morita citation and a term paper written by Michael Yanai, MBA. The author gratefully acknowledges these contributions.

Masara Ibuka who had already established a new company, Tokyo Tsushin Ken-kyusho or Tokyo Telecommunications Research Laboratories. Morita initially joined the company as a part-time moonlighting (see Chapter 11) researcher with the intention that he and Ibuka would launch a manufacturing company as soon as possible. They quickly did so, forming Tokyo Tsushin Kogyo or Tokyo Telecom-munications Engineering Company with initial capital of $500, supported by loans from Kyuzaemon Morita who backed his son's business ambitions. Shortly after-wards, the Allied Powers General Headquarters issued an edict that former Japanese-serving officers should be excluded from faculty teaching positions. Mor-ita had, by now, set his heart on a business rather than an academic career, so he cited this edict as an excuse for resigning his faculty position to devote himself full-time to the new company.

The U.S. Occupation forces had taken over the Japanese Broadcasting Company (NHK), the Japanese equivalent of the BBC. One of Ibuki and Morita's new compa-ny's early contracts was to manufacture electronic equipment for the NHK. While successfully fulfilling this contract, Ibuka saw his first tape recorder which had been imported from the United States. Like Morita, he had been fascinated by the wire recorder, but he too had been unable to make one with the materials available. They both immediately grasped the technical significance of what he had seen. A magne-tized tape (as opposed to a wire) has a much larger cross-sectional surface area, and so could be rotated past the recording head at a much slower speed when recording an audio signal. This meant that a tape (as opposed to a wire) recorder could make *longer* audio recordings of *higher quality*. Moreover, tape offers the added advan-tage that it can be easily spliced during the editing process. They agreed that a tape recorder was the very product that their fledgling company could manufacture and market to build a name for itself in postwar Japanese industry. They therefore secured a license to make tape recorders for the Japanese market.

The manufacturing and marketing of tape recorders proved to be much more difficult. Although the design and manufacture of the tape recorder itself proved to be well within their technical capabilities, making the magnetic tape proved to be much more difficult. Since it could not be imported from the United States (the only supply source at the time), until they could make the tape, they could not attempt to market the tape recorder! Furthermore, they recognized that, in the long term, the magnetic tape market could prove to be much larger than the tape recorder market; therefore, they did not wish to subcontract tape manufacturing to a supplier. Given that a suitable tape substrate could not be readily fabricated from the materials then available in Japan, as well as their ignorance of the underlying chemistry and physics involved, it took them some time to devise a viable manufacturing method for it. Nevertheless, after many unsuccessful efforts, they finally succeeded in this task—only to face a bigger obstacle. Nobody wanted to buy the product!

Morita had never lost his childhood enthusiasm for audio playing and recording, which he assumed he also shared with his fellow countrymen and women. Unfor-tunately, he was wrong. Although an audiotape recorder could now be made in substantial quantities, they were heavy and expensive. They weighed 75 pounds and manufacturing costs imposed a selling price of Y170,000, at a time when the average monthly salary for a young university graduate was about Y10,000. These barriers precluded the selling of tape recorders as consumer products, but fortunately, alternative industrial markets were found. First, because few had been trained during the War, there was an acute shortage of court stenographers to record trial proceedings at that time. An audiotape recorder provided an alternative to a stenographer, so sales for this purpose quickly grew. Second, the educational system was adopting American audio–visual teaching techniques, mainly based upon 16 mm films imported from the United States. Naturally, such films had English language soundtracks. The teaching and learning of English had been banned by the military regime immediately prior to and during World War II, so these soundtracks were incomprehensible to both students and teachers. Schools purchased tape recorders which played Japanese language soundtracks synchronized to run with the films as they were shown in classrooms. Perhaps of equal importance, this action also introduced a new generation of Japanese children to the audiotape recorder!

By 1952, the tape recorder provided a secure product base for Tokyo Tsushin Kogyo, and Ibuki and Morita were looking for further challenging new ideas that could promote corporate growth. The transistor had been invented in the Bell Laboratories four years earlier, and its underlying technology was being outlicensed by the Bell manufacturing arm, Western Electric. Although the transistors currently being manufactured could only be used in hearing aids, both Ibuki and Morita recognized that developing the new technology would provide the challenge needed. They therefore purchased a license and launched a research program to develop transistors that could be used in radio receivers. The outcome of that project was their simultaneous invention, with Texas Instruments, of the personal transistor radio in 1957 (see Chapter 4).

The Sony Corporation

Prior to the development and launch of the personal radio, the founders realized that they should change the name of their company. This was especially important if it was to grow into a global entity, since Tokyo Tsushin Kogyo was a cumbersome name in its original Japanese characters, let alone overseas. After some searching, they discovered the Latin word for sound, which is *sonus*. As well as its etymological attraction, this word possessed other image attributes. As modified in the English word "sonny," it was used by many GIs to describe the young and also sounded like another English word "sunny." Thus, it projected an image of youthfulness and happiness. Unfortunately, as spoken, it also means *to lose money* in Japanese! Therefore, to retain the Latin root and avoid this negative connotation, the founders chose *Sony* as both the new name for their company and to launch their personal radio.

The personal transistor radio (see Chapter 6) was but another of a succession of pioneering products that Sony had developed and launched since World War II. Following the precedent of the audiotape recorder, the company pioneered the development of the domestic videotape recorder in the early 1970s. Although suffering a reverse in this development, when Sony's superior *Betamax* format was rejected by the market in favor of the *Video Home System (VHS)* format, the company retained its dominant presence in the consumer electronics market,[11] and by the later 1970s, Akio Morita was seeking further new product ideas. This search was shared by the corporation's audio division.

The Walkman

By that time, a range of cassette audiotape was manufactured in the corporation's audio division, including radio-cassette recorders. In October 1978, the manufacture of radio-cassette recorders was transferred from the radio division, so the audio division was searching for a new product concept to replace its loss. The outcome of this search was the concept of a personal stereo-cassette player, analogous to the personal transistor radio.

One of the division's current products, a compact monaural recorder, was converted into a stereo player. First, the monoaural circuitry was replaced by stereo circuitry and a playback head installed. Then, the speaker was removed and replaced by a heavy stereo headset. All who listened, including Masara Ibuka and Akio Morita, were impressed with the quality of sound it produced. All, too, were less than impressed with the heavy headset. Coincidentally, the corporation's research laboratory was experimenting with new lightweight headphones, so the efforts of the two groups were combined.

When Akio Morita saw the tremendous potential of this new product, he gathered people from other Sony divisions to cover design, production, advertising, sales, and export, and to form a ten-person team under his leadership. The team drew up specifications for the new product, together with a targeted production output and schedule. Cost structure and pricing were also decided. The hardest part of the project was to bring down the size of the headphones without losing the quality of sound reproduction. Various promotional ideas and package designs were formulated, and presale consumer tests were conducted by the team. Morita insisted on a selling price of Y33,000 to the team, to celebrate the fact that the product was being developed during the 33rd year of the company's existence. This price was well under its estimated price, but Morita was correctly convinced that costs would later be brought below price from the scale economies of mass production. Upon production, each division supplying parts to the new product shaved its price in an attempt to bring down costs.

Along with the problem of reducing the production costs, the team was confronted with the problem of what to name the product. "Hot-Line" was proposed by the design division, but the sales force was not sure this would appeal to the general public in Japan. Then came another idea, "Stereo Walkie." It turned out that Toshiba had already registered the name "Walky" for one of its radios, so that name could not be used. But "Walky" appealed to the team members, and they particularly liked the preliminary logo design of a pair of legs sticking out from the bottom of the capital letter "A" in WALKIE. After more discussions, a team member made a suggestion to combine "walk" with "man." It became popular to those who understood the English language. It was a typical Japanese-made English word that would not sound right to the ears of an Englishman or an American. The team went on to design its logo with another pair of legs, this time sticking out from the bottom of the "A" in MAN. In the final stages of team discussion, a consensus took place, and "Walkman" was decided upon as the product name. The consensus took place in the absence of Morita, and he was furious to learn the name upon his return. However, he later admitted that it was a good choice. The Walkman's volume suppress switch was given the name "Hot-Line Switch." This was done to save the face of the design division who proposed the name "Hot-Line" for the product. Although other names were considered for different countries, it was finally decided that the *WALKMAN* should be a global brand label.

On March 24, 1979, five months after the tape recorder division lost its radio cassette business, Sony engineers completed its final design. Its development epitomized the fast innovation approach discussed in Chapter 8.

An innovative international warranty system was also introduced. Customers could receive after-sales service in countries different from their country of purchase. As the Walkman operated on two dry-cell batteries, customers did not have to worry about differences in voltage, either. All these factors encouraged the casual purchasing of the Walkman, regardless of where you were.

Promotion of the Sony Walkman began in a very modest way. Unsure about the success of the new product, the company did not allocate much money in its promotion. Initial advertising was in the form of television commercials and full-page newspaper ads in the two most influential newspapers in Japan. But perhaps the best promotion of the Walkman was afforded by its unusual launching in Yoyogi Park.

The press release announcing the Walkman launching was not in the usual fashion. Music and all necessary information were recorded on tape and sent out to magazines and newspapers instead of the conventional press release invitation cards. This increased the curiosity among reporters and contributed to the strong turnout attending the launching. The press first gathered at the Sony Building in Tokyo and were then transported to Yoyogi Park in central Tokyo. There, Sony arranged for a group of teenagers to rollerskate by with their Walkmans on. This unusual announcement was well rewarded by the massive publicity that the Walkman received in the newspapers the following day and in the magazines during the following few weeks.

In 1983, the company introduced two water-resistant Walkman models, one an AM/FM radio and another that also contained a stereo cassette player. The models were equipped with ultra-mini earphones, with suggested retail prices of $64.95 and $149.95, respectively. The radio-only model weighed just over 2 ounces, making it the lightest Walkman to date. Also in that year, Sony introduced the "Super Walkman," which was virtually the same size as the audio-cassette tape case, and in 1985, Sony put a new Super Walkman which used a rechargeable thin-plate nickel–cadmium battery on the market. In the summer of 1986, Sony introduced a solar-powered and waterproof Walkman in Britain. For times when the sun does not shine, the unit has a built-in 1.5 volt rechargeable battery. When the sun does shine brightly, the solar cell charges the battery as well as powers the Walkman. The suggested retail price was $284 at the time. During fiscal 1986, Sony's cumulative production of Walkman sets surpassed the 20 million mark. Although the Walkman craze has slowed down somewhat since 1982, the product is still a large contributor to the company's audio equipment sales. The product is so successful that it can even be found on dental patients to reduce their anxiety during dental work.

1.5 G. D. SEARLE AND NUTRASWEET™*

G. D. Searle began as a drug store opened by Gideon Daniel Searle in Fortville, Indiana in 1888.[12] By the time of his death in 1917, it had become G. D. Searle & Co. and was successfully manufacturing and selling elixirs and salves to doctors in the upper Midwest of the United States. His son, Claude Howard Searle, who succeeded him, recognized that pharmaceutical companies were becoming increasingly chemistry-based and that future products would be derived from substantive R&D efforts. This view was shared by Claude's son, John Gideon Searle, who assumed control of the firm in 1931 and expanded it into the national market, spearheaded by R&D-based new products. Pharmaceutical chemistry was pioneered in Germany, notably by the Bayer and I. G. Farben companies, so the two World Wars forced the United States to increase its pharmaceutical R&D efforts to compensate for the loss of German supply sources. Searle was an active participant in these efforts during and immediately after World War II, so that by the 1960s, it had entered international markets based upon successful products such as *Metamucil, Dramamine,* and *Bathine.*

Invention

Searle's R&D efforts were organized in clinical, biological, and chemistry groupings of researchers and, in 1965, William Schlatter was a member of the last group. The never-ending search for new drugs often involves the synthesis and examination of new chemical entities by the thousands (see next chapter), and Schlatter was synthesizing NCEs as part of a project on drugs for treating ulcers. One compound he synthesized was *aspartame,* a combination of two naturally occurring amino acids, aspartic acid or L-aspartyl and L-phenylalanine methyl ester or *L-phe.* Neither of these acids is sweet. The story is that one day he moistened his finger to turn the page of a book he was reading, experienced an amazingly strong sweet taste, and realized that it came from some aspartame that had accidently been left on his fingers. He reported his discovery to his colleagues including Daniel (Dan) Searle, a son of John Gideon Searle.

Dan, who was Harvard trained and succeeded his father as President of the firm in 1966, recognized the potential importance of the serendipitous discovery. The first step was to determine whether aspartame had already been discovered and

---

*This vignette is largely based upon the McCann citation and a term paper written by Betsy O'Neil, MBA. The author gratefully acknowledges these contributions.

patented. It emerged that aspartame had indeed already been discovered by the large British-based corporation Imperial Chemicals Industries (ICI), but that it had apparently not been patented for potential use as either a drug or sweetener. Searle's lawyers immediately filed a patent application in the United States and elsewhere for the use of aspartame *as a sweetener*. This decision proved to be a prudent one because their application just forestalled one planned to be made by the Ajimoto Company of Japan (see later). A new company, Searle Food Resources Inc., was launched to develop the product and it faced three formidable tasks. First, to secure approval of aspartame as a food additive through regulatory processes in the United States. Second, to formulate and implement mass-production protocols for its manufacture, which would also have to conform to regulatory requirements. Third, to formulate and implement a marketing strategy for selling it. Given the problems of manufacturing and marketing it, John Gideon (by now Chairman of the firm) was in favor of out-licensing aspartame and using the royalty income to finance more pharmaceutical R&D. Dan, however, saw aspartame as a diversification opportunity, so was keen to retain it as an internal product.

Regulatory Approval

The regulatory body in the United States is the Federal Food and Drug Administration (FDA), and initially, it was thought that securing regulatory approval for the use of aspartame would be straightforward. Since it was composed of two naturally occurring amino acids, it was hoped that it might enjoy a GRAS classification. This acronym stands for *generally recognized as safe* and can be applied to drugs, foods, and additives which have enjoyed a history of widespread safe use prior to the introduction of regulative legislation. Moreover, the only then currently approved artificial sweeteners were saccharin (with its well-known bitter aftertaste) and cyclamates. Searle made its first application for approval in 1970, just prior to the 1971 ban on cyclamates as potential carcinogens, so prospects for swift favorable approval looked to be good. In fact, securing this approval proved to be an, at times, harrowing "roller coaster" ride.

Initial approval for the use of aspartame in dry uncooked food was granted in July 1974 after the performance of 200 tests on 12,000 animals and 600 people. Approval was thus restricted because aspartame is unstable in acidic liquids and at high temperatures. Therefore, it was approved for sale as a sweetener for powdered drinks (which are less acidic than carbonated beverages) and cold breakfast cereals. This approval proved to be short-lived. First, it was claimed that aspartame might cause brain damage in young children. Second, Searle's test data collection procedures on two of its best-selling drugs were questioned in a Senate subcommittee televised hearing.

The upshot of these concerns was that the FDA initiated a formal audit of all the firm's regulatory tests from 1968 onwards, including those performed on aspartame, and its approval for use was suspended in December 1975.

The subsequent FDA investigations did identify flaws in the Searle's testing procedures.[13] However, the investigations were so lengthy and frustrating to the firm that Searle took the unprecedented step of suing the Agency for a ruling in October 1980. This suit was filed only hours before a two-year-long Public Board of Inquiry ruled that there was no evidence that aspartame was a health hazard, but further tests should be performed at lower dosage levels. The judge hearing Searle's suit ruled that the FDA should either make a decision or he would make a summary judgment. By now, Ajimoto had performed further animal tests which revealed no evidence of brain damage. The FDA therefore decided that further testing was unwarranted and the approval of aspartame was restored on July 15, 1981, although some critics are still concerned with its effects on a small minority of users.[13]

In the eyes of the firm, the six-year study of regulatory approval constituted an injustice. As with a drug, the U.S. patent ran for 17 years after it had been granted, so that the patent on aspartame was due to expire in 1986. The 17-year period is designated to give the first-to-market inventor of a new drug a period of monopoly profits to recover the massive costs of discovering, developing, and testing new drugs. The FDA's stay on its approval had reduced the period in which it could enjoy a monopoly on the U.S. sales of aspartame by a half, from twelve to six years. Searle argued that the patent should be extended to compensate the firm for the FDA's unreasonable delay in granting the approval of aspartame. A special amendment allowing such an extension was appended to a bill being considered by Congress in 1981. The amendment was drafted in a generic form, so as to apply theoretically to any drug subjected to undue regulatory delays, but with certain restrictions which limited is applicability to aspartame. Congress passed the amendment with the rest of the bill and the U.S. patent on aspartame was extended to 1992.

Manufacturing

Parallel with the pursuit of regulatory approval, protocols for the mass production of aspartame had to be developed. The development of effective and efficient manufacturing methods was particularly important because of the high cost of manufacturing the product. In 1985 the cost of manufacturing a pound of aspartame was about $90.00. Since it is 180 times as sweet as sugar by weight, this represents an equivalent cost of 50 cents per pound for sugar. The latter only cost 27 cents per pound, so aspartame effectively cost almost twice as much as sugar. Further, saccharin, which aspartame would have to replace as an artificial sweetener, only cost $2.90 per pound and, being 300 times as sweet as sugar by weight, had an equivalent cost of about one cent per pound when compared with sugar!

Also, although it is synthesized from two naturally occurring amino acids, its large-scale manufacture presented a significant technical challenge. L-aspartyl could be readily manufactured by the firm using a fermentation process based upon its existing knowhow and facilities, but Searle did not yet know how to make L-phe in large quantities. The only company with L-phe manufacturing expertise was the giant Japanese food company that had almost beaten Searle's application for the aspartame patent—the Ajimoto Company. This company was regarded as the world leader in amino acid research and manufacture and dominated the global monosodium glutamate (MSG) market. The Ajimoto Company possessed the knowhow and facilities for the large-scale manufacture of L-phe based upon chemical synthesis rather than fermentation technology, claiming that the latter approach would always be more expensive than chemical synthesis. The two companies therefore entered into a strategic alliance. Ajimoto provided Searle with the initial knowhow for the large-scale manufacture of aspartame in what what was called the *Z-process*. In return, Searle would pay Ajimoto manufacturing royalties and the two would pool their research results seeking process improvements to reduce manufacturing costs. Ajimoto and Searle would each hold an exclusive license to market aspartame in Japan and North America, respectively. They agreed to market it jointly in Europe and some North African countries, but to compete for markets in the rest of the world.

This alliance provided an excellent solution to the firm's manufacturing needs in the shorter term, but entailed longer-term risks, particularly after the aspartame patent protection ceased. Some people in the company felt that Searle should develop manufacturing alternatives to the Z-process to reduce any long-term dependence upon its powerful partner. One of these was a Searle process researcher named Lou Goldsmith. He believed that by taking advantage of recent advances in recombinant biotechnology, L-phe, like L-aspartyl, could be manufactured cheaper by fermentation than by chemical synthesis. Despite opposition from some of his superiors, Goldsmith researched the fermentation approach using funds and facilities bootlegged from his other research projects. He ultimately justified this belief. By 1988, using his manufacturing approach, L-phe was being sold back to Ajimoto at a lower price than the latter company could produce it with the improved version of the Z-process.

Marketing

The third challenge faced by Searle was to sell aspartame in markets with which, as a medium-sized pharmaceutical firm, it was totally unfamiliar. Moreover, since its primary markets would be as a sweetener additive, the primary customers for aspartame would be major firms in the highly competitive consumer food and beverage markets: large multinational corporations such as General Foods, Kraft, Nestle and Proctor & Gamble, The Coca-Cola Company, and PepsiCo. As stated earlier, regulatory approval was first granted to aspartame as a dry uncooked food additive and was marketed under two trademarks, *NutraSweet*™ for commercial preparations and *Equal*™ as a tabletop sweetener. General Foods became a major customer for NutraSweet™ as an additive and also helped promote its use in products such as *Kool-Aid*™. Searle also enlisted the sales force of Thomas J. Lipton Inc. and Diamond Crystal Salt to sell and distribute Equal™ to grocery outlets and food service chains, respectively. The latter included fast-food and hotel chains such as Burger King, McDonalds, and Holiday Inn, where consumers received free samples of Equal™ in order to discover aspartame's advantage over saccharin. Equal™ rapidly acquired and retained market share from both saccharin and sugar users.

The 1981 FDA approval only applied to the use of aspartame in dry goods. The key to really major sales was the huge soft drinks market dominated by Coca-Cola and PepsiCo. Because saccharin had been banned there in 1977, the Canadian Health Protection Branch approved the use of aspartame in soft drinks in Canada in 1981. Aspartame-based diet soft drinks proved to be popular and, what was equally important, despite its unquestioned instability in liquids, the drinks enjoyed a six to nine month shelf life. Sales of diet soft drinks increased from 2 to 21.3% of the total Canadian soft drink market demonstrating potential billion dollar U.S. sales. Searle therefore sought FDA approval for its use in the U.S. soft drinks industry. Approval was granted from August 1, 1983 and NutraSweet™ played a key role in the "Cola Wars" waging at that time.

An especially astute feature of Searle's overall approach to marketing aspartame was its "branded ingredient" strategy. Although such an approach had been used in marketing *Teflon*™ coatings on cooking appliances and *Dolby*™ Systems in stereo equipment, it had never been used in the food additive market. The development and promotion of the distinctive NutraSweet™ red and white swirl logo proved to be an inspired move. Searle spent $60 million over three years in direct advertising promoting NutraSweet™ as an artificial sweetener to generate consumer awareness and brand loyalty, and by 1984, when its aspartame sales exceeded a half billion dollars, most Americans recognized the name. Given its high cost, the soft drinks industry preferred to use a blend of aspartame and saccharin as an artificial sweetener, and found that such a blend containing no more than 15% aspartame created an acceptable aftertaste. However, Searle insisted that only products in which it was used exclusively as a sweetener could display the distinctive NutraSweet™ logo on their labels. This stance was implicitly supported in at least one state (New York) since saccharin could still be viewed as a potential health risk. It also helped entrench the brand image prior to the expiry of the patent in 1992.

Corporate Evolution

The aspartame story unfolded against a background of other major changes in G. D. Searle. It still remained a family business until well into the 1970s, with John Gideon's two sons and son-in-law, Dan, Bill, and Wes Dixon, respectively, forming a leadership troika. None of the three really wished to be CEO, so it was decided to find an outsider for this position who would bring new thinking into the firm. This was Don Rumsfeld, the Illinois politician and statesman who had held a series of senior appointments in the Nixon and Ford administrations, finally as Secretary of Defense, who was appointed CEO in 1977, with Dan as chairman. With the family blessing, Rumsfeld rapidly recruited other able outsiders, including two other lawyers, John Robson, a former chairman of the Civil Aeronautics Board and Undersecretary of Transportation in the White House, and Marty Hoffman, a former Secretary of the Army.

When aspartame finally won FDA approval in 1981, it was recognized that, to fulfill its commercial promise, it should be manufactured and marketed in an autonomous division, distinctly separated from Searle's mainstream pharmaceutical operations. The NutraSweet Company was created for this purpose, with Bob Shapiro as its first President and CEO. He proved to be an excellent choice. Being youthful, creative, and energetic, he was able to attract numerous others with these qualities to initiate a corporate start-up culture *"able to make ninety yard passes in tight situations."*[12] For example, he was credited with originating the very successful branded ingredient strategy, and his support enabled Lou Goldsmith to bootleg the L-phe fermentation process project until it too was proven successful.

During this time, Don Rumsfeld was pondering the future of G. D. Searle. Over the five plus years of his presidency, he had orchestrated its transition from a family business to a medium-sized, professionally managed corporation, still primarily based in the pharmaceutical industry. Currently, Searle was spending about $100 million a year on R&D and he recognized that this expenditure should ideally be tripled if the firm was to be a long-term major player in that industry. Both he and his senior colleagues saw that there were three options for G. D. Searle: First, stay at its present size, using profits from its aspartame sales to fund R&D; second, acquire or merge with another pharmaceutical to create the required economies-of-scale; third, sell the firm. These options were discussed at the corporate board level where the Searle family still held majority representation. The maintenance of the *status quo* was deemed undesirable. Acquisition or merger was considered, but the Searle family wished to reduce rather than increase its stockholdings in the firm, so it was felt that this option was not viable. G. D. Searle was therefore put up for sale.

It is a commonly held belief that Monsanto bought the firm in order to acquire the NutraSweet Company. At least in the first instance, this was not true. Monsanto was interested in buying G. D. Searle so that it could diversify from its own commodity chemicals base into pharmaceuticals. Initially, the presence of the NutraSweet Company was viewed as an impediment since Monsanto did not wish to diversify into the food additives industry, so it declined to make the purchase. Only later did it have second thoughts and Monsanto bought G. D. Searle for $2.8 billion in August 1985. By this time Monsanto recognized that the NutraSweet Company was the jewel in the crown that it had purchased. Promptly after the purchase, the Ajimoto Company offered to buy the NutraSweet Company from Monsanto, but this offer was declined.

1.6 WORLD INNOVATION OLYMPIC MARATHONS

These four vignettes share several common features, and we shall return to them at the end of Chapter 2, after we have identified some of the major characteristics of the innovation process. We conclude this chapter with some very brief comments. Reading these vignettes which span a hundred years, two features appear common to all of them. All four describe processes in a national and international setting. The first two have U.K. national settings, a country now past its prime, but still an important member of the European Union and world community of nations. The third is from Japan which, like the Sony Corporation, has emerged as a firebird from the ashes World War II to become an economic superpower and leader of the Asian world, based upon its imaginative creation and superb management of new technology. The fourth is from the United States still *the* world superpower, substantially based upon its technological and managerial innovations in the past century or so. However, despite their varied national settings, the key determinant of corporate success in exploiting the innovation was success in the U.S. market. Technological innovation is a global business, but companies rarely succeed without succeeding in the United States, or what is now the NAFTA market.

Jelinek and Schoonhoven[14] view the technological innovation process as a marathon race and the above vignettes all illustrate the appositeness of this metaphor. The technological innovation business may indeed be viewed as a globally-based Olympic marathon competition, in which preliminary heats and selection competitions take place throughout the world, but it is one in which the final race which determines the gold medal winner, almost invariably takes place in a U.S. Olympic arena. This is the reality of the technological innovation business as the twentieth century draws to its close. The rest of this book explores some aspects of this reality relevant to the needs of current or prospective major or minor runners in future worldwide innovation marathons.

FURTHER READING

1. A. Morito with E. M. Reingold and M. Shimomura, *Made in Japan: Akio Morita and Sony.* New York: NAL Penguin, 1986.
2. J. E. McCann, *Sweet Success: How NutraSweet Created a Billion Dollar Business.* Homewood, IL: Business One Irwin, 1990.

REFERENCES

1. C. Layton *et al., Ten Innovations.* London: Allen & Unwin, 1972.
2. J. Langrish *et al., Wealth from Knowledge: A Study of Innovation in Industry.* New York: Wiley, 1972.
3. L. A. B. Pilkington, "The Float Glass Process." *Proceedings of the Royal Society of London, Series A* **314**, 1–25 (1969).
4. "Living Case Histories in Innovation," *Research Management* **23**(6), (1980).
5. W. J. Baker, *A History of the Marconi Company.* London: Metheun, 1970.
6. F. Donaldson, *The Marconi Scandal.* New York: Harcourt Brace Jovanovich, 1962.
7. J. A. Powell, *Lubbock Lecture.* Oxford University, Oxford, 1977.
8. J. A. Powell, "Exploiting a Technological Breakthrough," Unpublished presentation. Copenhagen and Brussels: Top Management Forum, Management Centre Europe, September 20–22, 1978.
9. Thorn-EMI Press Release.
10. A. Morito, E. M. Reingold, and M. Shimomura, *Made in Japan: Akio Morita and Sony.* New York: NAL Penguin, 1986.
11. R. S. Rosenbloom and M. A. Cusumano, "Technological Pioneering and Competitive

Advantage: The Birth of the VCR Industry." *California Management Review* **24**(4), 51–76 (1987).

12. J. E. McCann, *Sweet Success: How NutraSweet Created a Billion Dollar Business.* Homewood, IL: Business One Irwin, 1990.

13. S. A. Farber, "The Price of Sweetness." *Technology Review* **92**, 46–53 (1990).

14. M. Jelinek and C. Schoonhoven, *The Innovation Marathon Lessons from High Technology Firms.* Cambridge, MA: Basil Blackwell, 1990.

____2
SOME FRAMEWORKS FOR VIEWING THE PROCESS

A technological innovation is like a river—its growth and development depending on its tributaries and on the conditions it encounters on the way. The tributaries to an innovation are inventions, technologies, and scientific discoveries; the conditions are the vagaries of the market place.

ERNST BRAUN AND STUART MACDONALD,
REVOLUTION IN MINATURE

2.1 INTRODUCTION

In Chapter 1 we examined examples to illustrate the contention that technological innovation is a complex socioeconomic and technological process which often extends over several decades or longer, requiring substantial front-end financial investment and not a little luck. In this chapter we will consider some conceptual frameworks for viewing this process. We begin with Bright's treatment of the process, which we then link first with Popper's and Kuhn's contrasting views of the evolution of scientific knowledge, and then with more recent work by Abernathy and Utterback, Sahal, and others on the evolution of high-technology industries. These considerations provide us with an overall framework for discussing the innovation process at the individual firm level.

2.2 BRIGHT'S TREATMENT OF THE INNOVATION PROCESS AND THE INNOVATION CHAIN EQUATION

The innovation process at the individual enterprise level is quite complex and has been described in varying detail by a number of writers (see, for example, Saren[1]). However, Bright[2] provides an eight-stage conceptual treatment of the innovation process which is sufficient for present purposes. He begins with the following definition:

The process of technological innovation embraces that sequence of activities by which technical knowledge is translated into a physical reality and becomes used on a scale having substantial societal impact. This definition includes more than the act of invention; it includes initiation of the technical idea, acquisition of necessary knowledge, its transformation into usable hardware or procedure, its introduction into society, and its diffusion and adoption to the point where its impact is "significant."

Bright's stages (slightly modified here) are as follows:

Stage 1: The innovation begins in one or both of two ways. One is by suggestion and/or discovery; that is from the speculations and/or discoveries of scientists, or possibly craftspeople in pursuing their activities. Another way is by the perception of an environmental market need or opportunity. Many commercially successful innovations arise, at least partially, from such perceptions, and this important factor will be discussed later in this chapter and in much more detail in Chapter 6.

Stage 2: This is the proposed theory or design concept; that is, the synthesizing of existing knowledge and techniques to provide the theoretical basis for the technical concept. This synthesis usually occurs after considerable trial and error.

Stage 3: The verification of the theory or design concept followed.

Stage 4: The laboratory demonstration of the applicability of the concept, such as the development of the "breadboard" model in electronics.

Stage 5: Alternative versions of the concept are evaluated and developed to be defined as the full-scale approach. At this stage, a prototype is developed and subjected to field trials. Alternatively, a pilot production plant produces small quantities of the new product which may be submitted to test markets or clinical trials.

Stage 6: The commercial introduction or initial operational use of the innovation.

Stage 7: The widespread adoption of the innovation when its scale and scope of usage are sufficient to generate substantial cash flows in the producing enterprises and significant societal impacts.

Stage 8: Proliferation, when either the generic product (e.g., radar equipment to detect speeding motorists) or the generic technology (e.g., microwave technology in cooking ovens) is adapted for use in newly defined markets.

Bright stresses that this generalized treatment simplifies a complex socioeconomic and technical process. Many innovations never survive to the eighth stage, while the progress of others is delayed by pursuing developments down technical or market "blind alleys." Thus, it is not a simple linear process because multiple feedback loops may be present as developments are recycled to earlier stages when unexpected difficulties arise. It may also incorporate feedforward loops since, in well-managed institutions, the potential for proliferation of the generic technology is evaluated at as early a stage as possible and certainly by Stage 4. The distinctions between the stages are ill-defined since several stages may be occurring simultaneously, and are better viewed as overlapping "phases" rather than distinct "stages," particularly if the innovation incorporates two or more scientific inventions. They will be described as phases from now on. The phases may perhaps be viewed as analogous to Shakespeare's "Seven Ages of Man" insofar as they represent the achievement a defined state of growth. Bright's application of his treatment to wireless telegraphy is shown in Table 2.1.

The details of technological innovation processes are obviously dependent upon their technological and industrial contexts. Clearly, there are process differences among (say) the electronics, aerospace, and pharmaceuticals industries. They also vary among corporate industries. They also vary among corporate settings, from small high-technology firms through large, divisionalized corporations and multinational organizations. Such differences will be implicitly explored in later chapters. Moreover, technological innovation more often than not requires the synthesis of knowhow and expertise from a diversity of fields, rather than narrow academic specialist knowledge.[3] For example, the impressive achievements of the semiconductor industry, particularly since the fabrication of the first integrated circuit (IC), is based upon the synthesis of knowhow and expertise in chemistry, physics, materials science, electronic engineering, computer science, and mathematics manifest in the research, development, and manufacturing domains.

Nevertheless, the essence of the technological innovation process may also be expressed succinctly, using a chemical analogy, as a chain process expressed in the innovation chain equation in Figure 2.1. Commercially successful innovations re-

TABLE 2.1   The Innovation of Wireless Telegraphy

Stage/ Process	Date	Individual	Activity
1	1846	Michael Faraday	Observation leading to scientific suggestion
2	1864	James Clerk Maxwell	Electromagnetic wave theory
3	1886	Heinrich Hertz	Experimental detection of electromagnetic waves.
	1892	Sir William Crookes	Suggests their use in wireless communication
4	1894	Oliver Lodge	Laboratory demonstration of use
5	1896	Guglielmo Marconi	First patent and field trial
6	1897	Guglielmo Marconi	Commercial introduction
7	1910–1912	Guglielmo Marconi	Increasing adoption—Crippen-Titanic effect
8	Later	Many	Proliferation—radio industry, radar, TV industry, etc.

Figure 2.1. The innovation chain equation.

quire the synthesis of scientific, engineering, entrepreneurial, and management skills, combined with a social need and a supportive sociopolitical environment, if a sustained chain reaction is to be achieved.

2.3 TIME AND COST CHARACTERISTICS OF THE INNOVATION PROCESS

Table 2.1 shows that it took 60 to 70 years for wireless telegraphy to complete the first seven phases of the innovation process. Bright cites other examples in which the total times to complete the innovation process range from 16 $\frac{1}{2}$ years for integrated circuits (ICs) to 87 years for the electron beam welder. These figures suggest (and this is confirmed by a historical analysis of other examples) that the process usually takes at least a decade and frequently much longer for major innovations, particularly in the aircraft and pharmaceutical industries. Figures 2.2*a* and 2.2*b* illustrate, in varying detail, the activities and timescales involved in two innovations, the Boeing 767 aircraft[4] and a new drug,[5] respectively. More minor innovations (we shall distinguish between major and minor innovations later, in Section 2.8) may take shorter lengths of time, but an organization must recognize that if it plans to participate in a major innovation at an early phase in the process, it may expect to invest corporate resources in this effort for at least a decade before receiving significant fiscal rewards for its efforts.

Some observers believe that the time scale of the process is shortening, but others disagree with this view, and examples of both decreases and increases can be cited.[6] Certainly, a rapid rate of innovation is manifest in the computer industry. Recall the rapid introduction of successive generations of floppy and Winchester disk drives in the 1980s, as well as the successive microprocessor chips that drive each new generation of microcomputers. More importantly, many companies are seeking to reduce this innovation time through improved management of the process. This important topic will be discussed further in Chapter 8. Nevertheless, the capacity to reduce the innovation time is limited by safety considerations in some industries and may be *increased* by changing government regulative requirements. For example, there is evidence that changes in U.S. federal regulations for the testing and introduction of new drugs has both lengthened innovation time and decreased the rate of innovation in the U.S. pharmaceutical industry.[7]

Although the rate of innovation may have fallen in the U.S. pharmaceutical

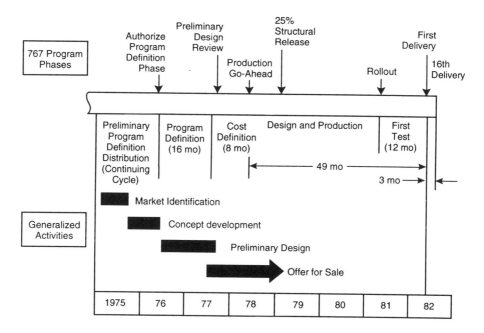

Figure 2.2a. The Boeing 767. *Source:* The Boeing Company.

industry, it has unquestionably increased in the computer industry, and this creates a further problem. Successive new systems are developed and marketed by the computer industry too rapidly for the users, so that the purchaser of an up-to-date computer installation finds that it is obsolescent and needs replacing before its high capital cost can be fully amortized. Thus, the substantial front-end financial investment that must be made by the innovator must be matched by the corresponding investments of the users, and this is a real concern for both manufacturers and users in that industry and in others.

The cash-flow profile associated with the innovation process is similar to that for any new product or process development, except that the time scale can be longer and both the technical and commercial uncertainties are typically greater. This profile is illustrated in Figure 2.3. The technological innovation process clearly acquires a substantial front-end financial investment before any positive cash flow accrues. Returning to the pharmaceutical industry, overall statistics suggest that it costs as much as $300 million to develop a successful new drug. Although this figure partially reflects the high costs involved in the successive stages of clinical trials properly required before a new drug is released for general use, significant costs are expended before any drug reaches this stage. An average of 1200 new chemical entities (NCEs) are synthesized and tested for every drug that reaches clinical trials and only 10% of those that enter clinical trials are finally marketed. That is, 12,000 NCEs are screened in some way for every drug that is marketed,

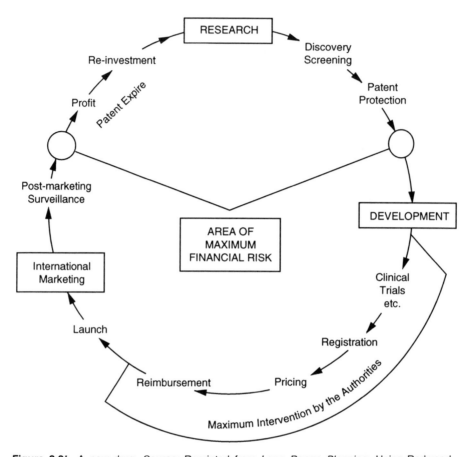

Figure 2.2*b*. A new drug. *Source:* Reprinted from *Long Range Planning,* Heinz Redwood, "Pharmaceuticals: The Price-Research Spiral," P. 19, copyright 1991, with kind permission from Pergamon Press Ltd., Headington Hill Hall, Oxford OX3 OBW, U.K.

successfully or otherwise. These considerations are reflected in the high R&D expenditures in the pharmaceuticals industry, which amounted to over $6 billion in 1988, with the largest prescription drug manufacturer, Merck & Co., spending $650 million.[8] These high costs are matched by high revenues from "winner" drugs. Merck's leading drug (*Vasotec,* used for treating high blood pressure and congestive heart failure) reached one billion dollars in annual sales in 1988 and the company had another 14 that each exceeded $100 million in annual sales. Squibb Corp. also achieved one billion dollars in annual sales with its own successful antihypertension drug, *Capoten.*

The breakdown of costs between the successive stages of the innovation process do, of course, show substantial industrial variations (see, for example, Souder[9]). Also, R&D costs often only represent 5–10% of this front-end investment. The balance is spent on pilot/prototype design and trials, production tooling and engi-

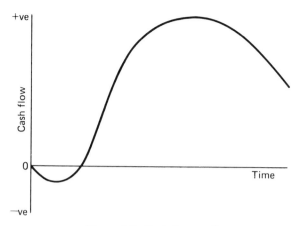

Figure 2.3. Cash flow profile.

neering, and production and marketing start-up operations. Naturally, an innovation will not be pursued unless it is anticipated that the revenue generated during its life will ensure an acceptable ROI before it becomes obsolete. Since R&D may require no more than 10% of the total cost of innovation and a good R&D laboratory may generate numerous promising inventions, some may be abandoned because the company lacks the financial resources to exploit them. This fact has important impacts on the behavior of individuals in high-technology enterprises, as will become apparent in later chapters.

At the end of the previous section we compared Bright's phases of the innovation process with Shakespeare's Seven Ages of Man. Readers familiar with Jaques words in *As You Like It* may recall that Man's seventh age is senility:

> Last scene of all,
> That ends this strange eventful history,
> Is second childishness and mere oblivion;
> Sans teeth, sans eyes, sans taste, sans every thing.

Although innovations (as opposed to some enterprises) do not become senile, to preserve the logic of this life-cycle approach to the innovation process, we must add a ninth-stage—namely, the mere oblivion of the innovation, which typically occurs when it is superceded by a new one. Most technological innovations offer an improved method of providing a given product, and hence supercede an earlier innovation. Therefore, we must expect any innovation to be superceded by a successor sooner or later. In fact, the innovation process in the evolution of technology may be viewed as analogous to the emergence of a new species in biological evolution, as suggested by, for example, Businaro[10] and Marchetti.[11] Both innovations and a new species proliferate if they identify social and ecological niches in which they display superior performance or survival characteristics. Both face the ultimate prospect of displacement through socioeconomic or natural selection.

Therefore, an individual innovation can be viewed in the context of the general evolution of technology, as discussed in the remainder of this chapter. Since much contemporary technology is rooted in contemporary science, we begin with Kuhn's and Popper's treatments of the growth of scientific knowledge.

2.4 REVOLUTIONS AND EVOLUTIONS: KUHN'S AND POPPER'S TREATMENTS OF THE GROWTH OF SCIENTIFIC KNOWLEDGE

The American philosopher and historian Thomas Kuhn and the Austrian-born English philosopher and mathematician Sir Karl Popper have developed to contrasting approaches to the accretion of scientific knowledge that can be applied to the evolution of technology.[12]

Kuhn's Paradigms

Kuhn provides a macroscopic treatment of the history and social-psychology of science, based upon his notion of *paradigm*.[13] Traditionally, scientists have used the terms *theory, law,* or *model* to describe their conceptual treatments of a given body of knowledge. Despite their value and acceptability, these terms tend to be limited operationally to the labeling of specific novel development (such as relativity or quantum theory) usually attributed to an individual scientist (such as Albert Einstein or Max Planck), rather than the generalized framework which develops from them (such as relativistic or quantum physics). Kuhn uses the term paradigm to describe both levels of scientific development. That is, it is used to describe both the novel achievement which can often be attributed to an individual scientist, and also the consequent novel framework of attitudes, assumptions, and approaches to scientific research which derive from it.

He suggests that there are two stages in the development of a given branch of science. The first is the *preparadigm stage* when there is no single generally accepted conceptual treatment of the phenomena in a field of study, but a number of competing schools of thought. The second is the *paradigm acquisition stage* which marks the achievement of a level of maturity of the science. A paradigm which provides the conceptual framework for further investigations in the field is proposed and generally accepted. The acquisition of a paradigm characterizes scientific maturity and the acceptance of an agreed conceptual framework or set of rules for the scientists in the field to plan and conduct their research activities. These rules remain in force until the paradigm is replaced by a new one. This evolution leads Kuhn to define two types of scientific activity in a mature science:

The first is *normal* or *puzzle-solving science*, which constitutes the majority of research activity. The universally accepted paradigm facilitates the conducting of specialist or esoteric "puzzle-solving" research (where the "puzzle" is defined within the framework of the paradigm), with research results communicated within the field through the specialist journals. In fact, the state of the art or body of knowledge is defined within the framework of the paradigm, so that expansion of this body of knowledge is achieved incrementally by puzzle-solving within this framework.

The second activity is *revolutionary science,* which occurs much more rarely. It occurs when a paradigm is first acquired (as in the paradigm acquisition stage cited earlier) or more frequently, when one paradigm is replaced with a new one (and, in this sense, a revolution occurs). The history of science is replete with examples of such revolutions, including the Copernican revolution in astronomy in the seventeenth century and the transition from classical to modern physics around the beginning of this century based upon relativity and quantum theories. Although Kuhn provides a quite realistic description of the social-psychology of scientific activity, the distinction between normal and revolutionary science is by no means clear-cut, and detailed study of scientific innovations suggests there is a continuous spectrum of puzzle-solving/paradigm-shift activities between these polar extremes. Moreover, his essentially macroscopic treatment does not deal explicitly with the puzzle-solving or normal activities of the vast majority of scientists. In contrast, Popper's evolutionary methodology of science focuses on the conduct of the individual scientist engaged in such activities.

Popper's Evolutionary Methodology

Popper summarizes his methodology in *The Rationality of Scientific Revolutions.*[14] A scientist always begins with a problem (P_1) which requires a solution (Figure 2.4). This may be a practical problem or an inconsistency between current theory and observation. The scientist then seeks to solve the problem or remove the inconsistency by postulating a new tentative theory (TT) or *conjecture.* The conventional view of scientific method (held by many scientists as well as laypersons) is that the scientist then seeks to confirm the tentative theory (hypothesis) by suitable experimentation or observation. Popper argues that a converse methodology should be applied. Rather than seeking to confirm his tentative theory, the scientist should devise experiments or observations that seek to disprove or *refute* it by an error elimination process (EE). Only after the tentative theory or conjecture has withstood the severest error elimination or refutation that can be devised does the scientist accept it. Critics of Popper's methodology argue that most scientists follow the conventional approach and seek to confirm rather than refute theory/conjecture by experiment. However, although this latter view may typify the conduct of the

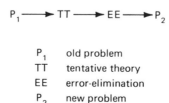

$$P_1 \longrightarrow TT \longrightarrow EE \longrightarrow P_2$$

P_1	old problem
TT	tentative theory
EE	error-elimination
P_2	new problem

Figure 2.4. Popper's evolutionary methodology of science.

individual scientist, Popper's methodology does more aptly describe the collective activities of scientists. A new theory/conjecture, particularly if it postulates unexpected ideas, is only likely to be generally accepted by the scientific community after it has withstood the severest attempts to refute it by other scientists in the field. Thus, a theory is accepted by a process of *conjecture* and *refutation* rather than confirmation of hypothesis by experiment or observation.* This process is exemplified in the furor aroused in the scientific community by Fleischman and Pons' cold fusion conjecture in 1989.

In Popper's methodology, a theory or conjecture is never confirmed or proved; it is, at best, as yet unrefuted. Thus, the highest acceptance a theory can enjoy is *an as yet unrefuted conjecture*. Once a theory is accepted, it generates a new problem (P_2), so Popper represents the process by the sequence of transformations illustrated in Figure 2.4. P_2 is attacked in the same manner, so scientific knowledge continuously evolves from old problems to new problems by means of the conjecture and refutation selection or trial and error elimination learning process analogous to natural selection in biological evolution.

2.5 REVOLUTIONS AND EVOLUTIONS: THE GROWTH OF TECHNOLOGICAL KNOWLEDGE AND INDUSTRIES

Popper and others have applied his treatment not only to science, but also to history and social change, art, and the "body–mind" problem in philosophy; that is, to both the biological and cultural evolution of humanity. What is relevant here, however, is that others have explicitly and implicitly employed both Kuhn's notion of revolutionary paradigm changes and Popper's evolutionary trial and error elimination learning to describe the growth of new technology-based industries.

The central role of innovation in both corporate and economic development was first propounded by Schumpeter in the 1930s.[15] Building upon his ideas, Nelson and Winter have developed the concept of the technological trajectory analogous to a Kuhnian technological (versus scientific) paradigm to describe the development path of a given technology, as well as evolutionary treatments analogous to Popper's to describe the innovation process into the theory of the individual enterprise and overall economic development.[16-18] Moreover, Dosi,[19] in applying and extending Nelson and Winter's models to the development of the semiconductor industry, explicitly discusses the trajectory of that industry as a Kuhnian technological paradigm. Members of the PREST Group in Manchester, England have also extended Nelson and Winter's approach in their evolutionary treatment based upon longitudinal studies of ten firm-technology segments in the United Kingdom,[20,21] while Hanan and Freeman[22] view an enterprise from an ecological perspective. However, although such efforts should yield an important understanding of the developmental economics of new technology-based industries, which is reflected in Moore's ecol-

---

*Moreover, the successive peer review processes that govern a person's advancement in an academic career involve successive evaluations of their performance to date, which implicitly invoke a critical refutation process.

ogy of competition,[23] other descriptive and empirical work is of more immediate interest.

2.6 THE ABERNATHY–UTTERBACK TREATMENT OF THE GROWTH OF HIGH-TECHNOLOGY INDUSTRIES

Abernathy and Utterback (A–U) provide a descriptive treatment of the evolution of high-technology industries generated by major technological innovations (such as the automobile or computer), which implicitly reflects the Kuhnian treatment.[24] They suggest that such industries evolve through three stages or states.

In the initial stages of development or *fluid state*, product designs (e.g., early automobiles and computers) are fluid, manufacturing processes are loosely and adaptively organized, and both product and process may be subject to major changes relatively frequently. It corresponds to the preparadigm stage in scientific evolution. The potential markets for the generic technology will still remain to be fully identified and the acceptability of the technology in these markets is, as yet, unknown.

After a period of fluidity marked by developments through trial and error, there emerges a *dominant design*. The Model T Ford and IBM 360 are examples of the dominant designs in the automobile and computer industries, respectively. These represent superior product designs that provide the standards for the industry to follow and correspond to the paradigm acquisition stage in scientific evolution. The establishment of a dominant design tends to stimulate, delineate, and coalesce the markets for the revolutionary technology so that market uncertainty is reduced. The company or companies which launch the dominant design on the market enjoy competitive advantages, but do not necessarily secure monopolistic or ologopolistic positions. In this *transition state*, emphasis is placed upon performance maximizing innovations, so another company with a strong enough technological base can reap good profits by developing incremental technological improvements in the dominant design in terms of performance, reliability, and after-sales servicing.

Manufacturing methods and product designs become standardized based on the dominant design. Outputs increase through increased sales in the larger markets created by the dominant design. Unit costs decrease both through the increases in manufacturing efficiency facilitated by scale economies of increased output, standardized procedures, and cost-reducing process (more than performance-improving product) innovations. The industry becomes more rigidly organized or in a *specific* or *mature* state. Once the industry develops into this mature state, the scope and sources of innovation change. For composite products, such as automobiles, product assembly is performed by an oligopolistic group of manufacturers (GM, Ford, etc.) using components supplied by vertically integrated subsidiaries or independent suppliers. At the product assembly level, further innovations are cost reducing and focus on production process rather than product improvements. It should be recognized, however, that such incremental cost reducing innovations can offer substantive profits because they apply over a large sales volume.

Also, once an industry has reached the mature state, there is a strong internal resistance to further substantive innovation. The industry is enjoying the benefits of its mature position on the technological learning curve (see Chapter 9), with little technological uncertainty, relatively low costs, and a large established market. The majority of the work force (from senior management to craft level) will have invested their personal career skills in the industry and will probably feel threatened by change. Capital investment in plant and equipment which might be rendered obsolete by change may be high. Thus, the investment commitment in technological, fiscal, and human capital present barriers to change.

This treatment implies that industries become technologically moribund on reaching a mature state. This does not usually happen in practice, but further substantial innovation may only be stimulated by external factors. Abernathy and Clark[25] extend the A–U treatment to incorporate the "dematuring" of industries through such changes, while Sahal provides a more extensive treatment based upon a larger sample of industries.[26] Again, his treatment implicitly reflects Kuhn's and Popper's ideas.

2.7 SAHAL'S TREATMENT OF TECHNOLOGICAL EVOLUTION

Sahal views a given technology as a self-organizing system which evolves by trial and error elimination learning with the following features:

Through the interplay of the self-organizing system formed by a given technology and its sustained application and development, there emerges a pattern of artifact design or *technological guidepost* which charts the course of further innovative activities. The concepts of dominant design and technological guidepost appear similar, both corresponding to a paradigm acquisition in Kuhnian scientific evolution. Before the emergence of this design or guidepost, the evolution of the technology can be subject to a wide variety of changes. It is, like putty, malleable. Once a design or guidepost is established, this malleability is lost and the technology will only be subject to minor changes. That is, it will be cast into a clay-like mold. Sahal describes this as the *putty-clay principle* of technological evolution.

However, thereafter, the A–U and Sahal treatments differ. Rather than progressing to a specific or mature state, Sahal argues that a technology continues to evolve through a succession of stepwise improvements in its capabilities. This stepwise growth is well documented in the literature (see Appendix 1 to Chapter 4). The successive computer generations reflect stepwise growths in computing capabilities, and we have witnessed four such generations in about 25 years. Other technologies have, however, remained on a plateau prior to a stepwise improvement for several decades. Sahal cites the Fordson farm tractor as a design which lasted for 25 years (circa 1920–1940). A stepwise improvement is typically consolidated with a new technological guidepost, and we shall discuss this further in the next section.

Two other related characteristics of the concept of dominant design or technological guidepost should be noted. First, two or more designs may share the shaping of further evolution. For example, the Boeing 707 and DC 8 were both successful competing designs in the evolution of jet airliners. Similarly, Apple and IBM have developed successive competing designs of personal computers. Second, the dominant design or technological guidepost is not necessarily the best realization of the technology. The standard QWERTY, etc. sequence became the dominant layout of letters on the mechanical typewriter keyboard about a century ago. It has so remained, despite the fact that it is not the most efficient layout ergonomically. Similarly, the successful IBM PC was not the most advanced personal computer that could have been designed at the time and, despite Sony's best efforts, BETAMAX lost out to the inferior VHS format in consumer VCRs.

Sahal emphasizes that there is one crucial difference between biological and technological evolution. Whereas distinct biological species cannot interbreed, stepwise technological growth is quite frequently achieved by the *creative symbiosis* of two (or more) previously unrelated technologies (e.g., the use of nuclear power in marine propulsion and solid-state electronics in numerical control systems). Further progress is then based upon a new coalition of technologies, possibly pioneered by invaders from other industries. We cited such an example at the end of the previous section since synthetic fibers are a product of the creative symbiosis of textile and chemical engineering technologies.

2.8 REVOLUTIONARY-RADICAL, MICRO-RADICAL, GENERATIONAL, AND INCREMENTAL INNOVATIONS

Kuhn describes a revolutionary paradigm in science as an achievement which possesses two characteristics: (1) it is sufficiently unprecedented to attract an enduring group of adherents from competing modes of scientific activities, and (2) it is sufficiently open-ended to leave all sorts of problems to solve.

If we substitute "scientific, engineering, entrepreneurial, and management activities" for "scientific activity" in the above, these words are equally applicable to technological (versus scientific) innovation. Based upon the Kuhn, A–U and Sahal treatments, we can define three categories of technological innovations.

Revolutionary-Radical Technological Innovations

These may be based upon major inventions which create a new industry (e.g., the transistor). They may also be associated with a creative symbiosis of previously unrelated technologies. They constitute *revolutionary discontinuities* in the technological evolution since they invoke new paradigmic frameworks for technological puzzle-solving expressed in the dominant design or technological guidepost. Such discontinuities are comparatively rare, but can be expected to induce revolutionary changes in market and industrial structures. Like their scientific counterparts, they readily attract an enduring group of adherents. This may take the form of invasions of the previously mature industries by "outsiders" and the creation of small entrepreneurial firms which subsequently establish themselves as major corporate entities. This is dramatically illustrated in the growth of Texas Instruments based upon the revolutionary-radical innovation of the transistor. Thus, top managements of companies in mature industries need to be continuously sensitive to the threats of revolutionary-radical innovations and make it desirable for them to engage in technology monitoring to detect the signals of technological change (see Appendix 1 to Chapter 4, Section A 4.1.3).

Paradoxically, revolutionary-radical technological innovations must also be approached circumspectly. They require a substantial front-end commitment of corporate resources and, as was EMI's experience with the CAT scanner, can lead to commercially unfortunate consequences. This is one reason why some companies prefer to follow defensive technological strategies (see Chapter 4, Section 4.8). It should also be noted that scientific revolutions and technological revolutions are not necessarily concomitant. Although the invention and application of transistors constituted a technological revolution, it did not constitute a scientific revolution. The theoretical explanation of transistor action, although an impressive scientific feat, was based upon the well-established band theory of solids in physics.[27] In contrast, the new biotechnology industry is based upon both revolutionary science (embodied in the double-helix and recombinant DNA) *and* revolutionary technology.

Micro-Radical and Generational Technological Innovations

The above innovations (whether scientific or technological) are revolutionary because they overturn the old paradigm and are *competence destroying*.[28] The transistor was a revolutionary innovation because it destroyed the prior competence in vacuum tube technology by rendering it obsolescent. However, there is another category of innovations corresponding to Sahal's step-wise improvements in capabilities. These constitute *nonrevolutionary discontinuities* in technological innovation, because they build upon existing capabilities and are *competence enhancing*. Durand describes them as *micro-radical innovations*.[29] They can also be perhaps best illustrated in the technological evolution of direct random access memories (DRAMs) and microprocessor chips in the microelectronics industry, from the 1970s onwards. Each new generation of DRAM (16K, 64K, 256K, etc.) or microprocessor (8088, 386, 286, etc.) introduces stepwise improvements in capabilities, which often offer substantial opportunities for new market applications while building upon existing competences, and so can be described as *generational innovations*.

It must be stressed that the distinction between the first two types of innovations is not always clearcut, and they may be best viewed as the polar extremes of a continuum. The distinction is sometimes dependent upon the perspectives of the enterprises involved. For example, in the eyes of Texas Instruments, the introduction of David Kilby's first IC into the market could definitely be viewed as a micro-radical innovation since it was based upon the enhancement of that firm's existing competence in silicon semiconductor technology. In contrast, although most of them made and marketed a limited range of silicon transistors, the competences of the other firms in the semiconductor industry at that time were mainly invested in germanium semiconductor technology, unsuited to IC fabrication. Therefore, for these firms, the first IC represented a more radical innovation since it forced them to transfer their existing competences in germanium technology into the enhancement of their more limited competences in silicon technology. Abernathy and Clark[25] define the term *transilience* as the capacity of an innovation to influence the firm's existing resources, skills, and knowledge. Clearly, the innovation of the IC represented less transilience to TI than it did to other companies in the semiconductor industry at that time. Interestingly, although the new biotechnology is based upon both revolutionary science and revolutionary technology, it is *less* transilient than was the transistor, which rendered vacuum-tube manufacturing technology obsolete. Although expected to be competence-destroying by providing product innovations which replace current products, the new biotechnology should also be competence-enhancing by providing process innovations which are integrated into existing manufacturing practices in the biotechnology and chemicals industries.

Incremental, Puzzle-Solving, or Normal Technological Innovations

These constitute the vast majority of innovations and correspond to puzzle-solving activities in Kuhnian normal science. They are typically incremental performance-improving and/or cost-reducing innovations conducted within the framework of the established technological paradigm, and facilitate the Popperean cumulative trial and error elimination learning described in Section 2.4 and Figure 2.4, but in technology, as opposed to science. Again, they lie on a continuum from very minor modifications to established products to, say, the new model automobile which contains no radically new technology. A successful feature of some Japanese enterprises is their policy of *kaizen* or the performing of R&D to create continuous incremental improvements in existing products that are appealing to customers.

2.9 THE "S"-SHAPED CURVE OF TECHNOLOGICAL EVOLUTION

The distinction among the above three types of innovations can be best seen in the "S"-shaped curves associated with the evolution of technological capabilities as illustrated in Figures 2.5–2.7. Sahal[26] and Foster[30] give numerous examples in their texts. The lower end of the Figure 2.5 curve corresponds to the A–U fluid state, when technologies, products, and markets are ill-defined. The steep middle part corresponds to the performance-maximizing transition state, after the definition of the dominant design or technological guidepost. The upper part corresponds to the cost-minimizing specific state, when the technology may be reaching the upper limit of its capability. Figure 2.6 illustrates New Revolutionary-Radical Technology B, replacing Old Technology A. Figure 2.7 illustrates the overall evolution of a technological capability based upon a succession of micro-radical innovations. The important role that the S-curve can play in the technological strategy of the enterprise is discussed in Chapter 4, Section 4.9 and in Appendix 1 to Chapter 4, Sections A 4.1.4–A 4.1.6.

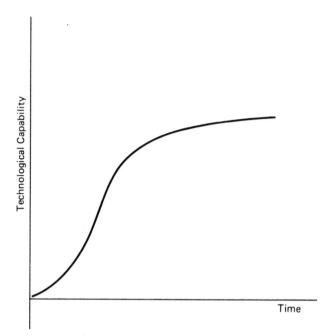

Figure 2.5. The "S"-shaped curve.

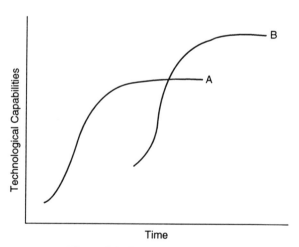

Figure 2.6. Technology substitution.

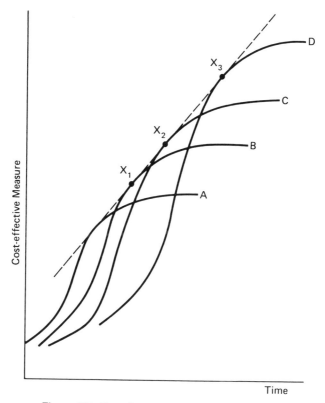

Figure 2.7. Overall evolution or envelope curve.

2.10   INNOVATIONS AS TECHNOLOGICAL MUTATIONS

In Stage 1 of his treatment of the innovation process (Section 2.2), Bright suggests that an innovation begins with either discovery or the perception of an environment or market need or opportunity. This underlying distinction is reflected in the innovation literature in the distinction between *technology-push* and *market-pull*. The former implies that a new invention is "pushed" through the R&D, production, and sales functions onto the market without proper consideration of whether or not it satisfies a user need—as shown in Figure 2.8*a*. In contrast, an innovation based upon market pull has been developed by the R&D function in response to an identified market need as shown in Figure 2.8*b*. We have just distinguished between revolutionary and other innovations. The former are major inventions and innovations (such as radio and the computer) for which there is no manifest need and which are created by technology-push or the visions and achievement drives of inventor–entrepreneurs such as Marconi. Once this essentially Kuhnian technological revolution occurs and the latent need becomes manifest through social recogni-

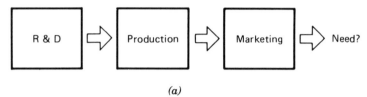

(a)

Figure 2.8a. Technology-Push.

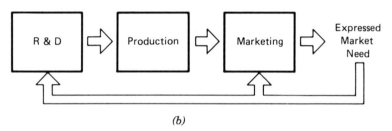

(b)

Figure 2.8b. Market-pull.

tion, market-pull stimulates the proliferation of incremental innovations to satisfy evolving specialist user needs, as reflected in the steep rise of the S-curve in Figure 2.5. Because most innovations are incremental, it is hardly surprising that there is some evidence that innovations based upon technology-push are less likely to be successful than those based upon market-pull.

This distinction between technology-push and market-pull can be best viewed in the context of the extension of the evolutionary treatment. The technology-push sequence illustrated in Figure 2.8a is similar to that for biological organism which has experienced a genetic mutation (except that the latter occurs randomly). The success or failure of each is determined by a trial and error selection process—the first in the marketplace and the second in the ecosphere. Li invokes the same analogy.[31] He compares the innovation process with the fundamental building block of biological evolution—the DNA double helix (see Figure 2.9). One spiral of the helix constitutes the technology-push or ensemble of scientific and engineering skills, while the other constitutes the market-pull or ensemble of entrepreneurial, managerial, and marketing skills which are required. The notion of technology–market synergy introduced above is synonymous with the judicous synthesis of Li's helices. Thus, once we view technological as well as scientific growth as a Lamarckian evolutionary process, we may view an invention which leads to a technological innovation as corresponding to a genetic mutation. Indeed, the eminent ethologist Konrad Lorenz made the same analogy when describing the genetic mutation process:

> The process whereby a large modern industrial company, such as a chemical firm, invests a considerable part of its profits in its laboratories in order to promote new

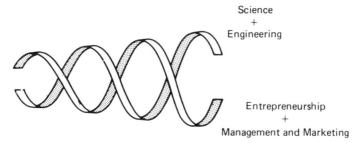

Science
+
Engineering

Entrepreneurship
+
Management and Marketing

Figure 2.9. The innovation double-helix.

discoveries and thus new sources of profit is not so much a model as a specific case, of the [genetic mutation]* process that is going on in all living systems.[32]

In employing this analogy, it is important to recognize that Herbert Spencer was mistaken in describing Darwinian biological evolution as the *survival of the fittest*. In fact, as the eminent biologist Kenneth Boulding has pointed out,[33] it is more appositely described as the *survival of the fitting*, since an organisms's survival depends upon how well it fits into its environmental niche rather upon by some undefined standard of fitness. The same can be said of a technological innovation. It does not have to be the *fittest* in the sense of incorporating the most up-to-date, powerful, and sophisticated features of the technology. Rather, it should be *fitting* in the sense that it should be designated to satisfy its end-user needs as closely as possible. For example, when IBM finally launched its own PC in the early 1980s, it did not incorporate the most advanced computer technology then available, so it could not be viewed as the fittest PC on the market. However, it did satisfy its end-users needs and, coupled with IBM's marketing abilities, service support, and reputation in the industry, quickly established the leading share of worldwide PC sales. The development of another very successful innovation, 3M's *Post-It Note,* also illustrates the appositeness of Boulding's aphorism to an innovation. Art Fry, the intrapreneur (see Chapter 12) responsible for its development, was frustrated with his scrap paper bookmark falling out of his hymnal as he sang in his church choir. He then recalled that one of his colleagues, Spencer Silver, had accidentally discovered a weak adhesive for which there was no apparent market need. Fry realized that a bookmark coated with a weak nonpermanent adhesive which would remain temporarily in place, but could then be removed without damaging the hymnal page, was what he needed. He recognized the value of Silver's adhesive, and the Post-It Note concept was created. The importance of end-user needs should be kept in mind by R&D managers. Several studies on successes and failures in developing new technology-based products have shown that the recognition and satisfaction of such needs is the most important determinant of success, as we shall see in Chapter 6.

Although both biological and technological evolution may be based upon the

*This author's addition.

survival of the fitting, Baruch makes a crucial methodological distinction between the two of them.[34] The former is linear Darwinian error elimination process, since the adaptive characteristics acquired by an organism in its life cannot be genetically transmitted to its offspring. In contrast, the development of an innovative new product is a conscious attempt by a company to adapt to its environment and past experience, and is thus a Lamarckian trial and error-elimination feedback learning process, using procedures for evaluating the innovation potential *fittingness* of inventions.

Effective high-technology companies will typically possess R&D functions which monitor and extend the state of the art in science and engineering to generate inventions with, as yet, undefined market needs. The companies will also possess the marketing capabilities and insights to identify both manifest and latent market opportunities for the evolving technological artifacts in the latter's changing economic, social–cultural, and political contexts. This observation that potential new technology must be judged in its contemporary socioeconomic and cultural context was also expressed by Graubar in his preface to a special use of Daedelus on the problems and opportunities inherent in modern technology.[35]

> Too many of us still tend . . . to think of technology in 19th Century terms. . . . We should think of technology (and technological innovation) not just as a collection of artifacts, however sophisticated and complex, but as a system whose social, cultural, intellectual, managerial and political components are seen as integral to it.

If we view technology and technological evolution in the manner Graubar suggests, we may conventionally label such a system as the *technosphere,* by analogy with the ecosphere in biological evolution. Now, a biological organism can only test the selective advantage of a genetic mutation in the ecosphere by a Darwinian process of successful trial or death. However, just as a scientist can allow his unsuccessful theories to die in his stead in the conjecture–refutation process, so a company can conduct a Lamarckian evaluation of an invention using prior learned experience and knowledge, prior to substantive resource commitment, to determine its selective advantage in the technosphere and its potential contribution to overall corporate objectives and goals. Companies will achieve success by possessing the judgment and resources to identify synergy among their technological, entrepreneurial, and managerial capabilities and this technosphere. This can alternatively be expressed by Schmookler's comparison with the blades of a pair of scissors.[36] Scissors will only cut with matching blades, and Schmookler suggests that a technological innovation and its market must match in a similar complementary manner.

This concept of synergy may be illustrated by an extension of Figure 2.8 in Figure 2.10 which views the company as an adaptive open system (consistent with Sahal's treatment) in the technosphere. The marketing function monitors this technosphere to identify new needs and opportunities and feeds information back to R&D and production. The R&D function monitors and interacts with the state of the art in the relevant science and engineering fields to identify new technological trends and opportunities. Information from these two sources may be combined to

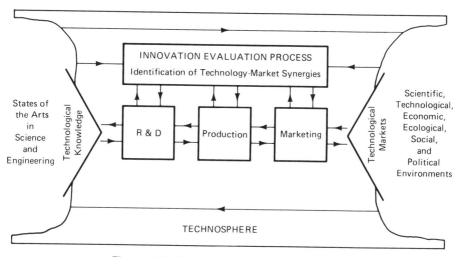

Figure 2.10. The innovation evaluation process.

generate new product ideas or inventions, or *technological mutations* which will meet evolving market needs or opportunities. Furthermore, according to Popper, the highest status of any scientific theory is *as yet unrefuted conjecture* since it always remains exposed to the possibility of being refuted by another theory or conjecture. Recalling also (Section 2.3) that the ultimate fate of any innovation is "mere oblivion" as it is superceded by a new one, in our Popperian treatment, the highest status of an innovation is an *as yet unrefuted product.* Sooner or later, in a free market economy, if the company does not do it, a competitor will "refute" or replace any product with a better one. Therefore, it is imperative that it maintains technology and market monitoring activities to ensure its long-term survival.

Although sufficient for present purposes, as Rothwell[37] points out, this view of the process needs to be modified in the light of innovation management practices introduced in the 1980s. First, in many companies, the R&D, production, and marketing functions now participate in the innovation process *simultaneously* rather than *sequentially* (as shown in Figure 2.10) in order to accelerate the process. This approach will be discussed in Chapter 8. Second, the innovation process is often shared between two or more companies. The role of partnering processes in innovation networks is discussed in Chapter 13.

2.11 CRITERIA FOR EVALUATING INNOVATIONS

As well as the notion of a technological mutation, the innovation chain equation (Figure 2.1) provides a fruitful analogy of the innovation process. All chains, whether physical, chemical, or abstract, are only as strong as their weakest links, and in Chapter 6, we shall cite evidence of innovations which failed through the weakness of one or more links in the chain. This analogy, coupled with the Popperian conjecture–refutation evolutionary methodology, provides a useful conceptual foundation for evaluating potential innovations since they may be invoked to postulate a series of challenging refutations, or hurdles which must be overcome, if an innovation is to achieve commercial success. They are also consistent with Graubar's comments, cited earlier.

1. As a technological conjecture, the invention must be able to withstand any scientific or engineering refutations with which it may be attacked. That is, it must be <u>demonstrably feasible technologically</u>. It also requires that the organization has the scientific and engineering resources to develop the invention.

2. It must be possible to produce and sell it profitably. This means it must satisfy a manifest or latent social or market need. If the need is latent rather than manifest, this need must be felt and accepted by potential users once the innovation is offered to society. That is, it must be <u>demonstrably feasible commercially</u>. It also requires that the organization has the entrepreneurial, managerial, and financial resources both to exploit the invention. Furthermore, remembering from the previous section that the highest status of any innovation is an "as yet unrefuted product," the longer-term product enhancement approaches need to be considered to withstand the competition from other organizations which may also be seeking to exploit it.

3. Whatever hazards or disadvantages might be expected from its continued and widespread use must be acceptable to individual users and society as a whole. That is, <u>any health, safety, and environmental impacts must be socially acceptable</u>, so these and other government regulative requirements must be satisfied.

4. Then, if its commercial success may be contingent upon regional, national, or international government policies, it should be ensured that it is compatible with these policies. That is, <u>relevant government policies should be supportive</u>.

5. As has already been stated, enterprises with productive R&D often create more inventions than they can support through the innovation process to be launched as new products. Therefore, a proposed innovation must be evaluated competitively with others to identify its <u>congruency with corporate objectives and goals</u>, and determine whether it offers a prudent investment of scarce corporate funds. Some companies hold an inventory of patents of their inventions which they are unable to exploit for this reason which others may purchase or license. As was also stated earlier, this rationing of corporate support for innovative opportunities has a significant impact on the behavior of individuals in technology-based companies, as will be demonstrated in later chapters.

2.12   THREE INNOVATIONS BRIEFLY REVISITED

It took Marconi about 20 years of sustained effort to overcome the hurdles (or refutations) to the exploitation of wireless telegraphy. It took until the 1910s and the "Crippen-Titanic effect" to convert that latent need for wireless telegraphy into a need recognized by both society in general and governments in particular. After that date, hurdles 2, 3, and 4 collapsed, and the Marconi companies were able to enjoy the rewards of their technological and entrepreneurial efforts.

In contrast, the latent need for computerized axial tomography was quickly converted into a manifest need by the presentation at the Radiological Congress in Chicago in 1972, and EMI appears to have quickly overcome hurdles 1 and 2 only to be defeated by hurdles 3 and 4. In light of subsequent events, it might also be argued that investment in scanner technology was not congruent with corporate objectives and goals, so insufficient consideration was given to hurdle 5. EMI's experience with its CAT scanner illustrates the perils of being an offensive innovator (see Chapter 4) and, with the benefits of 20:20 hindsight, a more prudent strategy would have been to license the technology and/or enter into a joint venture with a strong presence in the U.S. market (such as GE). Two other U.K. companies, Pilkington and Boots (see next chapter), used licensing as well as direct sales strategies to market float glass and *ibuprofen* in the United States. It is of interest to read Abell's description of the overall development of the scanner market and Li's account of the development of scanner technology at GE which, as events turned out, purchased the EMI operation.[38,39] GE appears implicitly to have adopted an effective defensive strategy in the face of a revolutionary technological innovation in its area of technological expertise. By the judicious acquisition of the EMI scanner operation and that company's error-elimination learning experiences, GE was able to secure a powerful position in the new market, facing competition from others, including Toshiba.

In contrast, both Sony and Searle appear to have shown good strategic judgment in developing and marketing their innovations. As with the personal transistor radio, Sony astutely recognized the existence of a latent market need for a personal stereo, developed a well-executed market launch, and a succession of product enhancements to meet subsequent imitative competition. Hurdles 3 and 4 represented a real challenge to Searle, but again, the company proved to be astute in surmounting them. Although there is evidence that aspartame may be a health risk to a small minority of users, it can be set against the health risks of the other artificial sweeteners and that of sugar to diabetics. Searle, like Sony, adopted an excellent approach to marketing aspartame as a branded product. Moreover, in contrast to EMI, by entering into a manufacturing and cross-licensing agreement with Ajimoto, they both ensured a raw material source and made an ally of a potentially formitable competitor. The company also avoided a long-term dependency on Ajimoto by developing its own L-phe fermentation process.

Having reviewed some of the characteristic features of the technological innovation process, we next study it in a corporate setting.

REFERENCES

1. M. A. Saren, "A Classification and Review of Models of the Intra-Firm Innovation Process." *R&D Management* **14**(1), 11–24 (1984).
2. J. R. Bright, "Some Management Lessons from Technological Innovation Research." *Long Range Planning* **2**(1), 36–41 (1969).

3. D. E. Kash, *Perpetual Innovation: The New World of Competition.* New York: Basic Books, 1989, Chapter 3.

4. W. G. Howard, Jr. and B. R. Guile (Eds.), *Profiting From Innovation.* New York: Free Press, 1992, p. 135.

5. H. Redwood, "Pharmaceuticals: The Price/Research Spiral." *Long Range Planning* **24**(2), 16–27 (1991).

6. J. E. S. Parker, *The Economics of Innovation,* 2nd ed. London: Longman Group, 1978, pp. 55–56.

7. R. Rothwell, "The Impact of Regulation on Innovation: Some US Data." *Technological Forecasting and Social Change* **17**(1), 7–34 (1980).

8. *Business Week,* January 9 (1989).

9. W. E. Souder, *Managing New Product Innovations.* Lexington, MA: Heath, 1987, Chapter 4.

10. U. L. Businaro, "Applying the Biological Evolution Metaphor to Technological Innovation." *Futures* **15**(6), 463–477 (1983).

11. C. Marchetti, "Swings, Cycles and the Global Economy." *New Scientist* 12–15 (1985).

12. M. J. C. Martin, "On Kuhn, Popper and Technical Innovation." *European Journal of Operational Research* **14**(3) (1983).

13. T. S. Kuhn, *The Structure of Scientific Revolutions,* 2nd ed. Chicago: University of Chicago Press, 1970.

14. Sir K. R. Popper, "The Rationality of Scientific Revolutions." In I. Hacking (Ed.) *Scientific Revolutions.* Oxford: Oxford University Press, 1981, pp. 80–106.

15. J. Schumpeter, *Capitalism, Socialism and Democracy.* New York: Harper & Row, 1942.

16. R. R. Nelson and S. G. Winter, "In Search of Useful Theory of Innovation." *Research Policy* **6,** 36–76 (1977).

17. R. R. Nelson and S. G. Winter, *An Evolution Theory of Economic Change.* Cambridge, MA: Harvard University Press, 1982.

18. R. R. Nelson, *Understanding Technical Change as an Evolutionary Process.* Amsterdam: Elsevier, 1987.

19. G. Dosi, "Technological Paradigms and Technological Trajectories," *Research Policy* **11**(3), 147–162 (1982); *Technical Change and Industrial Transformation.* London: Macmillan, 1984.

20. L. Georghio, M. Gibbons, and J. S. Metcalfe, "Staying the Distance—Technological Development and Competition." *International Journal of Technology Management* **1**(3/4), 425–438 (1986).

21. J. S. Metcalfe, and M. Gibbons, "Technological Variety and Organization: A Systematic Perspective on the Competition Process." In R. S. Rosenbloom (Ed.), *Research on Technology Management Organization and Policy.* Greenwich, CT: JAI Press, 1988, Vol. 4.

22. M. Hanan and J. Freeman, *Organizational Ecology.* Cambridge, MA: Harvard University Press, 1989.

23. J. F. Moore, "Predators and Prey: A New Ecology of Competition. *Harvard Business Review* **71**(3), 75–86 (1993).

24. W. J. Abernathy and J. M. Utterback, "Patterns of Industrial Innovation." *Technology*

Review **80**(7), 40–47 (1978); "A Dynamic Model of Product and Process Innovation." *Omega* **3**(6), 639–656 (1975).

25. W. J. Abernathy, and K. B. Clark, "Innovation: Mapping the Winds of Creative Destruction." *Research Policy* **14**, 2–22 (1985).

26. D. Sahal, *Patterns of Technological Innovation.*   Reading, MA: Addison-Wesley, 1981, "Technological Guideposts and Innovation Avenues." *Research Policy* **14,** 61–82 (1985).

27. W. F. Shockley, *Electrons and Holes in Semiconductors, with Applications to Transistor Electronics.*   New York: Van Nostrand, 1950.

28. P. Anderson and M. L. Tushman, "Managing Through Cycles of Technological Change." *Research Technology Management* **34**(3), 26–31 (1991).

29. T. Durand, "Dual Technological Trees: Assessing the Intensity and Strategic Significance of Technological Change." *Research Policy* **21,** 361–380 (1992).

30. R. N. Foster, *Innovation: The Attacker's Advantage.*   New York: Summit Books, 1986.

31. Y. T. Li, *Technological Innovation in Education and Industry.*   New York: Van Nostrand-Reinhold, 1980, pp. 81–83.

32. K. Lorenz, *Behind the Mirror: A Search for a Natural History of Human Knowledge.* New York: Harcourt Brace Jovanovich, 1977, p. 27.

33. K. J. Boulding, *Evolutionary Economics.* Beverly Hills, CA: Sage Publications, 1981.

34. J. J. Baruch, Foreword. *TIMS Special Studies in the Management Sciences* **15,** vii–ix (1980).

35. S. R. Graubard, Preface to "Modern Technology: Problem or Opportunity?" *Daedalus* **109**(1) (1980).

36. J. Schmookler, *Invention and Economic Growth.*   Cambridge, MA: Harvard University Press, 1966.

37. R. Rothwell, "Successful Industrial Innovation: Critical Factors for the 1990s." *R&D Management* **22**(3), 221–239 (1992).

38. D. F. Abell, *Defining the Business: The Starting Point of Strategic Planning.*   Englewood Cliffs, NJ: Prentice-Hall, 1980, Chapter 5.

39. Y. T. Li, *Technological Innovation in Education and Industry.* New York: Van Nostrand-Reinhold, 1980, Chapter 3.

PART II
THE CORPORATE SETTING

Part II begins by describing the technological innovation base of a firm in Chapter 3. It continues in Chapter 4 by discussing technology planning and the spectrum of alternative strategies available to it. Chapter 4 is supported by two appendices. Appendix 4.1 outlines the technological and social forecasting approaches which it may use to define and "invent" its future. Appendix 4.2 discusses "green" or natural environmental issues and reviews how firms incorporate technology assessment approaches and environmental concerns in their technology plans.

___3
THE CORPORATE TECHNOLOGICAL INNOVATION BASE

The primary purpose of any theory is to clarify concepts and ideas that have become, as it were, confused and entangled. Not until terms and concepts have been defined can one hope to make an progress in examining the questions clearly and simply and expect the reader to share one's views.

CARL VON CLAUSEWITZ, *ON WAR,* BOOK TWO, CHAPTER 1, "BRANCHES OF THE ART OF WAR"

3.1 INTRODUCTION

The purpose of this text is top provide guidelines for the management of the innovation process in high-technology firms, and in Chapters 1 and 2 we examined some of the features and underlying characteristics of this process. To reach commercial fruition, such innovations must typically occur within the framework of the extant technological capabilities or base of the firm. In this chapter we define the components of this technological base insofar as it applies to the process of technological innovation.

Traditionally, technological innovation appears to have been largely bypassed in defining the management structures of high-technology companies. Most companies build their structures around the traditional functions of finance, marketing, production, human resources, and R&D. Many also define an engineering function, concerned with advanced design and development or replication of existing technological capabilities. At first sight, it might appear appropriate to equate the technological innovation base with the R&D and engineering functions of the organization, but such an equation is too simplistic. Although the R&D and (if present) engineering functions provide the site for its initial stages, the technological innovation process continues and manifests itself in other functions of the organization. Porter illustrates this contention in the technology development activities of his value chain.[1] He provides a conceptually appealing approach towards integrating technology into the company, but to satisfy our immediate needs, an essentially descriptive approach suffices.

The technological innovation base needs to be defined broadly enough to encompass both R&D and those other technological activities which are performed in the other functional areas, but which contribute to the commercially successful outcome of the innovation process. A corollary of this diffuse concept is that the components of the technological innovation base should be reviewed holistically and should be designed to reflect the overall technological stance of the company. This notion will be made concrete when we examine technological strategies in the next chapter.

3.2 INNOVATION PROCESS AND TECHNOLOGICAL INNOVATION BASE

We have just argued that the technological innovation base manifests and defines itself through the enactment of the innovation process in the organization. Therefore, it is convenient to define such a base by sequentially examining the innovation process from a company viewpoint, rather than the Bright-individual innovation or the industry-wide approaches in Chapter 2.

Table 3.1 repeats the first seven phases of Bright's treatment of innovation, but expands it to include the ancilliary technology-related activities which may be required to promote it successfully. This sequence from research to full-scale pro-

TABLE 3.1   Innovation-Related Activities

Bright Process	R & D-Production-Related Activities		Market-Related Activities
1. Scientific suggestion. Perception of need	Nondirected fundamental research		Market monitoring
2. Theory/design concept	Directed applied research	C	
3. Verification	Primary development	o	Initial market definition
4. Laboratory demonstration	Secondary development	n	Patenting
	Design engineering	t	Development of market plan
	Further development and design engineering	r a c t	Design of user education and advisory services
5. Pilot/prototype to full–scale production	Pilot prototype to full-scale production. Design of production system. Quality and reliability engineering. Design of technical after-sales services	R & D	Test marketing Refining product-market concept
6. Commercial introduction	Provision of after-sales services		Market launch Provision of user education and advisory services
7. Widespread adoption	Incremental product improvements		Licensing

duction is elaborated further in Chapter 6 (Section 6.5) in the context of the innovation evaluation process, and so is briefly outlined here.

3.3 NONDIRECTED OR FUNDAMENTAL RESEARCH

As Bright indicates, the innovation process may begin with the invention of a new technological capability based upon a scientific and/or engineering discovery or suggestion. Such discoveries and suggestions occur, not infrequently, seren-

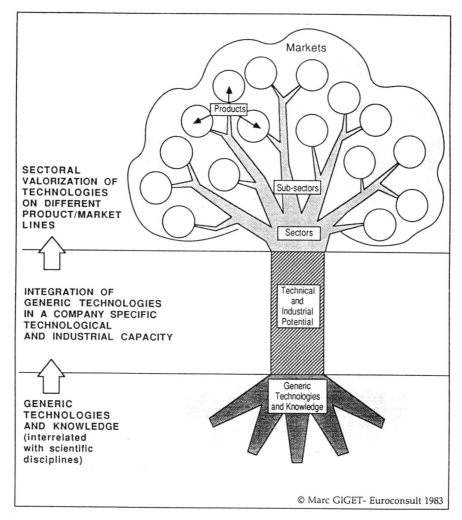

Figure 3.1. The tree of competences of a company. *Source:* This figure was first published in English *Futures,* Vol. 20, No. 2, April 1988, p. 148, and is reproduced here with the permission of Butterworth Heinemann, Oxford, U.K.

dipitously during fundamental research at the state of the art with the objective of expanding the body of scientific knowledge, rather than the development of new technological products. It can be argued that such fundamental research is the responsibility of the universities, government, and (possibly) industry-wide research laboratories rather than company R&D functions since it is not congruent with the latter's mission. However, some high-technology corporations, which wish to maintain a posture of technological leadership within their industry, believe it essential to participate actively in such research at the state-of-the-art level both to ensure that it maintains up-to-date scientific knowhow and to provide research results as an exchange currency with other in the field. It was the fundamental research of Carothers at Du Pont that created nylon.[2] As one senior manager and researcher with Xerox expresses it.[3]

For Xerox, the study of electron energy states and mobilities in photoconductors is an area of basic science with which Xerox needs to be conversant if it is to do its reprographics business knowledgeably. I believe that we and our customers are much better off, as we purvey copies containing photoreceptors, if we do so in a state of knowledge and understanding than if we are in a state of ignorance: It is better to do it "smart" than do it "dumb."

This belief is well illustrated by the bonzai tree metaphor[4-6] used by leading Japanese companies (Figure 3.1), including NEC, as described by Uenohara.[7] Fundamental research provides the root structure that enables the company to draw upon the scientific humus for the nutrients upon which its continued well-being depends, and we shall return to this compelling metaphor in later chapters. When compared with this western counterparts, Japanese companies also employ more staff, both to collaborate with basic researchers in universities and other organizations, and to analyze and sift scientific research outputs worldwide to provide knowledge for incorporation into their own applied research projects.

3.4   CONTRACT RESEARCH AND DEVELOPMENT

Many companies seek to subsidize their R&D efforts by performing contract R&D for outside agencies, notably the government. For example, GE spent $3 billion on R&D in 1987, of which slightly over half came from government contracts.[8] Quite often, it is possible to identify communalities between the innovations which the company is striving to develop and the R&D needs of outside agencies. Thus, the performance of contract R&D for an outside agency may directly or indirectly solve some of the problems faced in developing the in-house innovation, and the cost to the company of the latter may thereby be reduced. Historically, this was typical of the aerospace industry, where the knowhow discovered in the development of successive generations of military aircraft has been exploited in the development of successive generations of their commercial counterparts. The commercial spin-offs are frequently quoted by politicians and government servants to justify government R&D expenditure in industry. Contract R&D may itself lead to commercially successful innovations, and is often included in a company's R&D project portfolio.

3.5 TECHNOLOGY ACQUISITION AND IN-LICENSING SERVICES

Technology development at the company level is unquestionably a two-way street. Just as it may contract-in R&D as mentioned above, it may contract-out or buy technology in several ways.

First, it may contract-out R&D to another research organization, such as a university or other company. Alternatively, it may participate in precompetitive R&D consortia jointly with other firms in its industry, which often also include universities. The missions of many publicly funded universities such as the U.S. land-grant institutions, have historically involved R&D-based extension activities which have steadily grown throughout this century mainly from government grant and contract fundings. More recently, publicly financed universities, facing the specter of "creeping" privatization, have been eager to participate in collaborative R&D consortia with private industry. Thus, support for university-based research provides an effective means of supplementing corporate fundamental research and pushing stronger roots in the scientific humus of Figure 3.1. Cutler discusses alternative sources of external technology for U.S. firms,[9] while Sen and Rubenstein[10] discuss the factors influencing the integrating of such externally acquired technology with internal R&D activities. In Appendix 4.1 we discuss the notion of technology monitoring as a means of detecting and monitoring developments in emergent technologies that may impact upon a firm in the future. An obvious contemporary example is the future impact of recombinant DNA technology on the chemicals, pharmaceuticals, and food industries. Some mature companies in such industries are monitoring and accessing developments in this emergent technology by sponsoring research in the many small, new biotechnology firms (NBFs) that have been established in the last decade or so. Contracting-out R&D to an NBF also enables the mature firm to evaluate its technical capabilities as a preliminary to a closer relationship or outright purchase of the NBF. Technology acquisition will be discussed in Chapter 13.

Second, few (if any) inventions are converted into innovations as independent *technologically autonomous* new products. They typically embody other novel technical capabilities or *complementary technological innovations* in their design and/or manufacture. This is clearly true of, say, a new airliner, which will embody numerous inventions in engines and avionics, as well as a novel airframe design. However, even a new drug which, as a new chemical entity (NCE), is an invention consumed by the patient as the end customer, embodies other (possibly new) technologies in its chemical synthesis and manufacture. A company derives its innovative competitive advantage not solely from its capability to invent, but rather from its inventive ability to combine its own in-house capabilities with technology provided by suppliers, as illustrated in the IBM PC and Microsoft's DOS software. Thus, its technological innovation base needs to include the capability to evaluate and, if appropriate, to procure (through contracting-out work and purchasing products or licenses) the external technology it requires. This function is broader in scope than the traditional procurement or purchasing function, but must be present if a company is to derive the maximum value from its inherent technological capabilities.

3.6 DIRECTED OR APPLIED RESEARCH AND PRIMARY DEVELOPMENT

A firm's central R&D laboratory effort can normally be classified under this heading. Essentially, it represents phases 2 and 3 in Bright's treatment whereby a scientific discovery or suggestion is exploited to invent a novel technological capability which had the potential for development as a new product. Research on the potential capability is advanced to breadboard model or primary or experimental development stage when a laboratory scale demonstration of the technical viability of innovation is achieved. If cost and market analyses indicate that the potential innovation has at least some minimum potential innovation has at least some minimum potential for commercial success, work can proceed to the next phase in the process.

3.7 ADVANCED DEVELOPMENT AND DESIGN ENGINEERING

The project now moves from a very small-scale laboratory realization towards increases in scale and/or quantities produced, dependent upon the technology. It may involve two or more development steps in which increasing attention is given to cost and design considerations, both from the viewpoint of economy and ease of manufacturing/producing and market appeal. It may also be performed in a subsidiary rather than central R&D facility (see Chapter 5).

3.8 PILOT OR PROTOTYPE PRODUCTION

The project now moves from an R&D to a production environment. Production is scaled-up to the pilot run or initial prototype levels. The move to a profit-oriented production environment provides an improved test of the viability of the project against more demanding cost and time deadlines. As with development, it may incorporate two or more scaling steps in which increasing outputs are test marketed to selected customers and their responses evaluated.

3.9 FULL PRODUCTION

The final scale-up to full production is made. The technical feasibility of the project is proven and, given that the product is accepted by the market, it becomes an addition to the product lines of the company.

3.10 MARKET DEVELOPMENT

To ensure that the product satisfies a latent or manifest user need, parallel marketing activities should be undertaken. The market-product concept may be initially de-

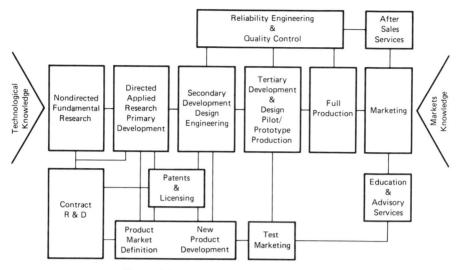

Figure 3.2. The technological innovation base.

fined at phase 1 and certainly by phase 4 of the innovation process. Thereafter, a technical marketing function will design and develop a marketing strategy in conjunction with the technical product development, design, and production. These activities are indicated in the third column of Table 3.1 and discussed later in Chapter 6 (Section 6.5).

The sequence "Nondirected Research to Full Production" represents what might be called the central sequence in the innovation process, with each stage in the sequence requiring varying technological needs. These stages may be viewed as the central components in the technological innovation base as shown in Figure 3.2, but for innovation to be successful, several other supporting components are required and these are now considered.

3.11 PATENTS, COPYRIGHTS, AND OUT-LICENSING SERVICES

The abilities of patents and copyrights to prevent the unlicensed copying of an invention is technology dependent. Patents are most effective in the chemical and pharmaceutical industries and the illicit copying of books and software is currently a concern in the publishing and computer industries.

Where the patenting and licensing of inventions is appropriate, their roles should be developed within the context of the overall technological strategy of the company. It is more or less standard practice for a patent application to be filed at phase 3 of the innovation process if and when "inventive merit" can be demonstrated. Patents may be used either to prevent competitors from marketing the same innovations or as bargaining counters in negotiations with such competitors as potential partners. For example, if a pharmaceutical company has discovered an NCE which offers promise as a potential new drug, it will typically patent a number of other

NCEs with closely related chemical structures to provide a barricade of patent protection for the chemical molecule that apparently possesses the desired pharmacological potency. Patent protection in the chemical, pharmaceutical, and food/beverage industries is also often supported by trade secrets, that is, the know-how that goes into the preparation of products and trademarks or brand labels. *Coca-Cola* and *Drambuie* (the liqueur scotch whiskey) are examples of world renowned products that are protected in this way. Trademarks are particularly useful as protection when the patent on an invention has lapsed, as illustrated in Searle's use of NutraSweet to describe aspartame.

The judicious use of licensing is illustrated in the manner the Boots Company of England marketed its successful antiathritic prescription drug *ibuprofen* in the U.S. market, which contrasts with EMI experiences in a quite similar situation described in Chapter 1.[11] Boots had no manufacturing or marketing operations in the United States, so it granted a nonexclusive license to Upjohn to sell the drug in the, as yet, unproven U.S. market. It proved to be a "winner," so Boots later set up its own U.S. company to manufactureand sell the drug. Ibuprofen also proved to be a potent general purpose anaelgesic. Hence, in 1987, when the FDA granted approval for its sale as an over-the-counter (OTC) or nonprescription anaelgesic in the United States, Boots again did wish to enter the very competitive OTC market in the United States, so it granted a license to American Home Products to sell the nonprescription version of the drug as *Advil*.* In- as well as out-licensing can play an important role in the development of new technology-based companies. Forrest and Martin studied a sample of 40 small U.S. NBFs and found that just over half of them engaged in this activity.[12] One company reported 80 in-licensing and 20 out-licensing agreements!

3.12   QUALITY, RELIABILITY, AND TECHNICAL AFTER-SALES SERVICES

Many high-technology products, even though they are manufactured to the highest quality and reliability standards, may require significant after-sales servicing. This factor must be considered in relation to manufacturing and product design as well as marketing and distribution decisions. For example, Motorola designed the circuitry of their domestic color TV sets in modules to enable the service repairmen to replace faulty modules easily, and then promoted sales of the product by emphasizing ease of repairs and servicing using the *works-in-a-drawer* advertising slogan. Similarly, Maytag emphasized the reliability of their washing machines on TV commercials which humorously focused on the boredom of their repairmen through lack of repair work.

Aspirin and *paracetemol* are the only other nonprescription anaelgesics approved for sale in the United States. Ibuprofen is also sold as an OTC anaelgesic elsewhere, including Britain (as *Nurofen*) and Canada (as *Actiprofen*).

3.13 EDUCATION, TRAINING, AND ADVISORY INFORMATION SERVICES

Ancilliary to after-sales services is the provision of before and after services which educate and advise the distributors and/or sales agents (if used), as well as customers, on the use of the product. The provision of such services can represent a substantial proportion of the product concept and its cost. This can be seen in the aerospace industry, where companies such as Boeing, Airbus, and McDonnell-Douglas must expend considerable effort in training airline staffs in the use of new generations of aircraft. It is manifest similarly in the computer and telecommunications industries where the changing requirements of the technology dictate that the manufacturers are required to provide very extensive ranges of training and advisory services.

3.14 COMPLEMENTARY ASSETS

The experiences of Boots and the biotechnology companies cited earlier illustrate another important point. Companies often lack the comprehensive range of capabilities to exploit an innovation. Boots lacked a U.S. facility, so it entered into agreements with first Upjohn and then American House Products with the distribution and marketing capabilities to sell ibuprofen. Also, 36 of the 42 NBFs in the above-cited study entered into manufacturing and/or marketing agreements to sell their products with larger firms, again because they lacked their own facilities. Companies enter into such agreements to gain access to Teece's complementary assets.[13] For example, computer hardware manufacturers collaborate with software houses when launching new products, as in the IBM and Microsoft relationship cited above. In some industries, suppliers and customers provide the complementary assets needed to design the innovation *jointly* with the company "producing" it, as von Hippel's work (to be outlined in Chapter 6) shows.[14] For example, Gardiner and Rothwell describe the producer–user collaboration between Boeing and Pan Am in the design and development of the 747 jumbo jet, which will again be outlined in Chapter 6.[15]

3.15 MANAGING THE TECHNOLOGICAL INNOVATION BASE

The above sections have outlined what maybe viewed as the components of the technological innovation base in the company. It can be seen that these are classified more or less independently of the conventional terminologies of finance, marketing, personnel, and production. Although these components may each lie within or impinge upon one or more of the above functions (as is illustrated in Figure 3.2), it is argued that they provide a more meaningful framework for managing the technological innovation process in a company. They might be viewed as the components which primarily sustain the dynamic process of technological change within the company and its markets, in contrast to the static manufacturing and marketing operation based upon unchanging technological capabilities.

Some (particularly small) companies, which may be either craft-based on exploring and serving relatively undemanding specialized or local markets, may be able to survive based upon a fairly static capability. However, most high-technology companies face a continuing and exacting demand to cope with technological change. Therefore, a holistic view of the dynamic technological innovation base should be taken if the company is to survive and compete in its changing technological environment. Also, as Adler and Shenbar argue, the manner of its competitiveness is reflected in the mode of management of the components defined above.[16] In Chapter 4, we define the spectrum of alternative technological strategies available to the company, and how they affect the structure and management of the components of its technological innovation base.

REFERENCES

1. M. E. Porter, "Technology and Competitive Advantage." *Journal of Business Strategy* **5**(3), 60–78 (1985).

2. D. A. Hounshell and J. K. Smith, Jr., *Science and Corporate Strategy: DuPont R&D, 1902–1982.* Cambridge, UK: Cambridge University Press, 1988.

3. G. E. Pake, "Research to Innovation at Xerox." In R. S. Rosenbloom (Ed.), *Research on Technological Innovation, Management and Policy.* Greenwich, CT: JAI Press, p. 21.

4. M. Giget, "The Bonsai Trees of Japanese Industry." *Futures* **20**(2), 147–154 (1988).

5. P. Abetti, *Linking Technology and Business Strategy.* New York: American Management Association, 1989.

6. P. S. Adler, "Technology Strategy: A Guide to the Literatures." In R. S. Rosenbloom (Ed.), *Research on Technological Innovation, Management and Policy.* Greenwich, CT: JAI Press.

7. M. Uenohara, "A Management View of Japanese Corporate R&D." *Research Technology Management* **34**(6), 17–23 (1991).

8. E. Koerner, "GE's High-Tech Strategy." *Long Range Planning* **22**(4), 11–19 (1989).

9. W. G. Cutler, "Acquiring Technology from Outside." *Research Technology Management* **34**(3), 11–18 (1991).

10. F. Sen and A. H. Rubenstein, "An Exploration of Factors Affecting the Integration of In-House R&D with External Technology Acquisition Strategies of a Firm." *IEEE Transactions on Engineering Management* **37**(4), 246–258 (1990).

11. M. J. C. Martin, "The Boots Company PLC: The Ibuprofen Project," Unpublished case and videocassette. Halifax, Can.: Centre for International Business Studies, Dalhousie University, 1988.

12. J. E. Forrest and M. J. C. Martin, "Strategic Alliances: Lessons from the Biotechnology Industry." *Engineering Management Journal* **2**(1), 13–21 (1990).

13. D. J. Teece, "Profiting from Technological Innovation: Implications for Integration, Collaboration, Licensing and Public Policy." In D. J. Teece (Ed.), *The Competitive Challenge: Strategies for Industrial Innovation and Renewal.* Cambridge, MA: Ballinger, 1987, pp. 185–220.

14. E. Von Hippel, *The Sources of Innovation.* New York: Oxford University Press, 1988.
15. P. Gardiner and R. Rothwell, "Tough Customers: Good Designs." *Design* **6**(1), 7–17 (1985).
16. P. S. Adler and A. Shenbar, "Adapting Your Technological Base: The Organizational Challenge." *Sloan Management Review* **25,** 25–36 (1990).

____4
TECHNOLOGICAL INNOVATION MANAGEMENT: PLANNING AND STRATEGIES

A prince or general can best demonstrate his genius by managing a campaign exactly to suit his objectives and his resources, doing neither too much nor too little. But the effects of genius show not so much in novel forms of action as in the ultimate success of the whole. What we should admire is the accurate fulfillment of the unspoken assumptions, the smooth harmony of the whole activity, which only become evident in final success.

CARL VON CLAUSEWITZ, *ON WAR*, BOOK THREE, CHAPTER 1, "STRATEGY"

4.1 INTRODUCTION

As we saw in Chapter 2, whether they are ultimately profitable or otherwise, all innovative products sooner or later "die," so a high-technology company must maintain a continued succession of new products if it is not to die with its obsolescent innovations. That is, quite simply, a company . . .

> . . . must "innovate or die." The process of innovation is fundamental to a healthy and viable organization. Those who do not innovate ultimately fail.[1]

These words were spoken by a senior manager in the food industry, and were later echoed by Paul Cook, the founder and CEO of technology-intensive Raychem Corporation:

> We get innovation at Raychem because our corporate strategy is premised on it. Without innovation we die.[2]

The process of technological innovation is central to the survival of the firm, but is, as we have seen, both complex and chaotic. As we indicated in Chapter 2, companies which view technological innovation as a simplistic linear process in which R&D inventions are PUSHED through to manufacturing and marketing are likely to

67

fail commercially. Rather, innovation should be viewed as a process whereby needs or unoccupied niches in the rapidly changing technosphere match the technological capabilities and aspirations of the organization—that is, the technology–market synergies are identified and exploited.

Ideally, this cybernetic open-systems process should be reflected in the organization's technological plans it formulates, the strategies it pursues, and the climate it seeks to create. The purpose of this chapter is to take an overview of the considerations required to formulate such plans and strategies and to embed them in the organization's technological innovation base and culture. The balance of the book is essentially concerned with examining many of these considerations in further detail.

4.2 GENERAL CONSIDERATIONS—THE INNOVATION MANAGEMENT FUNCTION

Schematically, we may extend the cash-flow profile of Figure 2.3 in time to incorporate a sequence or time series of innovations which maintain a steady positive cash-flow for the company, as shown in Figure 4.1. Obviously, this is an idealized and simplified treatment of how products are phased in and out of a company's portfolio in real life, but it suffices for the present purposes of exposition. Traditionally, we may view current production and marketing management as being involved in phase 5 of the innovation process onwards; that is, the production, distribution, selling, and servicing of a technologically (although not necessarily commercially) proven and adopted product. Equally traditionally, we may view the earlier stages of the innovation process (insofar as they occur within the company) as being within the purview of R&D management. The open-systems view of innovations as exemplified in Figure 2.10 clearly violates this linear managerial sequence. It suggests that we might benefit from open-systems management structures and "thinking" which reflect the nature of the innovation process; that is, that

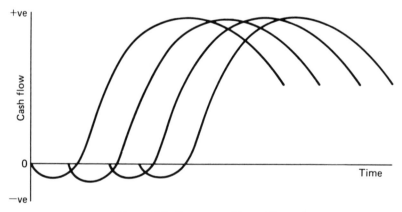

Figure 4.1. Sequence of innovations.

innovation should be a conscious explicit and *accountable* concern of all managers in the company and a permanent item on the agendas of intermanagerial and inter-departmental meetings.

In practice, such a viewpoint is utopian. Most, if not all, of the time and energy of many line managers are likely to be consumed in the day-to-day pressures of making and selling the company's current product range, leaving them little or no time to mediate on and develop new product ideas. Given that sustained technological innovations, implemented using an open-systems (as against technology-push) approach, is essential for corporate survival, it must be the *explicit and major concern* of *some* managers in the company. Reflecting the ideas of Wills,[3] we suggest that a high-technology company requires a dual structure reflecting its future–present or innovations–operations orientations, conceptually illustrated in Figure 4.2. Such dualism is reflected in the financial reporting conventions of some companies. For example, the Hewlett-Packard Annual Report identifies the share of current revenues that can be assigned to products introduced in *each* of the last five years.[4] It is therefore useful to invoke a nontraditional management function, called the *Innovation Management Function (IMF)*—in contrast to traditional operational functions such as production, marketing, finance, and personnel. This IMF can be closely identified with the R&D function and new product development teams, but as we shall see later, it needs to exercise a pervasive influence throughout the

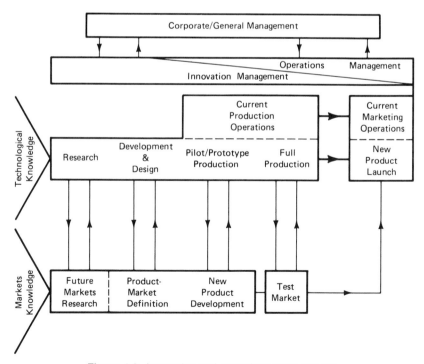

Figure 4.2. Innovation and operations management.

organization if a healthy innovation policy is to be sustained. Given its strategic importance, the function should report to a senior management group, which we call the *Innovation Steering Management Committee (IMSC)*. The IMSC should have a multifunctional membership (from R&D, production, marketing, etc.) and be chaired by the Chief Executive Officer (CEO) or Chief Technology Office (CTO) of the firm.

4.3   TECHNOLOGY PLANNING

It follows that technological innovation management should be viewed as central to corporate and business planning since, in high-technology firms, *business and corporate plans* are really synonymous with *technological innovation plans*. Despite the fact that technology is recognized as a potent source of competitive advantage, until recently, R&D, the prime source of new technology, has been managed in functional isolation rather than within a holistic framework integrating R&D and corporate or business goals. Roussel *et al.*[5] discuss the historical evolution of the changing status of the R&D function in the firm and argue in favor of *third generation R&D*. In first and second generation companies, R&D is treated as an expense in the corporate annual report. The R&D function is largely left "to do its own thing" and generate a smorgasbord of new knowledge and *inventions,* which may or may not be of value to the firm. In contrast, the third generation company has an organizational culture in which R&D planning is embedded in corporate and business planning, so that it generates *innovations* congruent with business and corporate strategies. That is, it is expanding the *core wealth-creating technology competencies* of the firm, rather than just expanding its knowledge base. As Matthews points out, the management of R&D is the process of *converting wealth into knowledge,* while the management of technology is that of *converting knowledge back into wealth.*[6] Clearly, the wealth to knowledge to further wealth creation spiral should be integrated if technology is to be a continued source of competitive advantage and growth for the firm.

Technology Planning at Business and Corporate Levels

High-technology firms range from the small start-up discussed in Chapter 11 to large organizations integrating numerous individual businesses or strategic business units (SBUs) into an overall corporate structure. In stressing the importance of integrating technology considerations in strategy and planning, one must distinguish between the roles of technology in planning at SBU and corporate levels in technologically diversified multibusiness corporations. As Steele (formerly Staff Executive for Corporate Technology Planning in GE) points out, an individual SBU competes in a market by offering superior value to customers that may be based upon superior technology.[7] Therefore, technology needs to be explicitly considered as a determinant of competitive advantage, in say the design of a new product. In contrast, he argues, a multibusiness corporation competes in the capitals market where it is valued by the return offered by its portfolio of SBUs. Therefore, at the corporate level, a portfolio analysis approach applies in which the financial appeal of technological possibilities and opportunities are considered, for example, when deciding whether to diversify into new areas and buy or sell businesses. Koerner reports that at GE, the evaluation of the potential competitive advantage to be derived from a particular technology is made at the individual business unit rather than the corporate level.[8] She quotes GE's vice president for corporate business development and planning as stating: *"The best context for a discussion on the impact of technology on a particularly competitive environment is the individual business."* This argument does not, of course, preclude substantial investment in corporate level R&D activities as a strategic resource to support the technology development activities of individual SBUs which share common core technologies, as is exemplified in GE's world renowned Schenectady Laboratories.[9]

The above argument may be applicable to technology planning in a multibusiness corporation with a diversified technological innovation base, lacking any common or core technologies. However, as Prahalad and Hamel argue, many corporations have grown by developing their core technologies and competencies across a number of SBUs.[10] 3M's growth has been fueled by its competencies in the technologies of coating materials coupled with its intrapreneurial culture which stimulates the creation and growth of SBUs (see Chapter 12). Canon, Honda, and NEC are three Japanese corporations with impressive growth records based upon their core competencies in microprocessor-controlled optical imaging, engines and power trains, and VSLI and systems integration, respectively. DuPont and Exxon are other examples of successful mature corporations that have grown through pursuing their largely chemistry-based core competencies, while Harris Corporation successfully managed a transition from mechanical to electronic core competencies over some 25 years from the 1950s to the 1970s.[11] Recalling the bonzai tree metaphor of Chapter 3, we may view the core technologies as the trunks of the tree, with SBUs being branches, and its leaves and fruits being its products.

It follows from this argument that technology planning needs to be performed at both corporate and SBU levels, with iterative cycling between the two of them. It also implies that IMSCs, as defined in Section 4.2, should exist at both individual SBU and corporate levels (perhaps with overlapping memberships), to oversee and participate in both the SBU and corporate technology planning processes. At the corporate level, the planning focuses on portfolio analyses to determine which technologies and SBUs are worthy of the corporate investments and the technological strategies that should be pursued with them. Logic, therefore, dictates that the technology planning process should be addressed at the SBU level before it is reviewed at the corporate level. Therefore, SBU technology planning will first be discussed in some detail before addressing it at the latter level.

4.4 SBU TECHNOLOGY PLANNING

The formulation of a technology plan and strategy can be viewed metaphorically as the merging of currents from both the external and internal environments of the SBU into a confluence or tide, which may be exploited through its technological navigation or steersmanship. These currents are illustrated on the left-hand side of Figure 4.3. At any time, the firm will be presented with a nexus of opportunities and threats implicit in its generic technological and industrial environment or technosphere. The extent to which it is able to meet and profitably exploit this nexus is as function of the stage reached in the industry life cycle, economic climate, and current societal values and concerns on the one hand and the firm's historically rooted technological, marketing, and other capabilities on the other. The notional matching of societal wants and needs to the SBU's historically evolved capabilities and aspirations generates its goals. These goals may be expressed in societal and economic rather than technological terms, but in practice, in a high-technology company they will be achieved largely through the pursuit of specific technology-based plans and strategies.

This exercise of technological entrepreneurship is essentially based upon matching evolving technological possibilities and capabilities to evolving market needs and opportunities—that is, identifying and exploiting technology–market synergies or occupying profitable niches or segments in the continually evolving technosphere. This consideration may be expressed in the two dimensions of a technology–market or technological opportunity matrix as shown on the right in Figure 4.3. The technology rows of the matrix distinguish between present product lines or technology, and a new technology continuum ranging from cost-reducing incremental innovations in present products for present markets to revolutionary new products for new introduction to the world markets (such as the microcomputer in the 1970s). The columns differentiate between present markets and a new markets continuum

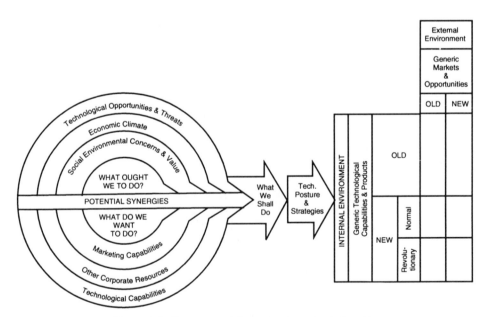

Figure 4.3. Strategy and the technology–market matrix.

ranging from new but already familiar to the firm to new to the world (again, such as the microcomputer in the 1970s).[12]

Technology planning can be seen as consisting of a set of exercises to:

1. Forecast evolving technological possibilities and capabilities together with evolving market needs and opportunities.
2. Disaggregate this technology–market matrix into its component submatrices and to assess the firm's present and future competitive strengths in order to identify potential future technology–market synergies or options.
3. Formulate a technological innovation mission or plan, based upon a selection from these options.

Given the need to integrate it into the business planning process, R&D or technology planning requires the joint participations of R&D staff and other functional managers. Such participation may be best achieved through the use of independent third parties (such as external consultants) who act as facilitators, with an overall framework.* Several writers suggest approaches to this exercise; see, for example, Burgelman *et al.*,[13] Adler *et al.*,[14] Steele,[15] and Contractor and Narayanan,[16] as well as Roussel *et al.*[17] In this context, it is vital to view all scientific and engineering R&D staff as line managers in that function and to invite their participation in the planning process. Given that technology planning should be overseen by the SBU's IMSC, it might be taken to imply that it should be entirely top-down, but this is not so. R&D and other functional staff at all levels who are monitoring the states of the arts in their respective technologies and markets should, and probably will wish to, participate in the process. This consideration is illustrated by arrows in Figure 4.4. Technology planning should involve top-down, bottom-up, and sideways participation. Both Steele and another senior technology manager, Mitchell (Director of Planning for GTE Laboratories), recognize that a major impediment to the effective integration of R&D inputs into business planning is the credibility gap between many R&D staff and managers.[15,18] R&D staff think in terms of technology inputs, that is, the knowledge and skills required to build the product, while managers think in terms of the commercial viability and attractiveness of the product itself. *Technology planning involves the recasting of technical terms and objectives into business terms and objectives.* Therefore, by its very nature, such an exercise helps to reduce the above credibility gap and contribute to the personal developments of R&D staff since it helps them translate their knowhows into business know-hows, an important concern that will be discussed in Chapter 13.

Forecasting Technological and Market Trends and Changes

This exercise provides a backdrop for technology planning in the form of a scenario of the social, economic, political, industrial, and technological trends and changes

---

*For example Pugh-Roberts Associates in Boston, MA act as facilitators in technology planning for client companies.

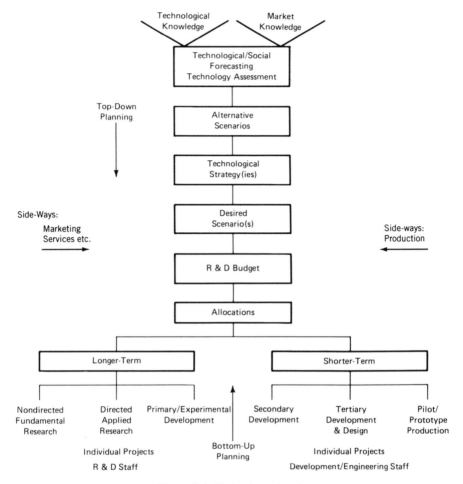

Figure 4.4. Technology planning.

that can be anticipated in the firm's future environment. The exercise may possibly extend over the next 25 years or longer, which is a realistic planning horizon in some industries. It will be performed by the technological forecasting function of the SBU or corporation if it exists. The approaches and techniques of technological forecasting are described in Appendix 4.1. If the firm is too small to justify the cost of establishing such a function, the exercise can be performed in an informal manner by the people cited in the previous paragraph since its requirements may well be satisfied by much more focused environmental study and shorter planning horizon than for a larger firm. The exercise should identify the levels of maturity and potencies of the core technologies of the SBU, as well as forecast changes in them over the future planning horizon.

Defining Technological Life Cycles and Potencies

To begin with, the technologies in which the business is rooted need to be identified and evaluated, both in terms of their maturities and potencies, as sources of competitive advantage. Arthur D. Little provided the following framework for this evaluation, based upon the technological trajectory and S-shaped curve (Chapter 2, Section 2.5 and Appendix 4.1, Section A4.1.4).[19] The S-shaped trajectory or life cycle of a given technology may be divided into the four stages, as shown in Figure 4.5, which offer decreasing risk and reward opportunities. They essentially correspond to the Abernathy–Utterback treatment, except that their mature state is divided into *mature* and *aging* stages. Wilkinson describes it in the *Curve of Maturity* and suggests that different strategic considerations apply in each of the stages.[20]

In its *embryonic* stage, knowledge of the technology and its potential market applications is limited, so that it faces an uncertain or murky future. R&D investment to reduce technological uncertainty can be viewed as speculative, but if successful, could yield substantial rewards. If successful, the technology moves into its rapid *growth* stage, when such rewards begin to be realized. As technological knowledge accumulates, approaching an inherent upper limit, the growth rate declines and the technology enters the *mature* stage. Now, rewards from R&D will be both much more predictable and modest. Finally, the technology reaches the upper limit of its potential in the *aging* stage and significant advances can no longer be expected. Roussel *et al.*[21] suggest that the technologies of the thermal pasteurization of milk and glass container manufacturing fall into this category. The role of the S-shaped technology maturity curve in strategy is discussed later in this chapter (Section 4.7) and in Appendix 4.1 (Section A4.11.6).

In practice, the technological innovation bases of firms may embrace several technologies of varying maturities, and we can categorize them in terms of their competitive values as follows:

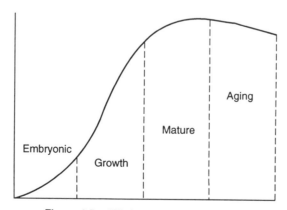

Figure 4.5. "S"-shaped maturity curve.

1. *Base Technologies.* This is commonly available, mature scientific and engineering knowledge that provides the business bedrock or foundation, such as the underlying science of silicon IC technology. It offers no competitive advantage because it is common knowledge.

2. *Key Technologies.* The knowhow that provides the immediate source of technical leverage and competitive advantage. This is the uncommon and often tacit inventive knowledge possessed by creative R&D, design, and manufacturing staff that generates the next new product, such as a new generation silicon DRAM. Note that it is not necessarily exclusively located in the R&D function. Jelinek and Schoonhoven[22] suggest that a key source of competitive advantage in the semiconductor industry nowadays lies in the linkages between development and production of new ICs and the manufacturing end-game.

3. *Pacing Technologies.* Those technologies entering the growth stages of their life cycles and beginning to replace the current key technology. The germanium transistor was the original key technology in the semiconductor industry of the 1950s. With the invention of the IC, the silicon transistor and IC became the pacing and then key technology in the 1960s. Gallium arsenide transistors and ICs are replacing silicon ones in some applications. The timing and scope of future impacts of pacing technologies are unknown. Gallium arsenide has been considered as a potential replacement for silicon for 30 years but, in 1992, had still not been substituted for silicon in most applications. Therefore, although it may not be necessary to invest in active R&D in a pacing technology, it is important to monitor its progress continuously to avoid being left behind by an unexpected breakthrough in the field.

4. *Emerging Technologies.* Those technologies in the embryonic stages of their life cycles that may have a revolutionary impact in the future. Molecular electronic devices may ultimately replace ICs but their adoption is not anticipated until the next century. Therefore, research in molecular electronics can be confined to universities, government, and a few large industrial laboratories, while other companies monitor progress informally since it is unlikely to have a commercial impact within the next 10 years. Others include nanotechnology, high-temperature superconductivity, and hot nuclear fusion which has been in the embryonic stage for the past 40 years. Because of its high cost, research into hot nuclear fusion is largely conducted by government-supported international consortia. It is justified by the potentially immense benefits, particularly if coupled with high-temperature superconducting materials, offered by the cheap electrical power generated and distributed by such means.

The preceding categorization is best performed in-house as part of the forecasting exercise outlined below since it should be company–market specific. Note that the maturity of the technology does not necessarily equate with its competitive potency. This, too, can be market specific. For example, the new biotechnology of rDNA and genetic engineering is still viewed as embryonic or emerging by some companies since it is as yet uncertain whether it will prove to be viable in their markets.

However, progress in developing new biotechnology-based products and processes has been sufficient to categorize it as a pacing technology in some markets and a possibly key technology in others.

Present SWOT Analysis: Defining Current Core Technologies, Products, Markets, and Competition

As many reachers may already know, the acronym SWOT stands for Strengths, Weaknesses, Opportunities, and Threats. The *present* technology–market matrix (Figure 4.3) may be disaggregated into a number of submatrices, as shown in Figure 4.6. The current products and markets can be defined, together with their supporting core technologies, manufacturing facilities and methods, product lines, market segments, and distribution/servicing channels. The features such as price/cost, features/performance, quality/reliability, etc. that may be viewed as sources of competitive advantage are identified. Present and potential future competitors are also identified and their strengths/weaknesses assessed and compared with the firm, to evaluate its present strengths, weaknesses, opportunities, and threats.

The above exercises, taken *in toto,* constitute a *technology audit* of the firm.

Future SWOT Analysis: Formulating the Technology Plan. By now, the participants from the various functional areas will have exchanged and examined a considerable range of ideas and should be sharing a common mind-set. That is, the recasting of technical terms and objectives into business terms and objectives, as suggested earlier, should have largely, if not totally, been accomplished. Therefore,

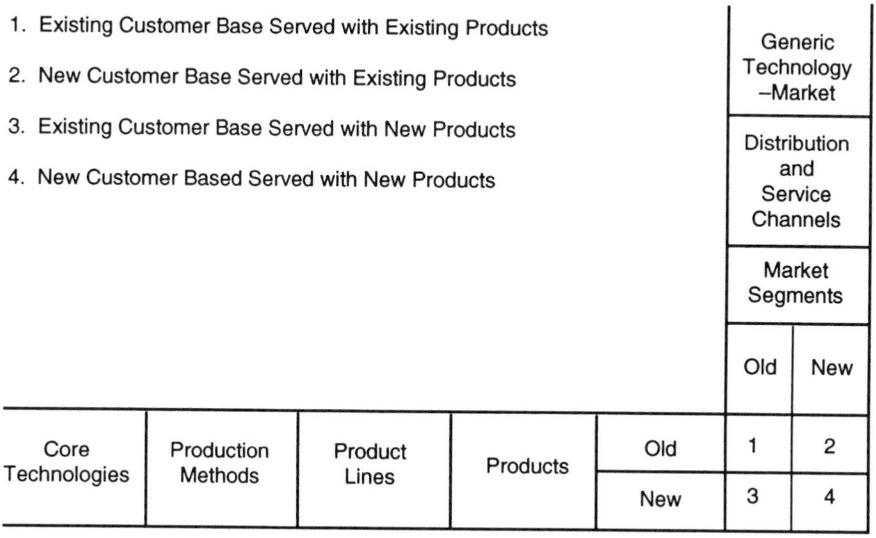

1. Existing Customer Base Served with Existing Products

2. New Customer Base Served with Existing Products

3. Existing Customer Base Served with New Products

4. New Customer Based Served with New Products

					Generic Technology–Market	
					Distribution and Service Channels	
					Market Segments	
					Old	New
Core Technologies	Production Methods	Product Lines	Products	Old	1	2
				New	3	4

Figure 4.6. Technology–market submatrices.

the above evaluation of the current situation, when combined with the forecasting exercises indicated above and based upon the techniques described in Appendix 4.1, should logically and psychologically stimulate a future SWOT analysis in which *alternative future scenarios* are explored and alternative technology options and target markets are evaluated over an appropriate planning horizon. This planning horizon may extend to 25 years in some industries, such as aerospace. A technology plan, based upon the options chosen and embedded in a business plan, can then be formulated. It will provide the basis for autonomous activity in the firm's internal R&D and technology acquisition and alliance strategies (to be discussed in a later chapter), as well as its business strategy. In the SBU of a larger corporation, it will be subject to review and possibly further iterative refinements at the corporate level. Werner Niefer (Chairman of the Managing Board, Mercedes-Benz AG., Germany) describes his corporation's approach to technology planning, reflecting many of the considerations discussed in this section.[23]

4.5 CORPORATE TECHNOLOGY PORTFOLIO ANALYSIS

As indicated at the end of Section 4.3, in a multibusiness corporation, the technology and business plans should obviously be reviewed at corporate level to ensure that they are congruent with corporate performance criteria, strategies, and goals. For the technologically diversified corporation, this evaluation may be primarily financial, as Steele suggests. For the corporation consisting of SBUs sharing common core technologies and perhaps markets, a more detailed review is required. First, it can be expected that any scenario generation exercises, as described in Appendix 4.1 (Section A4.1.18), will be performed by a future studies function located at corporate level, thereby providing a common base for the exercises performed in the SBUs. Therefore, it is likely that individual SBUs will have collaborated in developing their own technology and business plans, particularly if a corporate R&D function is a participant in the process. Thus, the corporate level process is likely to consist of a review of the individual SBU plans to ensure that they are mutually supportive (exploiting any potential synergies and economies of scale and scope), as well as ensuring that they are congruent with corporate goals.

A technique that is also used at the corporate level, for example, by Hitachi in evaluating research themes in its Central Research Laboratory, is *technology portfolio analysis*.[24] is essentially based on the approaches pioneered in corporate planning by the Boston Consulting Group, for which a useful tool is the *technology portfolio matrix* (Figure 4.7), which compares the future importance of a given technology with the corporation's strength in it relative to its competitors.[25,26] It is divided into four quadrants:

The *Bet* quadrant represents promising technologies in which the corporation is in a relatively strong competitive position. Clearly, the corporation should invest in SBUs developing this technology, following an offensive or defensive strategy (see next section).

The *Draw* quadrant represents promising technologies in which the corporation

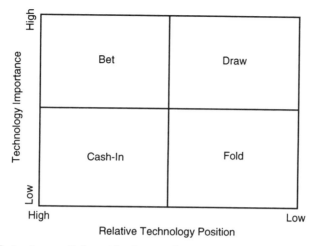

Figure 4.7. Technology portfolio matrix. *Source:* Reprinted from *Long Range Planning*, N. K. Sethi *et al.*, "Can Technology be Managed Strategically?" Pp. 96–97, Copyright 1985, with kind permission from Pergamon Press Ltd., Headington Hill Hall, Oxford OX3 OBW, U.K.

is in a relatively weak competitive position. Here, the corporation has two choices. Either it can invest substantially more in this technology, probably following an offensive strategy to strengthen its technology position, or divest its investment in it.

The *Cash-in* quadrant represents a technology in which the corporation is in a strong competitive position, but the market for it is small. This combination may reflect three alternative situations. First, an aging technology for which there is as declining market. Second, as rapidly evolving technology with a modest market and short product life cycles. Third, a niche market which may not warrant the attention of a large corporation. In all three situations, divestment may be preferable; otherwise, only modest investments should be made.

Finally, the *Fold* quadrant represents a weak position in an unimportant technology. Clearly, any investment in this quadrant should be divested as quickly as possible.

The technology investment matrix provides as means of identifying *technology investment priorities*, as shown in Figure 4.8, which is self-explanatory. Clearly, Technology A deserves further investment, Technologies B and C enjoy acceptable levels of investment, while Technology D should be abandoned, at minimum cost, as quickly as possible. 3M uses a very similar approach in evaluating its R&D programs.[27]

Some technology-based organizations face the challenge of diversification from a mature technology base and see a corporate research facility as a vehicle for assessing diversification opportunities. For example, to reduce its dependency upon steel-based products, Nippon Steel plans to diversify into new materials, chemicals, and biotechnology among other areas, from an R&D base.[28] Its central R&D organization also includes strategic planning, so technology portfolio analysis is a logical tool for use in evaluating its diversification opportunities. Isaac Barpal (Corporate

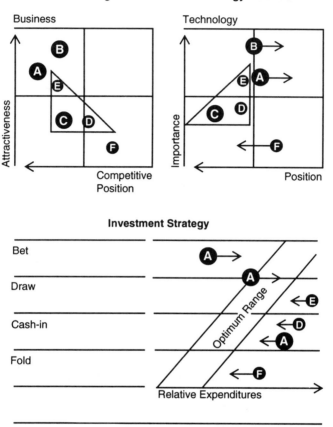

Figure 4.8. Technology–investment matrix. *Source:* Reprinted from *Long Range Planning,* N. K. Sethi *et al.,* "Can Technology be Managed Strategically?" Pp. 96–97, Copyright 1985, with kind permission from Pergamon Press Ltd., Headington Hill Hall, Oxford OX3 OBW, U.K.

Vice President for Science and Technology at Westinghouse) describes his corporation's use of the portfolio approach to provide a stronger technology base for corporate growth.[29] It led to the divestment of over 70 businesses and the acquisition of over 50 others, and a change in the role of the corporate R&D function from a technology-driven Research and Development Center to a business-driven Science and Technology Center.

Alternative scenarios and technology plans can be developed and evaluated at the corporate level, based upon an integration of those plans developed at SBU levels and the technology portfolio matrix, to constitute an overall corporate technology plan. Examples of corporations that have developed such plans are described by McDonald[30] and Klimstra and Raphael.[31]

4.6 TECHNOLOGY TIMING STRATEGIES

The technology-planning processes described in the previous two sections outline approaches to technology matrix and technology portfolio matrix mapping, but offer no explicit advice on the strategies to be employed in pursuing technology plans. The next few sections discuss alternative technological strategies in terms of the timing of technology entry into the market, together with the extent of its segmentation and specialization. The classification is based upon ideas developed by Ansoff and Stewart[32] and Freeman.[33] At the outset, it should be stressed that the classification scheme constitutes a continuum rather than a number of discrete types. Each strategy may shade into others, and a corporation or SBU may pursue one or more strategies simultaneously in different product–market areas. As Freeman suggests, the typology may be viewed as analogous to the famous psychologist Karl Gustave Jung's personality typology since a company, like an individual, is unlikely to conform to a single type, but rather be a blend of several types. In outlining them, we consider each strategy in turn in roughly decreasing order of technological and marketing challenge, relating it to the technology–market matrix, the third dimension of time, and the technological innovation base functions discussed in Chapter 3. We begin by describing the first of these—the offensive strategy, in some detail.

4.7 OFFENSIVE STRATEGY—LEADERSHIP OR "FIRST TO MARKET" POLICY

General Considerations

This is the most glamorous strategy, and can be identified with well-known successful high-technology firms such as IBM in computers, RCA in television, TI in semiconductors, and DuPont in chemicals. All these were the first to market the revolutionary innovations in their fields and could claim to be exercising technological leadership since they initiated industry life cycles. An example that is possibly less well known is Proctor and Gamble's leadership in performing the R&D that identified the role fluorides play in preventing tooth decay, despite the initial scepticism of the American Dental Association. By being first to market *Crest* toothpaste, P&G pioneered the addition of fluoride to toothpastes and water supplies to radically reduce the incidence of dental cavities in children in North America and elsewhere. Examples of unsuccessful attempts to achieve this leadership can also be quoted. The Comet airliner could be viewed as a revolutionary innovation which failed for technological reasons. In contrast, the Concorde appears to be technologically successful, but has failed for political, economic, and environmental reasons.

The introduction of a revolutionary new technology is a comparatively infrequent event and, having been first to market it, a company will almost certainly wish to maintain its technological and market leadership as the industry matures. It may thus seek to establish the dominant design in the industry and continue to be the first to market performance-maximizing (and later cost-reducing) incremental innovations. We saw that in the transition stage of the life cycle, the market is performance- rather than price-sensitive. It follows that companies can reap rich rewards by exploiting their technological virtuosity to produce a technologically superior product. The markets for the new technology are still not clearly defined, so these may be further explored. The ability to maintain such leadership and sustain a capability over some years demands a technological innovation management base which exemplifies *all around excellence*.

In Chapter 3 we defined the technological innovation management base of the company, represented schematically in Figure 3.2. We now discuss the implications that the pursuit of an offensive strategy has upon the structure and climate of this base.

R&D Requirements

As was suggested in Chapter 3 (Section 3.3), maintaining a nondirected research activity can be desirable if a company pursues this strategy since it provides the means to root itself in the scientific humus of its bonzai tree. Some companies that have unquestionably exercised a technological leadership role have also made outstanding contributions to fundamental knowledge, notably the Bell Laboratories of AT&T, until deregulation of the U.S. telecommunications industry.

Whether or not it performs basic research, a company can be expected to perform applied research, and this leads us to the important consideration of climate and communication in the R&D functions of offensive innovators. These are discussed more fully in the article cited earlier by Ansoff and Stewart, who suggest that such innovators should operate R&D functions which work in close proximity to the state of the art and are research- or R-intensive. These authors, quite correctly, distinguish between the states of the arts at different phases in the innovation process. They suggest that in research (corresponding to phases 1–3 in the innovation process), the state of the art denotes the frontier at which investigators seek to discover new phenomena or to devise new solutions to known problems. In contrast, they suggest that the state of the art in development (corresponding to phases 4 and 5 in the innovation process) denotes the situation where the validity of a theory or solution has already been proved, but has yet to yield a commercially successful application, thus focusing on both economics and technology. Thus, we may perceive a number of overlapping successive phases in the state of the art, corresponding to the successive phases of the innovation process.

Research in offensive innovators is characterized by scarcity of precedence and low stability and predictability. Working in situations of ignorance and uncertainty at the frontiers of knowledge means that neither R&D staff nor managers can readily rely on historically-based guidelines to assess the commercial viability of an innovation or to control activities. Also, such a company must anticipate rapid technical advances (either by itself or its competitors), leading to major improvements in the performances and/or reductions in costs of products. Thus, it must perpetually face the possibility of unpredictable opportunities to exploit, or threats to jeopardize its position in the marketplace.

The characteristics of an R-intensive organization are described by the above authors as follows:

1. *Nondirective Work Assignments and Indefinite Objectives which are Broadcast Widely.* At the early stages of the innovation process, the solution (and possibly the problem) is unknown, so alternative solutions must be searched for and evaluated. This implies nondirective work assignments and emphasis on individual contributions and scientific and technical insights, rather than on highly structured tasks and roles. It also implies the broadcasting of information on problems, market data, and possible solutions among technical staff to stimulate the generation of the widest range of possible solutions.

2. *Continuing Evaluations of Results and Swift Perception of Significant Out comes.* Given the fluidity and unpredictability of the work situation, alternative solutions can be expected to be continuously generated. At any time, an approach or project may be superceded by a superior one, arising either from within the group or possibly from a competitor's efforts. Technical management should be quick to perceive significant results and maintain a continuous review of project activities to permit quick switches in approaches in light of these results.

3. *Value Innovation Offers Efficiency.* Needless to say, because the objective is to generate new and markedly better products ahead of the competition, R&D efforts must be judged by their effectiveness rather than their efficiency.

What is clear from the remarks of these authors is that the R&D functions of offensive innovators should be staffed by highly able individuals who should be given considerable freedom to produce results. In other words, R&D management must practice *inspired adhocracy* rather than deadening bureaucracy! In fact, their recommendations, for what might be called Theory Y R&D management, are echoed in the seminal text by Burns and Stalker.[34] It is also illustrated in Kidder's engrossing description of the development of a new minicomputer in Data General Corporation.[35]

Pilot/Prototype Production

For the offensive strategy to be successful, this inspiration and momentum must be carried right through the innovation process, particularly given the increased emphasis of speeding up the process (see Chapter 8). If the company is to maintain its objective of being first to market an innovation, that innovation must be swept through the remaining stages as swiftly as effective problem solving allows. Although an offensive strategy implies an R-intensive company, this research emphasis is *not* at the expense of the development, design, manufacturing, and marketing functions of the firm. The company is R-intensive compared to companies following other strategies, as we shall see later in this chapter, but also requires equally intensive efforts to be made in successive stages of the process. As well as strengths in research and experimental development, it must have commensurate problem-solving talents in the design–engineering, pilot, and prototype development and testing stages and so on, often spread over a wide range of disciplines and skills. Because of the scarcity of precedence, problems may arise at, say, the pilot/prototype production stage which cannot be resolved by the "rule of thumb" methods. If the innovation is incorporating new science and technology, even at this stage problem-solving may require recourse to "scientific first principles." Therefore, the firm must have the ability to make this recourse. Examples of revolutionary innovations where this recourse is applied are Pilkington's development of the "float glass" process and IG Farben's development of PVC.[36] Scarcity of precedence also applies to craft (or technician) as well as science and engineering (or technologist) skills. The demand for welders who can work to the exacting standards imposed by deep-sea oil drilling programs is an example in this category.

Patents and Licensing

The same general considerations apply to what might be called the more peripheral functions in the technological innovation base. The company will probably wish to adopt a strong patent position to protect its technological leadership for as long as possible. This is not as self-seeking a characteristic as first appearances might indicate. A company pursuing an offensive strategy must be "technology-intensive" or T-intensive and invest a bigger proportion of its budget in R&D and related activities than the industry average. Despite its technological competence, it may also be expected to experience a high proportion of project failures, whether for technological or commercial reasons. Therefore, it will wish to protect its "winners" from the competition for as long as possible, to ensure that they accrue the maximum profits possible, which maybe invested in other projects. Recall the analogy with horse breeding, training, and racing suggested in Chapter 1. By adopting a strong patent position, the company is seeking to maximize the winnings of its successful "horses" so as to stay in the "breeding, training, and racing" business. Adopting a strong position implies filing a patent application as soon as feasibility allows, both for the primary invention and for any secondary and tertiary inventions which are developed as the innovation proceeds. These patents may then be used later as barriers or bargaining counters when competitors seek to enter the market. An astute licensing policy can also help maximize "winnings," particularly in an offshore market which the company does not wish to enter directly or where the patent protection law is weak.

Marketing, User Education, and Services

The roles of the user needs and services are equally, if not more, important. A revolutionary innovation, such as radio and the computer, can be viewed as a technology-push–market-pull synergy because it seeks to satisfy an unmanifest but, nevertheless, latent user need. Often, as with radio and the computer, the innovations are both technologically and socially revolutionary. Since the offensive innovator is the first to market such innovations, it means that considerable efforts must be made to ensure the reliability of the product in use and to educate users in their operation. A notable feature of the offensive innovators in the radio and computer industries is that they invested considerable efforts in setting up after-sales servicing networks and user training programs. Marconi and IBM are names associated with internationally famous schools for training wireless operators and computer programmers, respectively. Moreover, the process of training a user and servicing a piece of equipment leads to a more refined identification of user needs and of desirable incremental improvements in the equipment. During the performance-maximizing stage of the industry life cycle, the offensive innovator will be under pressure to maintain its technological and market leadership position by introducing incremental improvements in the innovation. Information feedback of user needs and problems to the R&D functions is an important role for the user needs and service functions, and training centers are usually located close to company R&D functions to encourage this communication.

This last sentence brings us to the last, but by no means least, point concerning the offensive strategy, which needs to be emphasized—what Ansoff and Stewart call the need for *high downstream coupling*. It is essentially identical with the technology-push versus market-pull distinction we made in Chapter 2. They argue that the implementation of an offensive strategy requires good coupling or communication among all the functions involved to ensure the swift identification and solution of problems, and they use a hydrological analogy to emphasize their point. Using our control engineering analogy, we express the same argument by suggesting that offensive innovation should be an open-system process replete with feedforward and feedback loops to ensure optimal goal-seeking behavior. That is, it is a cybernetic self-organizing system which has the ability to identify opportunities and threats in its environment, and to exploit and adapt to them before its competitors. As control engineers know, this responsiveness is a function both of the loop structures and of signal delays which occur in the system, since the system must exhibit a swift reaction time if it is to beat its competitors. Our earlier chemical process analogy or the innovation chain equation is also useful here. The offensive innovator seeks to complete the process before his competitors, so he needs to accelerate the chain reaction process which occurs. In chemical engineering, this could be achieved by raising the temperature and pressure. The offensive innovator need to achieve an analogous effect in the context of a managed institution, which is perhaps why such firms often have pressure-cooker climates.

Marketing Emergent Technologies

The above considerations are of general applicability to offensive innovators, but it is also worth noting the Japanese strategy used to launch as yet unproven emergent technologies.[37] North American and European companies attempt to apply such a technology to the *most difficult* problems and abandon it if it does not work. In contrast, Japanese companies first seek out *easy* applications in mass markets, such as consumer electronics. If successful, they enjoy large sales and the consequent manufacturing learning curve cost advantages enable them to drop prices faster than competitors. The strategy was used by Sony in marketing the first transistor radios (see Section 4.14) and the Walkman (Chapter 1). It was also used by Japanese companies (including Sony) in designing, manufacturing, and marketing low-cost domestic VCRs, after Ampex and RCA had pioneered the design and development of the professional product.[38]

Opportunistic or Maverick Innovators

Before concluding this section, a word must be said about one specialist type of innovator—the opportunistic–offensive *maverick* innovator. All technological innovation may be viewed as opportunistic since it involves seizing technico–business opportunities when they arise, and the successful technologically innovative company is one which effectively institutionalizes technological opportunism. However, not infrequently substantial, if not revolutionary, innovations are introduced by small companies set up specifically to seize the opportunity. Quite often, such companies are spun off from parent organizations, such as government, university laboratories, or large companies where the invention to be innovated was made. Such companies play an important role in the evolving technosphere since, if successful, they stimulate the adoption of the innovation, often against the reluctance of larger companies, and sometimes grow into large successful companies. The obvious recent example is Apple and the microcomputer. Another less well-known one from the 1950s is Wilkinson Sword Edge and the stainless steel razor blade. Wilkinson, a quite small British company that made stainless steel gardening tools, had also developed the capability of making stainless steel blades from its stainless steel cutting edges knowhow. By the 1950s, Gillette had the capability of developing a stainless steel razor blade, but declined the opportunity because the longer life of the stainless steep blade would reduce the total razor blade market. Since Wilkinson was not then in the razor blade business, it had nothing to lose from following a maverick strategy and invading this market with a stainless steep blade. Their new product was so enthusiastically received by shavers (including this author), who appreciated its superior performance every morning, that for a short time after its first introduction, supplies of these blades had to be rationed to retailers. The traditional manufacturers (Gillette and Schick) responded quickly with their own stainless steel blades, but by then, Wilkinson had established a permanent niche in the razor blade market through their maverick behavior. Wilkinson Razors has subsequently been purchased by Gillette.

4.8   DEFENSIVE STRATEGY—"FOLLOW-THE-LEADER" POLICY

The offensive strategy is fraught with risk and is rarely followed by a company indefinitely. Moreover, larger corporations are unlikely to encourage all their divisions to follow it at any one time, whether they are technologically homogeneous or otherwise. The technology life cycle model suggested that relatively good profit opportunities occur in the performance maximizing stage in the cycle when the innovation has been initially marketed, but the dominant design has yet to emerge. It also stressed that new product and company mortality rate can be high at this stage. These two considerations make a *defensive strategy* attractive.

The strengths of a defensive innovator is broadly identical to its offensive counterpart. Such an organization is likely to be equally technology intensive as its offensive counterpart, but with differences of emphasis. These differences are discussed here. The defensive innovator is averse to the risks of being first to market an innovation which is both technologically and commercially unproven. It reasons that if the innovation, when first marketed, looks like a "loser," it loses nothing. In contrast, if the innovation looks like a "winner," provided it swiftly follows the leader with its own (probably improved) version, it stands to win much of the spoils. The successful implementation of the strategy requires an organization with a technological innovation base which monitors technology–market opportunities actively and continuously, operates close to the state of the art in its successive phases, and is able to innovate swiftly. The differences of the defensive as opposed to the offensive innovator are as follows:

1. To maintain its close proximity to the fundamental research phase of the state of the art, the defensive innovator may undertake some nondirected fundamental research and, certainly, directed applied research. This research will be of a defensive rather than offensive character—that is, on topics which match the current research concerns of the offensive innovators. It ensures that the company has the autonomous scientific knowledge to exploit a new innovation once it appears to be successful.

2. Clearly, the defensive innovator must be strong in experimental development and design engineering and successive functions in the technology base since it needs to make up for lost time in marketing the innovation and manufacture a product with superior performance. The only area in which it may be able to place less emphasis is education, training, and advisory services. It maybe able to "piggy back" on the success of the offensive innovator in this area because the market will already have had some experience of using the innovation. Such services must still be given emphasis, however, since the offensive innovators will have penetrated only a relatively small part of the market, and substantial user education may still be required. Furthermore, dependent upon the technology, the first customers of the offensive innovator may have been locked into the latter's innovation, forcing the defensive innovator to seek out and educate different customers. This has been a notable feature of the computer industry.

3. One area whose function differs significantly between offensive and defensive innovators is patents and licensing. We have already argued that the offensive innovator will seek to establish a strong patent position to protect its technological dominance, a position which the defensive innovator must seek to subvert. The latter will therefore be required to develop its own patents wherever possible, and use them as bargaining counters to weaken the dominant position enjoyed by its

competitor. Needless to say, a cost of delayed market entry to the defensive innovator may be a much lower licensing revenue than its offensive counterpart.

In practice, as was stated at the beginning of this section, most of the larger multiproduct companies are likely to spread the risks endemic to innovation and enjoy economies of scale in R&D by following a mixed technological strategy— that is, be defensive in some areas and offensive in others. IBM has largely pursued an offensive strategy in the mainframe computer market, but followed an archetypal defensive strategy in its development of the IBM PC microcomputer in response to Apple's opportunistic–offensive strategy. Gillette also followed defensive strategies in response to the Japanese introduction of the fiber tip and, as already stated, the British introduction of the stainless steel razor blade into the U.S. market.

Historically, the French chemicals industry has followed a defensive strategy in deference to its German counterpart and, until recently, the European semiconductor industry largely followed a defensive strategy in deference to that of the United States. For example, after David Kilby of Texas Instruments made the first IC in the late 1950s, the major European companies quickly set up sections both to study the underlying chemistry, physics, and metallurgy of IC fabrication and to design and develop their own IC products. More recently, however, major European semiconductor manufacturers (Philips, Siemens, and Thomson) appeared to be pursuing offensive strategies by entering into joint ventures to develop the next generation of DRAMs.

The organizational structure and climate of a defensive innovator does not differ greatly from that of an offensive innovator. Both may operate with a comprehensive technological innovation base as defined in Chapter 3. We have already argued that the offensive innovator must possess high downstream coupling to secure and maintain its competitive position. If anything, this argument is *more applicable* to the defensive innovator from experimental development onwards, since success is dependent upon the ability to identify and produce improved versions of the innovation. This requirement places a high premium upon superior technological product development, marketing intelligence, and responsiveness.

4.9 DEFENSIVE VERSUS OFFENSIVE?
THE COUNTEROFFENSIVE STRATEGY

The merits of offensive over defensive strategies are discussed extensively by Foster.[39] He generally favors the offensive strategy, and cites numerous examples of firms successfully following such strategies in introducing new technology-based products into markets dominated by entrenched incumbents. However, Pavitt[40] argues that the choice of strategy should be dependent upon the firm's size and the nature of its accumulated technological competencies, with Wilkinson's[20] curve of maturity considerations also applicable. Teece[41] cites examples of successful and unsuccessful offensive strategies. Using EMI's experiences with CAT scanner as one of his examples, Teece suggests that a well-executed offensive or first-to-market strategy should be successful when:

1. The key inventive novelty embodied in the innovation can be effectively protected by patents, trade secrets, or other forms of tacit knowledge.
2. The offensive innovation constitutes the dominant design.
3. The offensive innovator has access to the complementary assets (such as competitive manufacturing, marketing, distribution, and services) needed to make and sell the product.

Even without the benefit of 20–20 hindsight, EMI's failure with its CAT scanner offensive strategy could have been anticipated. First, competitors were able to invent around its patents to produce rival machines. Second, the design paradigm remained fluid for some years as first brain and then total body scanners, with decreasing scan times, were developed and marketed. Therefore, EMI was unable to establish a dominant design. Third, in contrast to GE, EMI lacked all four of the complementary assets cited above. The IBM PC was also successful based upon a dominant design (as an alternative to the Apple IIE) and IBM's unrivaled complementary assets, whereas Osborne failed to maintain is position in the personal computer market, despite initial success.

The circumstances when a defensive strategy prevails over an offensive one are of particular interest to U.S. firms since, given the alluring size of their domestic market, they often have to defend it from the invasions of offshore firms following offensive strategies. Clearly, an offensive strategy should succeed when the firms launch new technology based upon impregnable patents, just as Polariod and Xerox did with their original cameras and document copiers, respectively. However, certainly outside of the chemicals/pharmaceuticals industries, such products are rarities. Foster includes as examples of successful offensive strategies Celanese's introduction of polyester as a tire cord against the entrenched position of DuPont's nylon, and Michelin's introduction of the radial ply tire into the U.S. markets against the entrenched position of the domestic cross ply tire makers. However, both of these successes can be at least partially attributed to the reluctance of the entrenched market incumbents to abandon their investments in the current technology. This reluctance is not unreasonable. As stated earlier, if the new technology proves to be a loser, the incumbent will lose nothing. If it proves to be a winner, provided it swiftly follows the leader with its own (probably improved) version, the incumbent can still gain much of the spoils. GE and IBM did precisely this, and followed very successful *counteroffensive strategies* in developing their own CAT scanning and PC capabilities, respectively. In contrast, both DuPont and the U.S. tire makers failed to mount a counterattack to meet the invasions into their established markets. This issue is also discussed briefly in Appendix 4.1 (Section A4.1.6), in the context of technological forecasting. The reasons why some firms are able to meet a technological threat with an effective response are probably managerial and cultural rather than technical, as discussed in Chapter 12.

4.10   IMITATIVE STRATEGY—"ME-TOO" POLICY

As an industry matures, a dominant design becomes established and the industry moves from a fluid to a transitional and then to a specific state, and other strategic options appear. As was stated earlier, the establishment of a dominant design tends to stimulate, delineate, and coalesce the market so that excellent opportunities are presented for incremental innovations or improvements in the dominant design, based more upon design, reliability, and cost considerations than upon radical technological differences.

This strategy is practiced in the pharmaceuticals industry by the generic drug firms which concentrate on manufacturing and marketing out-of-patent successful drugs or "winners" that are still extensively prescribed. Because they do not incur the high R&D expenditures of the major pharmaceutical firms associated with the search for and testing of effective new drugs, which includes many losers for every winner, generic firms can profitably manufacture such out-of-patent drugs at relatively low prices. IBM PC clone manufacturers also incur little R&D expense and often source components from low labor cost countries, so they are able to sell IBM-compatible microcomputers at a much lower price than the authentic product. The same strategy is pursued by others such as Amstrad, the U.K.-based company, in PCs and other market segments of the consumer electronics industry.

The imitative company will be development design, production, and service engineering intensive rather than R-intensive. It follows that its costs should be lower, except that it may well have to purchase the imitative technology through licenses and knowledge agreements with the primary innovators. This strategy can be particularly attractive to domestic companies in countries which traditionally lag behind the leading countries (such as the United States) in adopting new technology. If a U.S. primary innovator has no corporate presence in a given country, it may prefer to license the innovation to a domestic manufacturer rather than incur the costs, scarce resource investments, and risks associated with exploiting that market itself. The domestic manufacturer finds the arrangement equally appealing. Since it incurs no R&D expenditures, its direct manufacturing costs should be lower (dependent upon raw material, equipment, and labor costs in that country) and, particularly if it is protected by tariff barriers, profit opportunities may be excellent. Japanese companies followed this strategy very successfully after World War II, before some moved to defensive or even offensive strategies, via absorbent strategies (see below).

The imitative company possesses a truncated technological innovation base from design engineering onwards, as shown in Figure 4.9. If it does not enjoy a protected market, it will clearly have to be very efficient at this truncated operation. The primary innovators may still be able to produce technological improvements (particularly during the transitional as opposed to a specific state of industrial development), based upon their R&D capabilities. The imitative company can compete with only design improvements and lower manufacturing costs. This implies a design, production, and service engineering intensive company as was stated above. This difference in technological emphasis imposes different requirements on the manage-

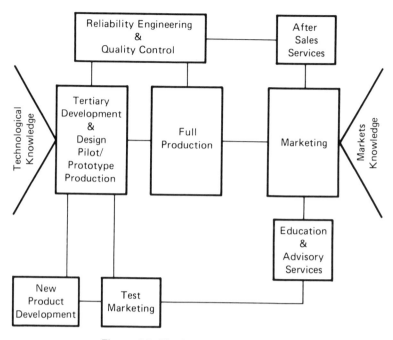

Figure 4.9. The imitative strategy base.

rial style and organizational climate of the company. Tasks are typically more clear-cut, since a major emphasis is placed upon the efficient enactment of a specific design concept. Supervision is more directive, individual and group tasks are structured to ensure the efficient dovetailing of the component activities, and established management techniques like PERT may be used. Thus, the organizational climates favor efficiency rather than innovativeness.

4.11   APPLICATIONS ENGINEERING OR INTERSTITIAL STRATEGY: TECHNOLOGICAL NICHEMANSHIP POLICY

Given the complex and kaleidoscopic nature of the technosphere, the primary innovators are unlikely to seek to satisfy every potential application or occupy every potential market niche for an innovation. A judicious analysis of the primary innovators' strengths, weaknesses, and strategies, combined with a search for unrealized applications will frequently identify specialty niches which can be profitably exploited. That is, other firms are able to identify and develop incremental innovations for new markets. Such companies are analogous to organisms which adapt to specialized niches in the ecosphere. Most of the firms in this category emphasize design and development. They can be properly described as applications engineers. However, there is as subcategory of firms which warrants special mention, which may invest substantial effort in primary development and follow an interstitial strategy. These companies also identify a niche or interstice in the market, which is too small to interest the larger primary innovators, but is still quite large and technologically demanding. One of the most notable examples of such a company is Control Data Corporation (CDC) which was able to market computer systems tailored to users whose needs could not be satisfied by systems available from the IBM product line. CDC was thus able to compete effectively in a U.S. computer market dominated by IBM without confronting the latter until later when it had built on the technology–market strength to do so. Its strategy at that time could be viewed as interstitial. Similar strategies have been followed by smaller companies in the aircraft industry which have been unwilling and unable to compete with Boeing, Douglas, and Lockheed. They may now be observed in the microelectronics industry, and are particularly attractive for companies located in countries with small domestic markets. Two Canadian examples are cited.

First, Nautical Electronics (Nautel), with plants in Maine and Nova Scotia, is a small company which was first to market all solid-state radio beacons in the early 1970s. Small radio beacons are needed at often remote and unmanned airstrips as homing and navigational aids to pilots. The company was founded in 1969 by several engineers who spun off from the subsidiary of a sonar equipment manufacture. At that time, all radio beacons used vacuum tubes, and the founders recognized that solid-state beacons would be much better because they would require less power and be more reliable, particularly important for remotely located equipment. They successfully designed and marketed the first solid-state radio beacons, and have become a worldwide leader in this $25 million/year niche market. Second, Gandalf in Ottawa started up in the early 1970s by being the first to market short-range modems.[42] At that time, only long-range modems were available, whereas many users required short-range versions to serve their local needs. Gandalf's founders recognized these unmet needs and successfully designed a short-range, low-cost modem to satisfy it. Some niche companies remain small, while others grow through expansion into neighboring niches or market segments to become quite large companies. Nautel remains small, still privately owned, and very profitable with sales of around $20 million per year. Gandalf has become a publicly traded company with worldwide marketing and manufacturing operations, and sales over ten times greater than Nautel's.

4.12 DEPENDENT STRATEGIES—"BRANCH PLANT" POLICY

The company following a dependent strategy is characterized, in the extreme case, by a technology base truncated to production and marketing and their ancilliary functions (see Figure 4.10). It is typically a subsidiary or specialist department of a larger firm or one of the latter's suppliers. Often, national subsidiaries of multinational corporations (MNCs) occupy such positions since this is one way for an MNC to exploit an innovation in offshore markets.

Figure 4.10. The dependent strategy base.

4.13 ABSORBANT STRATEGIES

An impressive feature of Japan's post World War II industrial achievement is the way that its companies have pursued absorbant strategies. An offshore company, probably with a truncated technology base, acquires a license or licenses from (typically) an offensive–defensive innovator to exploit innovations in its domestic market. However, rather than acting merely as a passive agent for the manufacturing and marketing of the innovations, the licensee uses the surplus cash-flow and market presence it thereby establishes to assimilate the technological knowhow and buildup its own R&D capability, to launch its own performance maximizing and cost-reducing incremental innovations in both its domestic and offshore markets. The sustained successful implementation of this strategy enables a company to extend its technology base from Figure 4.10 to first Figure 4.9 and then possibly Figure 3.2, to become an offensive–defensive innovator. It is implicitly illustrated in the Sony vignette of Chapter 1.

Tsuchiya describes the historical evolution of Japan's absorbant strategy, while Uenohara discusses its current emphasis on the shift to knowledge-intensive industries.[43,44] It could well continue since it appears that the country's inventive as well as innovative capacity is growing.[45,46] Although this strategy is especially associated with Japan, perhaps because the unique 1200-year-old ability of the Japanese culture to adopt, adapt, and improve offshore ideas, other technologically less developed countries and regions are seeking to replicate it.

We have now reviewed the repertoire of alternative technological strategies which may be pursued by a firm dependent upon its own strengths and weaknesses, the current stage of the industry in its life cycle, and the segments in the technology–market matrix the firm wishes to exploit. We have also seen that com-

TABLE 4.1   Strategies and Capabilities

Strategy Function	Offensive	Defensive	Initiative	Applied Engring./ Interstitial	Branch Plant
Nondirected Fundamental Research	M	M	N	N	N
Applied Directed Research	H	M	N	N	N
Experimental Development and Design	H	H	N	L/M	N
Advanced Development and Design	H	H	H	H	L
Pilot/Prototype to Full Production	M	M	H	M	M
Quality Control/Product Design	M	M	H	M	M
Patents and Licenses	H	M	L	M	N
"After-Sales" Services	H	M	H	H	M
Education and Advisory Services	H	M	L	H	M
Long-Term Planning	H	M	L	H	N

Key: N = None
　　 L = Low
　　 M = Medium
　　 H = High

panies must establish technological innovation bases congruent with their adopted strategies, and the profiles of these bases as a function of such strategies are summarized in Table 4.1. Before concluding this chapter, however, we will comment briefly on traditional and some other (nontechnologically) innovative strategies.

4.14   RENEWAL STRATEGIES

Much business activity, whether of low, medium, or high technology, encompasses the harvesting of profits from established products (or old technology) in established (or old) markets. It is largely characterized by companies in industries which are in the mature stage of their life cycle. One could place the farming and textile industries in this category. Both are very traditional, craft-based industries which have displayed a continued capacity to absorb technological changes in work methods. The post World War II farming industry is an interesting example because it is sandwiched between suppliers and customers who have both introduced quite radical technological innovations. Pesticides and fertilizers have increased productivities for farmers, while the development of freeze-drying techniques represents as radical innovation by food processors. The most obvious example is the automobile industry's contemporary efforts to adopt computer integrated manufacturing technologies.

4.15   OTHER (NONTECHNOLOGICALLY) INNOVATIVE STRATEGIES

In Chapter 2 (Section 2.10) we pointed out that potential technological innovations must compete for support with other innovative opportunities which a firm may perceive. These other opportunities may not necessarily be based on new technology but may still be entrepreneurally rewarding and may profitably extend the duration of the mature product life cycle since they often require relatively little front-end investment in new manufacturing plant, marketing, distribution, and servicing efforts.[47] Moreover, as Drucker comments, it is social rather than specific technological innovation that have the largest impacts.[48] As one very successful technological entrepreneur, Simon Ramo, implies, the successful management of high-technology companies requires as prudent balance between the introduction of innovations and the maintenance and extension of ongoing operations.[49] Thus, a company may have numerous opportunities for innovation, *not all* of which are technological, so it is useful to look at examples of these.

First, the sales of existing products may be increased by promotion innovations, distributing innovations, financial innovations, etc., as Ford and Ryan suggest. Many such innovations are technologically *cosmetic,* involving only changes in product presentation to enhance its customer appeal, while the products remain unchanged. Cosmetic innovations are vital competitive tools in consumer expendable industries such as detergents, personal toiletries, food, etc., where significant long-term technological innovation does occur (Proctor and Gamble's development of *Crest* toothpaste incorporating a fluoride, for example), but less frequently than the short-term cosmetic ones. It is significant to note that Telfer, in his remarks quoted at the beginning of this chapter, did not distinguish between technological and nontechnological innovation. Since he is a senior executive in a corporation in the highly competitive food industry (Maple Leaf of Canada), it could be reasonably conjectured that his remarks reflect the views of enlightened management in that industry. Second, many commercially successful innovations achieve this success through the judicious synthesis of technologically and nontechnologically innovative features in which the nature a role of innovative technology may vary. Two U.S. companies, McDonalds and Benihana of Tokyo, have achieved commercial success through marrying well-established and essentially noninnovative manufacturing management principles (but technologically innovative in their *own* industries) with innovative marketing approaches in the fast food and restaurant industries, respectively.[50] Third, the invasion of new markets with old technology can be highly profitable, provided the company understands the new market and has the necessary access into it. Quite often, a company perceives an opportunity for its own generic technological capabilities when technological and economic changes occur in the new market. For example, the Dowty Group, a developer and manufacturer of aircraft undercarriages, recognized that the introduction of face mechanization in the U.K. coal mining industry created the need for mechanized roof support systems to replace pit-props. They also recognized that the problems endemic to the design of mechanized roof support systems were generically similar to those endemic to the design of aircraft undercarriages. They therefore applied their generic technological expertise to the new problems, with considerable commercial success.

The examples of technological invasion cited so far illustrate situations where the generic capability has been developed in one industry and later transferred to another. We have already stressed the pivotal importance of identifying the potential synergies between the technology and the market place during the innovation process. This frequently involves identifying a number of potential markets for a proposed new product. Quite frequently, the market finally chosen for the introduction of new product was not one envisaged at the beginning of the innovation process. In this situation, it is difficult to distinguish between "old" and "new" markets since the introduction of an innovation may well trigger the invasion of one industry by another. This is illustrated by the invasion of the textile industry by the chemicals industry with the introduction of synthetic fibers and, more recently, the invasion of the watch by the electronics industry with the introduction of ICs.

Finally, we cite two examples illustrating that innovative thinking is not just confined to the technological rows of this matrix. Xerox, in introducing a revolutionary innovation (xerography) into the marketplace, enhanced its market appeal and its profitability to the company through an innovative pricing approach. Rather than selling or leasing its copiers as fairly expensive capital cost units, it marketed them on a charge-per-copy basis. This overcame potential customer resistance since it enabled users to experience the performance benefits of the new technology without a substantial capital outlay. Xerox thereby achieved more unit sales, and in many cases, a higher saleprice per unit, because the net revenue of copying charges derived from a unit exceeded the price that would have been charged in a direct sale. Both Texas Instruments and Sony rather similarly exploited the introduction of transistor radios. Because transistors are smaller, cheaper, and more reliable than vacuum tubes, the transistor as opposed to the vacuum tube radio could be made smaller, cheaper, and more reliable (requiring little or no servicing). This meant that it could be sold as a "new" product (the personal as opposed to the portable radio) in a new market (as a pocket radio to individuals, notably teenagers, rather than to households). Furthermore, its improved reliability rendered the traditional dealer–retailing network of the radio and TV industry redundant since negligible after-sales servicing was required and transistor radios could be distributed more cheaply through traditional nonspecialist retailing outlets such as drugstores. Similar thinking was also applied in the development of the Walkman personal cassette player. In these examples, the commercial benefits to be exploited from the introductions of revolutionary technologies were enhanced by novel approaches to market definition, distribution, and pricing. That is, they were especially successful because they constituted social, as well as technological, innovations.

All the examples cited in this section reinforce the point that technological innovations should be judged in the context of all the innovative opportunities available at any given time, and by business entrepreneurial *as well* as technological entrepreneurial yardsticks. A point to be kept in mind as we review the management of the R&D function in rather more detail in the next few chapters.

FURTHER READING

P. Abetti, *Linking Technology and Business Strategy.* New York: American Management Association, 1989.

L. W. Steele, *Managing Technology: The Strategic View.* New York: McGraw-Hill, 1989.

REFERENCES

1. J. A. Telfer, Senior Vice President, *"Strategic Planning at the Innovation Canada— 1976 Conference,"* Unpublished presentation. Domestic Operations, Maple Leaf Mills Ltd.

2. W. Taylor, "The Business of Innovation: An Interview with Paul Cook." *Harvard Business Review* **68**(2), 97–106 (1990).

3. G. S. C. Wills, "The Preparation and Deployment of Technological Forecasts." *Long Range Planning* **2**(3), 44–52 (1970).

4. T. Peters, "PART ONE: Get Innovative or Get Dead." *California Management Review* **33**, 9–26 (1990).

5. P. A. Roussel, K. N. Saad, and T. J. Erickson, *Third Generation R&D: Managing the Link to Corporate Strategy.* Boston: Harvard University Press, 1991.

6. W. H. Matthews, "Conceptual Framework for Integrating Technology into Business Strategy," Unpublished presentation. La Hulpe, Belg.: International Forum on Technology Management, July 1989.

7. L. W. Steele, *Managing Technology The Strategic View.* New York: McGraw-Hill, 1989.

8. E. Koerner, "GE's High-tech Strategy." *Long Range Planning* **22**(4), 11–19 (1989).

9. W. L. Robb, "How Good is our Research?." *Research Technology Management* **34**(2), 16–21 (1991).

10. C. K. Prahalad and G. Hamel, "The Core Competence of the Corporation." *Harvard Business Review* **68**(3), 79–91 (1990).

11. N. S. Langowitz, "Managing a Major Technological Change." *Long Range Planning* **25**(3), 79–85 (1992).

12. E. B. Roberts and C. A. Berry, *Sloan Management Review* **26**(3), 3–17 (1985).

13. R. A. Burgelman, T. J. Kosnik, and M. van den Poel, "Toward an Innovative Capabilities Audit Framework." In R. A. Burgelman and M. A. Maidique (Eds.), *Strategic Management of Technology and Innovation.* Homewood, IL: Irwin, 1988.

14. P. S. Adler, D. W. McDonald, and F. MacDonald, "Strategic Management of Technical Functions." *Sloan Management Review* **33**(2), 19–37 (1992).

15. L. W. Steele, *Managing Technology: The Strategic View.* New York: McGraw-Hill, 1989, Chapter 10.

16. F. J. Contractor and V. K. Narayanan, "Technology Development in the Multinational Firm: A Framework for Planning and Strategy." *R&D Management* **20**(4), 305–322 (1990).

17. P. A. Roussel, K. N. Saad, and T. J. Erickson, *Third Generation R&D: Managing the Link to Corporate Strategy.* Boston: Harvard University Press, 1991, Chapters 5 and 6.

18. G. R. Mitchell, "New Approaches for the Strategic Management of Technology." *Technology in Society* **7**, 227–229 (1985).

19. Arthur D. Little, "The Strategic Management of Technology." *European Management Forum* (1981).

20. A. Wilkinson, "Corporate strategy and the buying selling of knowhow." *R&D Management* **15**(4), 261–270 (1985).

21. P. A. Roussel, K. N. Saad, and T. J. Erickson, *Third Generation R&D: Managing the Link to Corporate Strategy.* Boston: Harvard University Press, 1991, pp. 60–61.

22. M. Jelinek and C. B. Schoonhoven, *The Innovation Marathon Lessons from High Technology Firms.* Cambridge, MA: Basil/Blackwell, 1990, Chapter 10.

23. W. Niefer, "Technological Expertise and International Competitiveness: Why New Strategies are Needed." *Siemens Review* **4**, 4–9 (1990).

24. Y. Kuwahara, O. Okada, and H. Horikoshi, "Planning Research and Development at Hitachi." *Long Range Planning* **22**(3), 54–63 (1989).

25. N. K. Sethi, B. Movsesian, and K. D. Hickey, "Can Technology be Managed Strategically." *Long Range Planning* **18**(4), 89–99 (1985).

26. J. M. Harris, R. W. Shaw, Jr., and W. P. Summers, "The Strategic Management of Technology." *Planning Review* **11**, 26–35 (1983).

27. L. C. Krogh *et al.*, "How 3M Evaluates its R&D Programs." *Research Technology Management* **31**(6), 10–14 (1988).

28. T. Kimura and M. Tezuka, "Managing R&D at Nippon Steel." *Research Technology Management* **35**(2), 21–25 (1992).

29. I. R. Barpal, "Business-Driven Technology for a Technology-Based Firm." *Research Technology Management* **33**(4), 27–30 (1990).

30. D. W. McDonald, "Strategic Management of R&D: Linking Technology and Business Planning," Unpublished presentation from a seminar on *Technology: The Challenge.* Toronto: Management of Technology Institute, McMaster University, September 1987.

31. P. D. Klimstra and A. T. Raphael, "Integrating R&D and Business Strategy." *Research Technology Management* **34**(5), 22–28 (1991).

32. H. I. Ansoff and J. M. Stewart, "Strategies for a Technology-Based Business." *Harvard Business Review* **45**(6), 71–83 (1967).

33. C. Freeman, *The Economics of Industrial Innovation,* 2nd ed. London: Francis Pinter, 1982, Chapter 8.

34. T. Burns and C. M. Stalker, *The Management of Innovation.* London: Tavistock Publications, 1961.

35. T. Kidder, *The Soul of a New Machine.* New York: Little, Brown, 1981.

36. C. Freeman, *The Economics of the Industrial Innovation,* 2nd ed. London: Francis Pinter, 1982, p. 175.

37. E. Herbert, "How Japanese Companies Set R&D Directions." *Research Technology Management* **33**(5), 28–37 (1990).

38. R. S. Rosenbloom and M. A. Cusumano, "Technological Pioneering and Competitive Advantage: The Birth of the VCR Industry." *California Management Review* **29**(4), 51–76 (1987).

39. R. N. Foster, *Innovation: The Attacker's Advantage.* New York: Summit Books, 1986.

40. K. Pavitt, "What We Know about the Strategic Management of Technology." *California Management Review* **32**(2), 17–26 (1990).

41. D. J. Teece, "Profiting from Technological Innovation: Implications for Integration, Collaborations, Licensing and Public Policy." In D. J. Teece (Ed.), *The Competitive Challenge: Strategies for Industrial Innovation and Renewal.* Cambridge, MA: Ballinger, 1987.

42. M. J. C. Martin and P. J. Rosson, *Further Cases on the Management of Technological Innovation and Entrepreneurship.* Ottawa, Ont.: Department of Industrial and Regional Expansion, 1986.

43. M. Tsuchiya, "From Process Innovation to Product Innovation: The Experience of Japanese Manufacturing Industries." *Japan Update* Winter, pp. 27–30 (1989).

44. M. Uenohar, "A Management View of Japanese Corporate R&D." *Research Technology Management* **34**(6), 17–23 (1991).

45. "Eyes on the Prize" Japan Challenges America's Reputation for Creativity and Innovation." *Time* March 21, pp. 48–49 (1988).

46. "Thinking Ahead: A Survey of Japanese Technology." *The Economist, Supplement* December 2 (1989).

47. D. Ford and C. Ryan, "Taking Technology to Market." *Harvard Business Review* **59**(2), 117–126 (1981).

48. P. F. Drucker, *Innovation and Entrepreneurship: Practice and Principles.* New York: Harper & Row, 1985.

49. S. Ramo, *The Management of Innovative Technological Corporations.* New York: Wiley, 1980.

50. T. Levitt, "Product Line Approach to Service." *Harvard Business Review* **50**(5), 41–52 (1972).

APPENDIX 4.1
TECHNOLOGY FORECASTING

Any firm operates within a spectrum of technological and market possibilities arising from the growth of world science and the world market . . . Its survival and growth depend upon its capacity to adapt to this rapidly changing external environment and to change it.

CHRISTOPHER FREEMAN,
THE ECONOMICS OF INDUSTRIAL INNOVATION

A4.1.1 INTRODUCTION

In Chapter 2 the innovation chain equation and double-helix metaphors were used to show that commercially successful technological innovation is dependent upon the occurrence of a particular chain of events which reflect the matching of a technological capability to an acceptable social need by entrepreneurial and management actions. We also saw that the time scale of this innovation process can extend over a number of decades. Although this innovation chain was represented in isolation, it does, of course, occur in the context of evolving technological and social milieux. The evolving technological milieu or technosphere reflects a continuous process of scientific and engineering inventions and developments embodied into economically competitive new products which satisfy social needs. The technology planning and strategy of the company constitutes an overall framework for evaluating, selecting, and managing the individual innovations in this context. If it to be effective, this framework must match the time scale of such innovations which, as we have seen, can stretch over decades. Since the 1960s, various approaches and techniques for perceiving and predicting such future environments have been developed under the general label of *technological and social forecasting* and, in this Appendix, we will review some of these techniques which the company may employ to pursue its goals. Because some readers may be either unfamiliar with, or have misconceptions about, the nature of technological forecasting (TF), we will begin with an outline of its scope and development to date.

A4.1.2 THE EMERGENCE OF TECHNOLOGICAL FORECASTING: THE CONCEPT OF FUTURE HISTORY

TF or, more generally, futures studies (FS) is an approach to the study of the future that has burgeoned over the past 30 years although, as with many new management aids, its origins can be traced back much further in time. Both Leonardo da Vinci and Jules Verne, for example, speculated as to future technologies, and H. G. Wells proposed in 1902 that it should be feasible to develop a systematic study of the future, comparable to the sciences.[1]

> And I am venturing to suggest to you that, along certain lines and with certain qualifications and limitations, a working knowledge of things in the future is a possible and practicable thing. . . . I must confess that I believe quite firmly that an inductive knowledge of a great number of things in the future is becoming a human possibility. I believe that the time is drawing near when it will be possible to suggest a systematic exploration of the future. . . But suppose the laws of social and political development, for example, were given as many brains, were given as much attention, criticism, and discussion as we have given to the laws of chemical combination during the last 50 years—what might we not expect? . . .

During the interwar years, political scientists, sociologists, and economists published materials (particularly those warning of the dangers of the Nazi regime in Germany) that could be labeled futures research. However, it was not until after World War II that widespread interest was shown in the idea, initially for military reasons. The East–West cold war and arms race based upon nuclear weapons technology made it imperative to explore possible future situations in which a war could breakout, in order both to develop cost-effective weapons and (more importantly) to try and minimize the risk of a nuclear holocaust. These considerations dictated a requirement to explore the political and socioeconomic impacts arising from the development of alternative new weapons systems, as well as to predict developments in military technology itself. There thus existed a practical need to develop methods of technological and social forecasting. By the 1960s, future thinking had been applied to other aspects of human affairs and became more visible through several notable publications. In 1964 (but based upon work which began there in 1948), Gordon and Helmer[2] of the Rand Corp. published their development of the DELPHI method of forecasting (see Section A4.1.10). Also in 1964, Gabor,[3] who won a Nobel prize for his invention of holography, published his *Inventing the Future,* while in 1967 Kahn and Wiener published *The Year 2000.*[4] A body of techniques for systematically studying the future was identified by Jantsch in *Technological Forecasting in Perspective,*[5] and De Jouvenal in *The Art of Conjecture*[6] sought to provide an epistemological basis for the study of the future. This interest was reinforced by public concern for the harmful impacts of technological innovations on the natural ecosystem. This environmental concern led the Club of Rome to sponsor a global study, using systems dynamics techniques (see Section A4.1.9), which forecasted worldwide food and other resource shortages, pollution buildups, population imbalances, and so forth, over the next 100 years. The controversial results of this study were published under the title of *The Limits to Growth.*[7]

At first sight, it might be supposed that TF or FS really consists of traditional business forecasting methods in a new guise. Economists and statisticians have been forecasting the future behavior of important variables and parameters in the economic environment of the firm for a number of decades, and industrial operations research groups frequently perform similar exercises. Furthermore, technological forecasters do use statistical, econometric, and OR techniques to extrapolate technological and economic parameters into the future (as the Club of Rome study cited above illustrates), so TF does exploit techniques common to other disciplines. Despite these commonalities, it is broader in both scope and intent than traditional forecasting methods for the following reasons. First, traditional forecasting methods are primarily concerned with forecasting key variables and parameters over the short and medium term—say, from a few months to five years or so. Because of the time scale of the innovation process, FS must be concerned with forecasting the environment of the firm over a much larger term—typically 3 to 25 years, and maybe even longer. Second, traditional methods are primarily based upon statistical forecasting techniques which extrapolate historical data into the future, assuming no underlying economic, social, and technological change—an assumption which may be valid for forecasts up to five years, but is unlikely to be valid beyond that. Technological forecasters do use similar techniques when forecasting incremental and changes-of-scale in technological capabilities, but they are also concerned with forecasting the technological, socioeconomic, and political impacts of technological changes, either individually or in combination, which disturb the status quo, thus making an extrapolative approach invalid.

These requirements dictate that FS must first adopt a truly multidisciplinary approach (embracing expertize in the social and behavioral sciences, as well as the physical and life sciences, engineering, and management) if it is to forecast future innovative opportunities. Second, because it seeks to predict technological changes based upon, as yet, undiscovered inventions or ideas, FS must be able to assess and evaluate technological trends creatively, to identify the characteristics, timings, and probabilities of occurrence of future inventions and innovations. These requirements are perhaps best described by using the term *future history* as a synonym for future studies or technological and social forecasting. Since the *Concise Oxford Dictionary* defines "future" as "*of time to come, . . . describing events yet to happen*" and "history" as "*study of past events, especially human affairs,*" we must clearly justify the use of this apparent oxymoron!

Historians, after careful research and scholarship, present their interpretations of past events by some form of descriptive analysis. The factual data and other evidence available to them are clearly limited by the gaps and biases inherent in the historical recording process, so that their analyses must inevitably be based upon partial information. Moreover, historians are judicious and selective in the use of these facts in developing and propounding their interpretations of the past. As Carr puts it: "*The facts speak only when the historian calls on them; it is he who decides to which facts to give the floor, and in what order or context.*"[8] Since history is a seamless web of unfolding events and trends, any competent historical analysis of the recent past should be amenable to extrapolation and informed speculative projection into the future. Such forecasts or projections will lack the formal rigor of the statistical methods analysis which is used in short-term forecasting and, because they must often accommodate the occurrence (or otherwise) of uncertain future events, they can rarely be expressed in terms of statistical confidence intervals. Nevertheless, the historical processes of technological, social, and political change can be projected into the future to provide a forecast of the future technological, social, and political environment for the company, and the opportunities and threats it may face in pursuing its corporate goals. It is in this sense that the term "future history" is used. It provides a "picture" or "scenario" of the future environment for the company, or the future technology–market matrices which the company may wish to exploit and, as such, is a required element in corporate planning. Any corporate planning activity must, either implicity or explicitly, plan on the basis of some scenario of the future environment for the company. FS is an approach to building up this scenario in as critical and logically consistent manner as the inherent uncertainties in the situation allow, rather than relying on an intuitive and impressionistic view of the future or assuming it will be like the present.

For example, an intelligent observer of the semiconductor industry in the early 1960s could have predicted that there was a high probability that within 10–20 years, integrated circuit technology would provide a cheaper, reliable, and more accurate replacement for the traditional mechanical watch. Had the Swiss watch industry studied its future in a critical systematic manner, it would have discovered this picture and would have taken appropriate measures earlier to meet the threat of invasion of its traditional markets by the electronics industry. The development of scenarios is discussed in some detail later (Sections A4.1.15 and A4.1.16). This example illustrates an important observation on the purpose of business forecasting in general made by Peter Drucker.[9] As he pointed out, business forecasting is concerned not with the future itself, but with the *futurity of present decisions* taken by management *today*. It is in this spirit that a high-technology company should explore its futures and apply the TF techniques which we now discuss.

A4.1.3 TECHNOLOGY MONITORING

One simple approach is to monitor *signals of technological change* as suggested by Bright.[10] Throughout the 1950s and 1960s, NCR failed to appreciate the implications of technological change in the transition from mechanical to electronic cash registers and accounting machines.[11] Similarly, had the watch industry done so, it could have anticipated the development of the electronic watch. It is an approach that can readily be used in small companies that cannot afford to employ a full-time forecasting staff. It may also constitute a first step in the establishment of a TF group which employs such staff in larger companies. It is particularly suited to the technological and market gatekeeping roles described in Chapter 7.

Technology monitoring involves reading or scanning a selection of publications which can be expected to provide signals of change. Martino[12] gives a detailed description of the approach and stresses that all relevant environmental aspects (including sociocultural, political, and ecological factors, as well as technological and economic ones) should be monitored, and Bright[10] recommends maintaining a journal for this purpose. The same essential idea was popularized by Naisbitt in his best-seller *Megatrends*.[13] The major difficulty with this approach is that if too many publications are monitored, continually updating the journal becomes too onerous an exercise, and the signals of change remain undetected in the large amount of published material collected. Sources relevant to corporate needs should therefore be selected judiciously. For example, Schwartz suggests that the magazines *New Scientist* and *Scientific American* together provide excellent sources for emergent scientific ideas.[14] He suggests that if an idea first appears in the *New Scientist* and then *Scientific American,* it will become important. Photocopies of extracts from relevant articles, etc. may be collected in a scrapbook which is periodically pruned to discard redundant material. Significantly, it is probably Japanese companies that have pioneered the use of competitive intelligence gathering.[15]

A4.1.4   THE "S"-SHAPED LOGISTIC CURVE

As discussed in Chapter 2, the growth in a new technological capability typically follows an S-shaped curve (Figure 2.5) which can be roughly divided into three stages. The first stage is slow initial growth as the new technology has to prove its superiority over existing technologies. Then, once this superiority is demonstrated, a period of rapid or explosive growth follows. Finally, its growth is limited by technological or socioeconomic factors and levels off towards some upper limit. The commercially successful exploitation of technology often depends upon the astute perception and exploitation of this growth, so forecasters have paid significant attention to extrapolating the growth of the S-shaped curve of a technological capability at some relatively early stage of its life. In so doing, they have used mathematical functions or models which were originally developed to describe the growth of biological organisms. Thus, they have explicitly exploited the conceptual similarity between biological and technological evolutions discussed in Chapter 2.

One of these is the Pearl function (named after the American biologist and demographer Raymond Pearl, who extensively studied the growth patterns of organisms and populations):

$$y = \frac{L}{1 + ae^{-bt}}$$

where y is the dependent variable whose growth is to be forecasted, L is the upper limit to growth, and a and b are parameters.

The Pearl function may be reduced to a linear form by the appropriate logarithmic transformation. If, at an intermediate stage in the growth of the technological capability, we have a historical record of y values for different years t, we may plot a least-squares regression to determine the parameters a and b. The Pearl function is then fully determined and may be extrapolated to forecast y.

An alternative function which generates a S-shaped curve with a longer upper tail is called the Gompertz function (named after an English actuary and mathematician):

$$y = L \, e^{-be^{-kt}}$$

The Gompertz function may also be converted to a linear form by a double logarithmic transformation, and forecasted values of y determined by least-squares regression.

Both Pearl and Gompertz functions have been applied quite extensively in TF work, and many computer centers have standard software packages to perform the required logarithmic transformations and regression analyses. Thus, provided a reliable database is extant, the goodness of fit of the data to each model can be readily tested. Apart from ensuring that one is working with a homogeneous database (see Section A4.1.9), the practical utility of these curve-fitting techniques for generating forecasts is limited by two factors. There is difficulty in setting a value to the upper limit L, particularly when the technological capability under examination is a synthesis of several subtechnologies. Clearly, this is often the case. Martino[16] points out that aeroengine technology is a synthesis of several subtechnologies, so that developing an accurate performance forecast for a given engine type is difficult. In addition, the inherent scatter in the data points may place relatively wide confidence limits on the parameter estimates and subsequent extrapolated forecasts.

A4.1.5 ENVELOPE CURVES AND TREND EXTRAPOLATION

Technological evolutions typically progress through successive substitutions of new approaches. Whereas individual technological capabilities (A,B,C, etc.) follow the S-shaped curve, each capability is superceded by a technology superior successor (A by B, By C, etc.) so that overall functional performance continues to rise along an *envelope curve* generated by successive technologies (see Figure 4.7). The changeover points (X1, X2, . . .) may occur only when the performance of the new innovation is *significantly* better than the old one. This is because of the economic, social, and political factors which tend to resist technological change. An industry may have so much financial and human capital invested in a given technology that it is unwilling to change to a new one, even when the latter is markedly superior. Indeed, sometimes such a change is imposed by a company outside the industry. Some of the issues implicit in S-curve transitions were discussed in Chapter 4 (Section 4.7), and we shall return to them in the next section. However, for the moment, we are concerned with the overall envelope curve.

A plot of the envelope curve over a lengthy period of time may yield a statistical trend which again may be extrapolated forwards to obtain a forecast.

Martino[17] cites numerous examples of logarithmic measures of given capabilities against time including:

1. Efficiency of light sources since 1850.

$$y = \log \text{ (lumens/watt) against time } t.$$

$$y = -128.71511 + 0.06851t$$

2. Top speed of U.S. combat aircraft since 1909 (bombers & fighters)

$$y = \log \text{ (speed m.p.h.) against time } t.$$

$$y = -118.30568 + 0.0640t$$

Innovations occurring were closed cockpit, monoplane, all-metal airframe, and the jet engine.

3. Capacity of Random Access Storage/Random Access Time for computer since 1951.

$$y = \log \text{ (random access storage/access time, bits/microsec)}.$$

$$y = -1080.81958 + 0.55125t$$

Innovations occurring were vacuum tubes, transistors and ferrite cores, and integrated circuits.

The above results plus the others given in Martino show remarkedly high correlation coefficients (over 0.95), so it is hardly surprising that trend extrapolation has been used quite extensively to derive a performance forecast. However, we stress that trend extrapolation must be used intelligently and not naively. This caveat can be illustrated with a fairly obvious example.

As we shall see in the next section, the log-linear trend to combat aircraft speed (example 2 above) is replicated in passenger aircraft transport cruising speeds and a naive extrapolation of trends in the 1960s would have forecast fleets of SSTs, flying in the 1970s. This forecast would have been proven invalid, for the following reasons:

1. Environmental pollution concerns made the SST socially and politically unacceptable.

2. The energy crisis and escalating costs of aviation fuel made it less attractive commercially.

3. Cruising speed is not an appropriate performance measure for passenger air transports. Passengers are primarily interested in getting from A to B as quickly and comfortably as possible. Clearly, for the shorter haul flights, any reduction in flying time will be masked by the time to access and egress the airplane from/to a downtown location. Thus, total trip time is a more appropriate performance measure and an SST will only offer a worthwhile reduction in this measure over the longer transcontinental and the intercontinental routes.

A4.1.6 MANAGING S-CURVE TRANSITIONS

The S-curve has been familiar to forecasters since at least the 1960s, but it was the publication of Richard Foster's best seller *Innovation: The Attacker's Advantage* in 1986 that focused management attention upon it. As already discussed in Chapter 4 (Section 4.7) and mentioned in the previous section, perceiving and reacting appropriately to S-curve transitions is the important management issue discussed in Foster's book.[18]

To anticipate such transitions, it is important to note that although the S-curve of a given technology is typically easily tracked after the event (with the benefits of 20–20 hindsight), it is much more difficult to do so during its evolution. This situation is schematically illustrated in Figure A4.1.1. At the "Present Time" (point A in the figure), the technology is entering the steep part of its S-curve, but its future trajectory may be difficult to extrapolate, because its upper-limit and the steepness of its slope are unknown. The naive extrapolation of the existing curve based upon the scant extant data to point A is likely to yield a projected trajectory with wide statistical confidence limits. It may follow a shallow growth to a modest upper limit defined by path A–B–C. Alternatively, it may grow rapidly to a much higher upper limit defined by path A–D–E. Although such extrapolations may be useful, they must be supported with technical and management insights if S-transitions are to be anticipated.

First, the technical upper limit, where possible, should be identified. For example, the upper limit to the number of transistors that may be fabricated on a single semiconductor chip is on the order of a billion for a number of technical reasons. This limit should be reached in the first decade of the next century. Further technical progress in microminiaturization beyond then will probably require a transition to molecular level electronics. Large companies in the microelectronics industries, such as IBM, are already performing research in molecular electronics in anticipation of the possibility of this transition. The design and fabrication of molecular electronic devices will require expertise in chemistry and biotechnology (perhaps including genetic engineering) currently reposing mainly in the chemical and pharmaceutical industries. Thus, companies in the present semiconductor-based microelectronics industries face the threat of invasion from companies in the chemical and

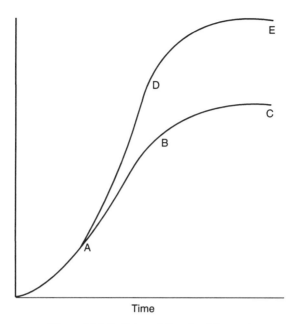

Figure A4.1.1. Extrapolating the "s"-curve.

pharmaceutical industries, if and when this transition occurs. By performing such research today, one to two decades in advance of when the knowhow could be required, IBM ensures that it keeps up with the state of the art in molecular electronics, to be able to pursue an offensive or counteroffensive strategy, should the transition occur in early the next century.

Second, even if the limits of the current S-curve cannot or have not been explicitly identified, there are other management signals of technical change. Whether the S-curve is represented by a Pearl, Gompertz, or some other function, by definition it approximates to the shape shown in Figure 2.5. Although, in that figure the two axes represented Time and Performance, respectively, they can equally well represent Technical Effort and Progress, as shown in Figure A4.1.2a. Without delving into calculus, it is intuitively apparent that the first derivative or slope of this curve is represented in Figure A4.1.2b. What does this curve measure? The answer is the progress per unit (say dollar) of effort invested in improving the technology—that is, it is a measure of R&D productivity. It shows us that R&D productivity attains its maximum value around the middle of slope of the S-curve and declines thereafter.

This observation is hardly surprising in light of the Abernathy–Utterback treatment discussed in Chapter 2 (Section 2.5). The steep part of the S-curve corresponds to the performance-maximizing product innovations emphasis in that treatment, when high payoffs from R&D can be expected. Once the curve starts to level off, corresponding to the move into the mature or specific state, emphasis shifts towards cost-minimizing process innovations and a law of diminishing returns takes over, so that declining R&D payoffs can be expected. A declining R&D performance, partic-

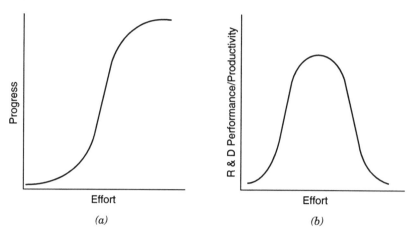

Figure A4.1.2. (*a*) The cumulative technical effort curve. (*b*) The incremental technical effort curve.

ularly when coupled with a shift in emphasis from product to process innovations, should be accepted by management as a signal of a potential S-curve transition. Prudent management will view this as signaling an opportunity as well as a threat, and decide whether to invest R&D efforts into alternative successor technologies and S-curves. As Pogany argues, the final decision whether to abandon an entrenched technology in favor of a new one depends upon company-, industry-, and technology-specific factors that are expressed in an S-curve.[19] Whether to pursue offensive or counteroffensive strategies, if and when the transition occurs, depends upon other factors and the considerations discussed in Section 4.7.

A4.1.7 PRECURSOR TRENDS

We have just referred to the fact that both combat and passenger aircraft speeds follow log-linear trends, as is illustrated in Figure A4.1.3. More importantly, the passenger aircraft trend appears to parallel its combat aircraft counterpart, but displaced about 10 years in time. That is, passenger aircraft performance matches combat aircraft performance 10 years later. The reason for this cross-correlation between the two trends is fairly obvious. Combat aircraft performance is achieved by the aerospace industry based upon R&D funded by government. This technological knowhow can be readily transferred to the development of civilian transport aircraft, so two parallel envelope curves are produced.

Although the application of precursor trends to forecasting civilian aircraft speeds broke down in the 1970s for the reasons cited earlier, it is an approach which can be used in other contexts. Schmidt and Smith[20] found similar associations between aircraft and marine technology, while Kovac[21] used the performance pa-

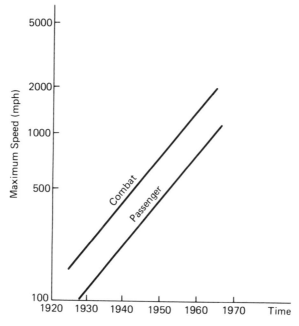

Figure A4.1.3. Precursor trends.

rameters of racing car tires to forecast those for passenger car types. Precursor trends can also be used to forecast the adoption of new geographical as well as application areas. In many instances, U.S.-based multinational businesses develop and sell new products in their domestic U.S. markets before selling and possibly manufacturing them offshore. Mumford found a consistent lag between the growth of sales of aerosol products in the United States and corresponding sales in the United Kingdom which could be used to forecast the latter.[22]

A4.1.8 FISHER–PRY SUBSTITUTION MODEL

While the S-shaped curve describes the growth of an individual technological capability and the envelope curve the growth of an overall performance through a succession of technologies, when an S-curve transition is occurring, it is often of considerable commercial interest to focus on the specific rate of substitution of technology A by technology B in Figure 2.6 over time—that is, the rate and total extent of the adoption of technology B and its diffusion in various product and geographical applications.

Rogers[23] has led research on the adoption of innovations by individuals more than organizations, but his conclusions are of sufficient interest to warrant a brief discussion here since, like innovations, such adoptions are often pioneered by individual champions. Because technological evolution depends upon learning by using, the exact path of the technological capability S-curve for technology B is dependent upon the rate of adoption of the new technology by potential users. Figure A4.1.4 is a related S-curve showing the percentage of the total population of potential users that have abandoned technology A in favor technology B over time. Logically, we can expect a strong positive correlation between the parameters of these two S-curves. If technological capability B progresses rapidly as shown in a relatively steep S-curve, we can expect potential users to adopt it more quickly, so that the percentage of adopters will also rise more steeply. This, in turn, should stimulate further technological improvements and, thence, further adoptions to create a virtuous circle which is the basis of learning or experience curve strategy used to advantage in manufacturing and marketing by TI and other companies.

Figure A4.1.4 is, in fact, the cumulative distribution of the population of adopters over time. Based upon considerable research, Rogers suggests five categories of adopters:

Innovators: Venturesome, representing the first 2.5% of the adopters (adopting technology B more than two standard deviations earlier than the mean adoption time). These are typically individuals who are ventursome even to the point of rashness, but are able to understand and apply any new technical knowledge required to adopt the new technology. They like to try out new things, and have the money to absorb the costs and risks of doing so.

Early Adopters: Respectable, representing the 13.5% adopting the technology between one and two standard deviations earlier than the mean time. Whereas the

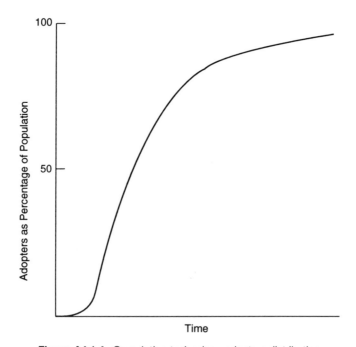

Figure A4.1.4. Cumulative technology adopters distribution.

innovator may be viewed as an unreliable maverick by his peers, early adopters are typically peer group opinion leaders. They are more judicious than innovators in adopting new ideas and, for this reason, once they have done so, other more cautious individuals who view them as role models will follow their examples.

Early Majority: Deliberate, representing the 34% up to one standard deviation below the mean. They do not wish to be opinion leaders, but prefer to deliberate for some time before adopting novelty. As Rogers puts it, they value the maxim, "*Be not the last to lay the old aside, nor the first by which the new is tried.*" They resemble defensive innovators in attitudes.

Late Majority: Skeptical, representing the 34% up to one standard deviation above the mean. These wait until perhaps the economic and social pressure of the majority having adopted the innovation forces them to do so also.

Laggards: Traditional, representing the last 16% of the population. These are the diehard rearview mirror drivers of society, who are suspicious of all novelty and view everything in the context of the past.

Rogers' user typology implies that companies launching technology B should develop a marketing strategy that concentrates on the early adopters (if not the innovators) since, once converted, they will encourage their more cautious peers to adopt the new technology. Such individuals (or organizations) are described by others as *lead-users,* and their important role will be discussed further in Chapter 6 (Section 6.7). However, the key point to recognize here is that a well-executed market launch targeted on Rogers' innovators should maximize the steepness of Figure A4.1.4*a* S-curve or the substitution rate of technology A by technology B. As stated at the beginning of this section, knowing this rate is very important since it strongly influences the plans for manufacturing and marketing the new technology-based products.

Many studies proposing models for measuring substitution rates are described in the literature, and a useful review is provided by Linstone and Sahal.[24] Because technology B substitutes for technology A by a Darwinian competitive process, a number of workers have suggested explicit models for generating these curves, and the one due to Fisher and Pry[25] has received the most widespread acceptance. From the evolutionary treatment, we can argue that technology B is a new and superior mutation of technology A, and that B invades and occupies A's environmental niche(s), systematically "killing" A in the process. Fisher and Pry start with an assumption consistent with this analogy—that the rate of adoption of technology B is proportional to the fraction of the market (environmental niche) still using technology A.

Thus, if f is the fraction of the market *already* captured by Technology B, t is time, and b is a constant, we have:

$$\frac{1}{f}\frac{df}{dt} = b(1 - f)$$

where df/dt is the first derivative of f with respect to t.

Now, if t is the time at which technology B has been substituted in half a market (that is when $f = 1/2$), the above first-order differential equation yields the following solution:

$$f = \frac{1}{1 + e^{-b(t-t_0)}}$$

Bearing in mind that if $f = 1/2$ corresponds to $y/2$ (since the upper limit to f is 1) and writing $a = e^{bt_0}$, we find that the Fisher–Pry model yields the Pearl Function of Section A4.1.4.

The Fisher–Pry model and its extensions are possibly the most widely used of the extrapolative TF techniques. Also, it can be combined with precursor trend identification to produce useful forecasts. For example, Jones and Twiss[26] report that a Swedish TV manufacturer successfully used this substitution model together with an analysis of the lag between U.S. and Swedish adoptions of TV innovations to predict the changeover from 21″ to 27″ tubes in their market. Their forecasts enabled them to avoid the overproduction and excessive stocking experienced by their competitors.

A4.1.9   SOME PITFALLS OF EXTRAPOLATIVE TECHNIQUES

The extrapolative techniques described above and others described in the now volumninous TF literature have proven useful to technological planners in both government and business. Indeed, as Sahal argues, they may well reflect some underlying laws of technological evolutions.[27] There are, however, some common pitfalls to be avoided if they are to be used effectively. We now briefly discuss these before moving on to other techniques.

It is important to identify and clarify a common stage in the innovation process when measures are observed. Historical data may well be drawn from a variety of sources which may report performance measures at different phases of the innovation process. One source, for example, might report the speed of a passenger air transport at the prototype trials stage when it is being tested for air worthiness, while another may report the speed of another air transport when it is in widespread use. The performance specifications of new capabilities are, of course, quite frequently changed during the successive phases of the innovation process, in light of the experience gained, so that it is important to use the same innovation phase for all the observations. Also, a forecaster must test the robustness or sensitivity of his forecasts to changes in model parameter values. As we indicated in Section A4.1.4, the extrapolated values obtained using the above models are typically quite sensitive to the parameter values which, in turn, may be subject to quite large statistical standard errors. Thus, the forecaster should develop a good feel for the robustness and limits of his forecasted values. Needless to say, as with other extrapolation techniques, the data should not be extrapolated further forward than the total time base warrants. A lengthy extrapolation increases the statistically based error, but more important, it is unlikely that the underlying core assumptions will be valid over a longer period of time. This important issue will be discussed next.

This pitfall was illustrated above when we referred to SST forecasts. Implicit in all extrapolative techniques is the *ceteris paribus,* other things being equal, assumption. Clearly, the forward extrapolation of historical trends is only valid as long as the underlying assumptions continue to hold. Apart from the fact that higher air speeds do not produce commensurate reductions in total journey time, the trend forecast of speeds in the 1970s proved invalid because the underlying assumptions broke down. The forecaster must never apply extrapolative techniques in a mechanical unthinking manner, but must always examine the underlying assumptions to ensure that they remain valid, or modify the approach in light of any new exogenous factors which must be considered. Indeed, Ascher,[28] in an extensive review of the accuracy of many technological and social forecasting studies, argues that many such studies have proved inaccurate through the choice of wrong underlying assumptions rather than the use of wrong techniques. He argues that the choice of a correct set of core assumptions is crucial to the forecasting task.

A4.1.10 SYSTEMS DYNAMICS MODELS

The crucial importance of developing forecasting models based upon valid core assumptions suggests that better forecasts might be obtained from models which explicity simulate the behavior underlying the innovation process. Given the complexity of this process, such simulations must be performed on computers.

Such models seek to represent and quantify the interconnections or causal paths among the various technological, economic, and sociopolitical factors in the innovative process. This typically involves developing a system or set of equations describing the process, and identifying critical parameters and variables and their values. The equations and underlying logic of the model, together with the deterministic and stochastic properties of the parameters and variables (including time), are incorporated into a computer program. The program may then be run repeatedly under different assumptions to simulate the dynamic behavior of the process under different conditions. Computer simulations may be performed sequentially over successive time periods (incorporating past–present–future) to generate a forecast.

The most common of these approaches was originally pioneered by Professor Jay Forrester of MIT, and was known originally as Industrial Dynamics and now more generally as Systems Dynamics. Forrester's approach is to view any socioeconomic system as a hierarchy of closed-loop feedback control systems (he views the feedback control loop as the atomic building brick of socioeconomic systems) which may be described as a series of mathematical equations. From this basis, he developed simulation programs first for individual companies—*Industrial Dynamics,* and then for cities—*Urban Dynamics,* and finally for the Earth as a whole in *World Dynamics.*[29]

Systems Dynamics leaped to prominence in the TF context through the *Limits to*

Growth study performed by Meadows *et al.*[7] for the Club of Rome. These authors used Forrester's approach to stimulate and forecast the impact of population growth, resource consumption, and pollution over the next 100 years. Their forecasts, which predict a virtual breakdown of human society within the next century, aroused widespread controversy in the early 1970s. Much of the *Limits to Growth* controversy centered upon the validity of the core assumptions made in the study, and, particularly as we are concerned with TF in the corporate context, need not concern us here. Systems Dynamics has, however, also been applied to TF at the corporate as well as the global level.[30]

Computer simulation approaches not using Forrester's systems dynamics paradigm have been developed by other workers. Some have developed models for specific high-technology industries. For example, Allenstein and Probert[31] describe a long-range planning model of the telecommunications business developed for the British Post Office. Although this is a corporate planning rather than TF model, one module within the model is explicitly concerned with technological forecasts, since microelectronics and fiber optics or optronics will have a radical impact on the telecommunications business over the next few decades.

A4.1.11 DELPHI METHOD

Some of the most widely used of the TF approaches are those based upon the Delphi method and its extensions. It was originally developed by Helmer and coworkers for the Rand Corporation.[32,33] Since the original description of the method, it has enjoyed widespread use and been subjected to considerable extension and modification. It derives its name from the famous Oracle of Delphi who was the prime source of future forecasts in Classical Greece. The Delphi method is based upon the not unreasonable premise that the best sources of technological forecasts are the opinions of experts in the given technology—that is, the simplest way of making a forecast is to ask the experts in the field to do it. It is undesirable to base a forecast on a single oracle or expert, however distinguished, so the opinion of a sample or committee of experts is sought. The considered judgment or consensus of a committee of experts provides a viable approach to deriving a technological forecast, but suffers the disadvantage that this concensus may be biased to the opinions of its dominant members. The Delphi approach avoids this disadvantage by requiring members to participate anonymously.

The approach is iterative, with each iteration called a round. In each round, a panel of experts is interrogated individually and confidentially (usually by a form containing a series of questions) for their views on the likelihoods and timings of the occurrences of certain hypothetical future events. Interactions among panel members is forbidden. The whole process typically takes four rounds. Thus, the structured interactions have the following characteristics: anonymity between panel members, iteration with controlled feedback, and statistical group response. The Delphi procedure which we now outline is usually conducted by one individual (with whatever support staff that is required) known as the director.

Selection of Panel

The selection of panel experts must usually be performed on a pragmatic basis. Factors that must be considered follow:

1. Individual expertise in area of study. This may require organizational as well as technical knowledge.
2. Commercial and military security. This may limit the organization(s) from which panel members may be selected.
3. Individual availability and willingness to participate. Delphi can be very time consuming.
4. Avoidance of bias.

Bearing in mind these factors, panel members can usually be selected from peer judgments, literature citations, honors and awards, patents, and professional society status. Typically, a panel will have 10–50 members. We now discuss the rounds in turn.

Round 1

The first questionnaire is as unstructured and open-ended as possible. It requires panelists to provide a forecast of developments over the period and within the area under study. Panelists may respond in whatever form they think fit, such as by suggesting a sequence and chronology of key events or by providing a narrative or scenario (see Section A4.1.16).

Round 2

From the results of the first round, the director compiles a list of key events which are presented to the panelists as the Round 2 questionnaire. They are asked to suggest their estimated dates when each event will occur (including never or after a given date). They are asked to provide their justification for each estimate. From these results, the director prepares a consolidated statistical summary of the distribution of dates given for each event. The median, lower, and upper quartile dates for each event are calculated.

Round 3

The third questionnaire presented to the panel includes the above statistical data plus the reasons advanced for the extreme event dates, and after reviewing this information, panelists are asked to provide revised event dates. Specifically, if they suggest dates outside the interquartile range (that is, outside the group consensus), they are asked to provide the arguments for doing so. Thus, they are required to cite facts or factors that they think the other panelists may be neglecting or of which they are unaware. The director then repeats the consolidation of Round 2, producing revised statistical distributions and arguments, particularly the pros and cons of extreme dates.

Round 4

This repeats Round 3, and panelists provide their *final* estimates of event dates, with full knowledge of the pros and cons that have been advanced. The director then consolidates the results of this final round which constitute the reported forecast. (More than four rounds may be required before opinions have converged sufficiently to constitute consensus, but usually this number is sufficient.) The panel's forecasts are usually presented in the form of the median date and interquartile range for each of the events considered. Figure A4.1.5 presents a typical output from one study.[34]

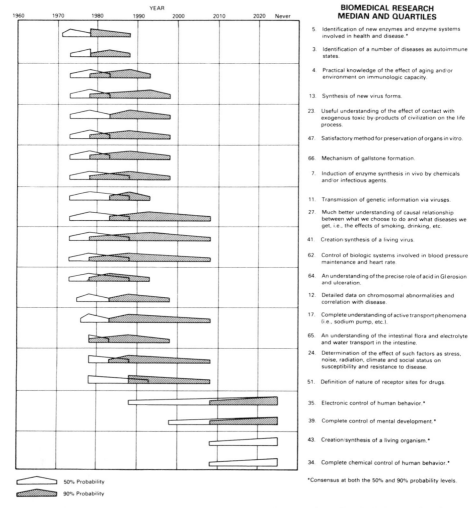

Figure A4.1.5. Sample delphi output. *Source:* M. J. Cetron and C. A. Ralph. *Industrial Applications of Technological Forecasting* (New York: John Wiley & Sons, 1971). p. 115. Reprinted with permission.

Although the Delphi Method has been used extensively since its introduction in the 1960s, there is little evidence that it produces better judgments than simpler expert opinion solicitation techniques. Two reviews of evaluations of the method, which included extensive literature citations, suggest this conclusion.[35,36] However, Rowe *et al.* in their review[35] do stress that many of the evaluations reported were based on laboratory Delphi studies using nonexpert subjects, which did not provide valid comparisons with realistic applications of the method. They also conclude that Delphi does have potential as a judgment-aiding technique, but that a better understanding of the psychology of judgment changes in groups is needed if this promise is to be realized.

A4.1.12 CREATIVITY STIMULATING TECHNIQUES

The techniques considered so far for forecasting technological futures have essentially been *passive*. That is, they have sought to estimate the dates of occurrence of future technological events either by the objective analytical extrapolation of the past into the future or by the subjective performance of a similar exercise by experts. In none of these approaches has the forecaster sought *actively* to influence the occurrence of the innovation under consideration. In practice, the realization of a new technological capability (such as the successful completion of a lunar mission with astronauts) requires the sequential completion of a number of inventive problem solving steps which often may be approached in a number of different ways. Clearly, the time taken to achieve a new technological capability is a function (among other factors) of the efficacy of attack on this problem solving sequence. Thus, another approach to making a technological forecast is to identify the time scale of the sequence of alternative problem solutions to be derived to achieve the technological capability to be forecasted. That is, the forecasting process is viewed as a creative problem solving process, and techniques which can equally well be used to stimulate group creativity (see Chapter 7 Appendix, Section A7.5) can be used in TF. We now consider two of the more important of these techniques.

A4.1.13 MORPHOLOGICAL ANALYSIS AND MAPPING

Like Delphi, Morphological Analysis is a TF approach which derives its name from the classical Greek world and *morphé* or the Greek work for "form." The origins of the approach can be traced back to Plato and Aristotle, particularly in descriptions of the animal kingdom; in the late eighteenth century, Goethe continued this tradition when he described the study of the structural interrelationships between living things as *morphology*. Darwinian biological evolutionary theory essentially continues this theme. Biological evolutionism is based upon the continued improvement of a species through the retention of advantageous genetic mutations which offer survival advantages to the organism as a whole. Since a technological innovation is achieved in the context of a comparable evolutionary process, it is hardly surprising that the systematic examination of the morphology (or form) of a technological or functional capability can also identify new mutations which offer potential performance improvements in the capability. Like an advanced biological organism, a given capability (for example, an airplane) often consists of a synthesis of a number of separate subcapabilities (engines, fuselage, wings, and tailplane), each of which may be realized in a number of different ways. Morphological analysis performs a systematic exhaustive categorization and evaluation of the possible alternative combinations of subcapabilities which may be integrated to provide a given functional capability.

We use the illustrative example of building bricks based upon one developed by Wills.[37] For purposes of illustration, we assume that building bricks may be manufactured by various combinations of material and process technologies, producing bricks of differing properties and shapes. In this example we have five subcapabilities, each of which may be realized in up to four different ways (see Table A4.1.1): Given the number of alternative realizations of each of the five subcapabilities, we have $4 \times 3 \times 3 \times 4 \times 5 = 720$ different combinations or realizations of building bricks. Thus, a relatively simple product will readily generate quite a large number of alternative realizations. More complex technological capabilities rapidly generate large numbers of combinations of realizations. By viewing Table A4.1.1 as a two-dimensional matrix, we may define a given realization as a sequence of matrix cells: Thus, A1–B2–C1–D1–E1 is a conventional house brick. In the United Kingdom, British Coal manufacturers house bricks from *discard* or noncarbonaceous waste materials, and this realization is represented by A4–B2–C1–D1–E1. The thermal insulation brick is represented by A1–B1–C1–D2–E1. These are three examples of known realizations of technologies. Configurations which have yet to be achieved in practice constitute technological suggestions, such as the combination A3–B2–C2–D1–E2—an opaque spherical plastic brick.

Zwicky applied the approach of an evaluation to the jet engine.[38] He defines 11 subcapabilities or parameters, each of which could be realized in two, three, or four different ways. He thus found that there were 36,864 possible combinations. Not all of these combinations are technologically compatible or feasible, but there re-

TABLE A4.1.1 Morphological Analysis of Building Bricks

	1	2	3	4	5
A Material	Natural Clay	Metal	Plastic	Waste Materials	
B Forming Process	Extrude	Mold	Press		
C Bonding Process	Heat	Chemical	Molecular		
D Properties	Opacity	Thermal Insulation	Elasticity	Aesthetic	
E Form	Rectangular	Spherical	Interlocking	Cubical	Aesthetic

mained 25,344 technologically feasible potential realizations. Not all of these have yet been tried out or invented.

By exhaustively examining all possible combinations, morphological analysis ensures that all feasible (as well as some infeasible) approaches to producing the capability are considered. Also, it often reveals promising unexpected combinations or new inventions which otherwise would have remained undiscovered. Wills[22] outlines an extension of the approach in the concept of morphological mapping and explores its potential applications in marketing to identify unsatisfied needs (or unoccupied niches in the technology–market matrix).

A4.1.14 RELEVANCE TREES

At the beginning of Section A4.1.12, we pointed out that the realization of a new technological capability requires the sequential completion of a number of inventive problem solving steps which can often be tackled in a number of alternative ways. The relevance tree provides another conceptualization of the sequence which can complement morphological analysis. This approach essentially involves the drawing of one or more tree diagrams which structure the sequence of technological problems which must be solved in order to reach some overall objective.

The basic concept is very simple, and its application to the domestic automobile is shown in Figures A4.1.6a and b. The form and detail of a tree are dependent upon the objective(s) sought. Figure A4.1.6a is a descriptive tree which identifies the components comprising an automobile. On the other hand, Figure A4.1.6b represents a solution tree in which, in this particular example, some of the possible alternative engine systems are shown. Corresponding solution trees could be developed for other components.

A relevance tree is drawn up in the context of the pursuit of some mission or objective which in turn requires the identification and solution of specific technological problems or tasks. As with morphological analysis, when developing a relevance tree care must be taken to ensure that:

1. The overall objective/mission is clearly defined and rigorously broken down into submissions and tasks.
2. All possible ways of performing submissions and tasks are considered exhaustively.
3. A detailed hierarchical tree is developed which rigorously relates tasks and submissions to the overall objective. It is vital to perform this analysis in a rigorous and exhaustive manner to ensure that no tree levels and possible problem-solving approaches are excluded.

Honeywell Inc. has extensively applied relevance trees to identify future health care needs and objectives. Given that specific objectives have been set, resources must be allocated to the branches to achieve them. Because such resources (particularly in health care) will almost certainly be scarce, the question arises as to how

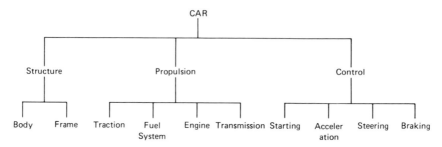

Figure A4.1.6a. Descriptive tree.

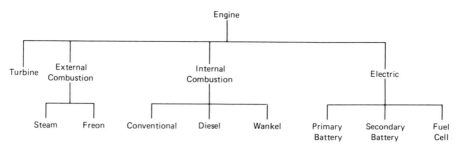

Figure A4.1.6b. Solution tree.

best to allocate them between alternative solutions at different levels. Although this allocation may ultimately be done subjectively, it can be facilitated by the development of relevance numbers of weighting factors which may be assigned to the alternative tasks and approaches required to achieve objectives. Relevance numbers is one way one can weight the relative importance of R&D projects, so it is of potential value in R&D project selection. Also, the comparison of relevance trees with Critical Path Method/Program Evaluation and Review Technique (CPM/ PERT) networks and the notion of predecessor or precursor events lead naturally to a role for them in R&D management and control.

At the end of the previous section, we indicated that morphological analysis is valuable in exploring for potential synergies in the technology-matrix. The relevance tree approach can also be used to integrate R&D planning and marketing, and indeed overall corporate operations through developing *binary relevance trees*. Rickards[39] describes such an approach to integrating a company's marketing strategy into its overall policy, and Hubert[40] describes its application to relate corporate and R&D activities in Unilever.

A4.1.15   CROSS-IMPACT MATRICES

Until now, we have considered the forecasting of capabilities individually. In practice, the probability of development of a given capability may well be dependent on developments in contiguous technological areas. When developing relevance trees, it is readily evident that communalities can exist between technological solutions of realizations developed in different branches of the tree. The process of rigorously and exhaustively developing relevance trees usually involves developing relevance matrices which identify the commonalities or interactions between activities in different branches of the tree. Beastall[41] developed such matrices to identify the commonalities between differing subsystems and activities in the British Post Office, and Hubert also developed similar matrices in the Unilever example cited above.

Gordon (who helped pioneer the Delphi method) and Haywood[42] recognized that such cross-impacts should be accommodated in Delphi exercises, and they independently developed the concept of cross-impact matrices for this purposes. Cross-impact matrices are similar to relevance matrices, but seek to identify the impact that one technological innovation may have on the probability and time of occurrence of another.

Technological interactions may be direct and of the first order or indirect and of second or higher order (two forecasted events may interact through others rather than directly).

The interactions between two forecasts can occur in several ways:

Mode of interaction. One event may enhance or diminish the likelihood of another event. It may advance or delay the second event. It may necessitate or obviate the second event. It may enable or prevent the second event.

Force of interaction. Strong . . . weak.

Time lag. Immediacy of influence, length of influence.

A cross-impact matrix (Figure A4.1.7) may be constructed as follows:

1. The events under consideration (E_1, E_2, E_3) are arranged in their expected chronological order in both rows and columns.
2. Consider cell E_1 E_2. This represents the impact of the *occurrence* of E_1 on the likelihood of occurrence of E_2.
3. Consider cell E_2E_1. This represents the impact of the *nonoccurrence* of E_1 on the likelihood of occurrence of E_2. (Since events are arranged in expected chronological order, E_2 cannot impact on E_1.)

Each cell includes the mode (positive or negative), strength, and lag of the interaction. For example, the entries in cell E_2 and E_1 indicate the nonoccurrence of event E_1 on the likelihood of occurrence of E_2. They indicate that the impact will be five years later and will diminish the probability of E_2 occurring by 20%. It is also of interest to compare E_2 E_3 and E_3 E_2. Comparing these entries indicates that E_2 is largely a precursor event to E_3.

Martino describes some conceptual and practical difficulties, as well as applications, of the method.[43] Cross-impact matrices must be developed with the same rigor and coherence as other approaches if they are to aid the development of reliable forecasts. Even quite a small list of events can generate a large number of

Event occurrence ↓	*Impact on:*		
	E_1	E_2	E_3
E_1		+ .20 Immediate	+ .70 5 years
E_2	− .30 3 years		0 Immediate
E_3	− .20 Immediate	− .90 Immediate	

Figure A4.1.7. Cross-impact matrix.

interactions. A five-event matrix (E_1, E_2, . . . E_5) can generate 20 cells (excluding the leading diagonal cells) of first-order interactions apart from any second order (say the interaction of event E_1 on event E_3 through event E_2) or higher order interactions. Clearly, it is very difficult to evaluate mentally all possible cross-impacts with even a quite small number of events, and formulae for doing this have been suggested.[44] They have also been employed fairly widely in Delphi and other TF studies. They can be incorporated into computer simulation models as part of a scenario generation process (see the next section). Jones and Twiss report a variety of cross-impact studies ranging from the original one by Gordon and Haywood used to evaluate the Minuteman missile system through studies on the future development of Canada to specific forecasts of farm tractor and industrial chemical technology.

A4.1.16 COMPOSITE FORECASTS—SCENARIOS

Cross-impact matrices represent one way of evaluating the interactions between events when combing forecasts. One may also wish to make a composite set of forecasts of numerous diverse events over a wide-ranging field. This situation first arose in social rather than technological forecasting and is particularly associated with the work of Herman Kahn of the Hudson Institute. In *The Year 2000* (coauthored with Wiener), the authors suggest a number of scenarios for the post-industrial society of the turn of the millennium.[4]

In Section A4.1.2 we introduced the notion of future history and, in their book, Kahn and Wiener were seeking to write a future history for the turn of the century. As we stated, historians present their interpretations of the past events by descriptive narrative of word picture, and forecasters may also present their conjectures and interpretations in a similar manner. In forecasting, this form of presentation of compositie forecasts is known as a *scenario* or "*a word picture of some future time, possibly including a discussion of the events which lead to the situation depicted.*"[45] To readers who question this reasoning, it is pointed out that the reliability or accuracy of a historical interpretation of past composite events or a historical scenario is clearly dependent upon the extant data and information on the period concerned. Even if the extant data are information on the period concerned. Even if the extant data are ample, historians frequently disagree in their individual interpretations or scenarios. However, when one reads interpretations of prewritten and prehuman history (that is, as history shades into anthropology and biology), one cannot fail to be amazed at the extrapolations and interpolations made by some workers on the sparse extant data to produce interpretations of scenarios of the distant past. This suggests that the epistemological credence for a future history is at least as valid as that of some past history. Bearing in mind that *no attempt* to forecast the future is, in effect, an implicit forecast that the *status quo* will continue, it is, of course, impossible to avoid either explicit or implicit forecasting. Therefore, it can be argued that a careful and rigorous composite forecast is the best picture of the future available. It also opens up the possibility of inventing the future, as we shall see shortly.

We have already stated that the pioneering work of Kahn and Wiener was concerned with broad socioeconomic and political forecasting at the international level, so it does not directly concern us here. However, their concept and approach of scenario generation was quickly adapted by others to produce forecasts of value to individual firms. Christine MacNulty, who has developed scenarios for numerous corporations in both North America and Europe, is one of the most experienced forecasters in this field.[46] Another is Pierre Wack who, in 1968, forecasted the 1973 OPEC oil price rise for his employer, the Royal Dutch Shell Group, thereby pioneering the use of scenarios in the Group.[47,48] Linneman and Klein[49,50] report that, in 1977, 150 of the Fortune 1000 industrial corporations were using multiple scenario analyses in their planning.

The basic approach is to develop several rather than only one scenario. These can be:

1. The *surprise-free* scenario, which we will call scenario A. This essentially involves the extrapolation of current trends based upon the assumption that they will continue indefinitely into the future, without major disturbance or turbulent events occurring. This can be viewed as the standard world for the organization.

2. *Alternative* scenarios, which are then developed using a set of internally consistent assumptions, will generate possible plausible outcomes will extreme variations in either direction from the surprise-free scenario. We will label these scenarios B and C, and they represent two extreme alternative futures for the organization.

3. In practice, it is unlikely that either of the extreme alternative futures will occur. On the other hand, because of the ubiquity and accelerating rate of technological and social change, it is likely that future actualities will progressively differ from the surprise-free scenario towards one or other of the alternatives as time progresses. Therefore, forecasters may generate a number of variations (known as canonical variations) upon the surprise free scenario;

that is, alternative futures which incorporate elements of scenario A with scenario B or scenario C in different combinations. This is also illustrated in Figure A4.1.8.

Schwartz is another writer with considerable experience developing scenarios for major organizations (including Royal Dutch Shell, Volvo, Nissan, and the International Stock Exchange).[14] He warns of an inherent danger of the above ABC approach; that is, two extreme variations about the surprise-free scenario. He argues that managers will view them as most likely (A), optimistic, and pessimistic (B and C) cases and automatically favor Scenario A. He also argues against assigning overall probabilities to scenarios since, again, managers may plump for the one with the highest probability and ignore the others. His view is that scenarios are vehicles for helping managers to learn and to change their perceptions or mindsets. Therefore, scenarios should not merely extrapolate present trends, but should present alternative images of the future. Schwartz's views echo Peter Drucker's maxim (quoted in Section A4.1.2) that forecasting is concerned with the futurity of present decisions—that is, the organization can make decisions in light of these scenarios and test their robustness against the alternative images of the future which have been generated.

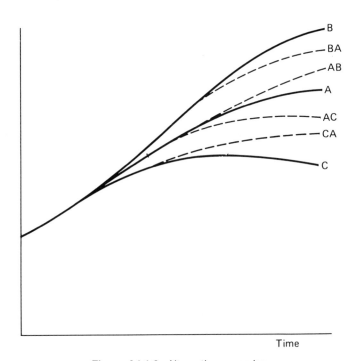

Time

Figure A4.1.8. Alternative scenarios.

A4.1.17 GENERATING SCENARIOS

As was stated in the previous section, Schwartz has considerable experience in generating scenarios for individual corporations, and his overall approach is summarized here. It is broken down into a number of steps.

Step 1: Identify Focal Decision or Issue. If scenarios are to change management perceptions, they should focus on a specific decision or issue of current concern and with long-term implications for the organization and impacts on its fortunes. Also, these impacts will depend upon the overall environment that the organization faces over the same period. The design might be whether to enter a new market, build a new plant, or adopt a new technological strategy.

Step 2: Identify Key Factors in the Local Environment. Once the decision and focus has been established, the key factors and considerations local to the organization (for example, its customers, suppliers, and competitors) which will influence decision outcomes can be identified.

Step 3: Discern Driving Forces. The next step, which is the most research-intensive, is to discern the driving trends in global environment of the organization that will also influence factors and therefore outcomes identified in Step 2. These may well include demographic, social, economic, political, ecological, and technological forces and trends.

Step 4: Establish Rank by Importance and Uncertainty. Each of the key factors and driving forces identified in Steps 2 and 3 are now ranked by two separate criteria:

a) Their degree of influence on the outcome of the focal decision.

b) The degree of uncertainty associated with their values.

From this step, usually a few factors and forces that have large influences and uncertainties are identified.

Step 5: Select Scenario Logics. It is from these few factors and forces that the axes or dimensions by which alternative scenarios are defined, so this definition is critical to the whole effort. They are chosen to yield a few scenarios and scenario logics that clarify the issues basic to the success of the focal decision. To take a very simplified example, suppose the focal decision faced by a U.S. automaker is whether to manufacture and market a new well-designed, large, gas-guzzling car. Suppose it is determined that the two most important factors or forces are fuel prices and the extent of economic protectionism. There can then be four scenario logics, based upon the four combinations of high or low fuel prices and highly protectionist or globally free economies. Clearly, the new gas guzzler could be successful in the two scenarios with low fuel costs, but would be unsuccessful in the other two.

Step 6: Flesh Out the Scenarios. Once the scenario logics have been selected, they are plotted into narrative form (like their theatrical counterparts), all incorporating the key factors and forces in an appropriate manner, how each scenario could unfold, and what events would be necessary to make its end point plausible. Schwartz also suggests that the scenarios should be given evocative names, such as Shell's *World of Internal Contradictions* which was continually updated and used for 10 years in the company. They should preferably be visually presentable in a matrix form.

Step 7: Determine Implications. Once the scenarios have been developed, then the decision focus has to be played out in each of them to determine the rewards, risks, and vulnerabilities in each of them. If a decision looks good for all scenarios or is very robust, then the work may be terminated. It is more likely that the decision choice will be dependent on what develops in reality, so a final step is needed.

Step 8: Select Leading Indicators and Signposts. The step requires the identification of signals of change which should be regularly monitored as signposts indicating which of the above scenarios is unfolding in reality. The selection of indicators and signposts should not be unduly difficult since it should be based upon each of the scenario logics originally developed. Decisions can then be made based upon the unfolding reality.

In his text, Schwartz emphasizes that scenarios can be developed in small as well as large organizations, and he gives two quite detailed examples. He describes three scenario logics for a small start-up company planning to market highly durable, quality garden tools for the global economy in 2005. He claims that the generation of good scenarios is dependent upon the choice of a focal issue or decision of pressing concern for managers (as indicated in Step 1 above) and the right selection of the scenario development team. This selection should be guided by three major considerations. First, it involves support and participation from top management. Second, team membership should embrace a broad range of functions and divisions of the organization. Third, team members should be people with imagination and open minds who work well together. As he puts it:

> You can tell you have good scenarios when they are both plausible and surprising; when they have the power to break old stereotypes; and when the makers assume ownership of them and put them to work. Scenario making is intensely participatory, or it fails.

A4.1.18 SCENARIOS AND INVENTING THE FUTURE

The last sentence of the previous section implied a passive strategic attitude towards the future. That is, a range of scenarios is developed which embraces the range of future possibilities, then organizational plans and decisions are modified in light of the future that actually unfolds. In Section A4.1.2, we pointed out that one of the books which stimulated an interest in future studies was Denis Gabor's *Inventing the Future* in which he postulated a proactive stance towards the future.[3] He introduced the notion of alternative futures or scenarios for a social group from which it should select or invent the future it desired. Once a firm has developed a number of alternative scenarios, it can pursue technological plans or strategies that it seeks to create or invent the alternative future or scenario that it prefers. This notion is illustrated conceptually in Figure A4.1.9.

The present state and environment is depicted on the left-hand side of this figure, with the alternative scenarios or futures depicted on the right-hand side. The continuous and dotted lines notationally represent the alternative paths through time which the firm must follow in order to realize each of these alternative futures. These futures may have common paths initially, but as time progresses, nodes are reached where they diverge. These nodes represent decision points where the firm must choose which path it wishes to follow or which future it wishes to invent. If the firm chooses Scenario AB, that is, invent that future, it must follow one of the continuous paths shown in the figure and make the appropriate decision at the shaded nodes to ensure that it does so. On each of the continuous paths, the sequence of decisions made at the shaded nodes can be viewed as a technological plan or strategy. Note that we have drawn two paths (I and II) which the company may follow to invent its preferred future, that is, it has two alternative technological strategies I and II. The strategy chosen will depend upon the company's estimates of the probabilities of each of the scenarios occurring, the expected benefits which each offers, and the company's attitude towards risk.

Readers may object to this view of inventing the future on the grounds that it is both naive and utopian. Clearly, most companies can exercise only a very limited influence on the future environment in which they operate, which is continuously subjected to uncertainties and turbulences over which they have little control. Running most businesses is more like sailing a small boat, in which continuous adjust-

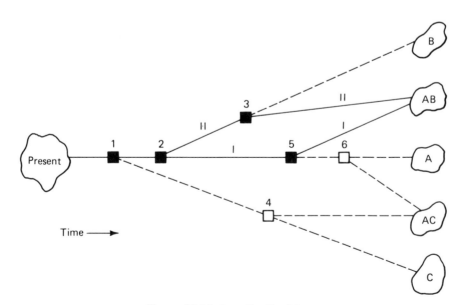

Figure A4.1.9. Inventing the future.

ments and changes may have to be made in light of changing sea and wind conditions, rather than steering a large ocean-going ship. However, provided the
organizational and environmental variables are judiciously selected in the process of
scenario development described in the previous section, the company may define a
comparative few which are relevant to its operations—and it is the alternative states
of these which constitute its alternative scenarios. (Rather like a biological organism
which evolves a set of senses which are relevant to its behavior and survival in the
environmental niche it has selected.) It may then be able to exercise a sufficient
influence on these variables significantly to influence future developments in the
direction it seeks. Most readers will accept that Texas Instruments has been one of
the most successful high-technology organizations since World War II, and that the
corporation provides a graphic illustration of the notion of inventing the future. In
the 1950s when the semiconductor industry grew rapidly based upon the manufacture of successive generations of discrete semiconductor devices, most companies
concentrated on germanium technology. Although it has inferior properties to silicon, most companies chose to develop and manufacture germanium devices because
to do so was technologically easier. Texas Instruments, however, committed itself to
silicon technology because its technological management believed the advantages of
silicon outweighed its disadvantages. Thus, they sought to invent a silicon future for
the company. The developments in integrated circuit technology from the late 1950s
onwards did produce a semiconductor industry primarily based upon silicon.

A4.1.19   THE VALIDITY OF TECHNOLOGICAL FORECASTS

We have now completed our review of the more frequently employed TF techniques. This overview does not claim to be exhaustive, and it is recommended that
readers see the specialist texts and journals on the subject if they wish to pursue it in
more depth. However, all readers will be concerned with the accuracy and efficacy
of TF efforts to date. We have already done so in Sections A4.1.9 and A4.1.11, but
it is worthwhile briefly to comment further on this matter.

First of all, it is important to distinguish between efficacy and accuracy in TF.
The purpose of TF is to aid technological planning and decision making, and it
should be viewed in that context. Its ultimate purpose is to promote efficacious
forecasts in the sense of promoting good technological decision making, rather than
accurate forecasts. Thus, it is possible to distinguish between what can be labeled
passive and active forecasting studies:

1. In a passive study, the criteria of efficacy and accuracy coincide. Its purpose
is to forecast the behavior of a technological function or capability as accurately as
the situation allows.

2. In an active study, the forecasting process seeks to promote a self-fulfilling or
self-defeating prophesy. Ideally, a self-fulfilling forecast will stimulate a sequence
of technological decisions which will make the forecast come true. (Particularly if it
is performed as an attempt to invent the future.) In contrast, a self-defeating forecast

will stimulate a sequence of technological decisions which will ensure that the forecast fails to come true. In both situations, the value of the forecast must be judged by its efficaciousness in triggering the desired sequence of technological decisions. The controversial *Limits to Growth* study, cited in Section A4.1.10, quite dramatically illustrates this argument. The assumptions and methodologies used in that study have been criticized so that its accuracy is, to say the least, questionable. It could, however, be viewed as an efficacious example of self-defeating forecasting. The prestige of its sponsoring body—The Club of Rome—and the dramatic doomsday scenario published, focused worldwide attention on the dangers of untrameled energy consumption, population growth, and pollution and helped make them political issues. The public concern which it aroused has stimulated governments to be more actively concerned with these dangers than they would have been had it not been performed and published.

Having pointed out that the criterion for evaluation is dependent upon its contextual use, it must also be pointed out that credibility of a forecast (whether extrapolative, self-fulfilling, or self-defeating) in the eyes of the technological decision makers is dependent upon the credibility of the methodologies and techniques used to generate it. One of Ascher's overall conclusions is that a major determinant of forecast inaccuracies is the validity of the core assumptions of the forecasters. That is, the validity of their insights into the underlying social, economic, political, and technological factors and trends which are interacting to produce the outcomes being forecast.[28] He suggests that forecasts based upon Delphi exercises, which identify the consensus of expert opinion, may be superior because experts are, by definition, capable of predicting from the best set of core assumptions.

A4.1.20 ORGANIZATIONAL LOCATION OF THE TF FUNCTION

At the beginning of this Appendix, it was suggested that a TF function should perform the technological futures studies required for corporate technological planning in light of the characteristics of the innovation process and the alternative technological strategies discussed in Chapter 4. It was also pointed out earlier that TF was a significant corporate activity in 1966 when Jantsch reported that 500–600 of the largest U.S. firms were engaged in such activities. The effective deployment of TF in corporate technological planning depends upon the way it is set up and its location in the organization, so we make some very general comments here. Fusfield and Spital cite GE, Monsanto, Whirlpool, Cincinnati Milacron, and Goodyear as firms which have successfully developed technology forecasting and planning programs.[51] They suggest three requirements for effective technology planning and forecasting, which are similar to those specified by Schwartz in Section A4.1.17 as the requirements for generating scenarios.

The first is a formal integrating mechanism. TF requires the part-time participation of diverse members of the corporate technological base, and some such individuals may assign it a low priority as compared to their primary work roles. It is vital, therefore, to assign specific responsibility and accountability for planning/fore-casting to an integrative group. This group may well constitute the nucleus of the innovation management function discussed in Chapter 4. The second requirement is that TF must be designed and geared to support planning. This implies three distinct, but interrelated, requirements. Forecasts must be derived and presented in a format which satisfies planning needs. The success of Monsanto's planning efforts is partially attributed to their forecasters' ability to identify potential planning problems and to provide forecasts that explicity address these problems. At Whirlpool, results are presented in the form of *"what this means to Whirlpool."* The cycle time of forecasting must be synchronized with the cycle time of the planning process which, in turn, is dependent on the lead time of the innovation process. This planning horizon can, of course, vary from six months for companies following an applications engineering strategy to a quarter century or longer for companies in some industries. Also, in industries characterized by a very rapid rate of technological change, forecasting and planning efforts may be required over several increasing time cycles, from a year upwards, to match this rate to ensure that such companies can respond rapidly to environmental changes. Also, forecasting must be comprehensive and supportive to planning efforts, and include assessments of competitors' future technological capabilities. The third requirement is that TF must enjoy top management support. The successful applications of TF in Monsanto, Whirlpool, Cincinnati Milacron, and Goodyear have been particularly attributed to support

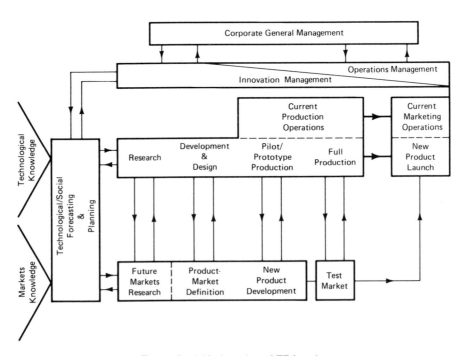

Figure A4.1.10. Location of TF function.

from top management. This is best achieved by involving senior management in the forecasting process. Obviously, the above remarks apply primarily to medium-sized or large organizations. However, as was indicated in Section A4.1.17, TF can also be used in small companies when its smallness should ensure good participation throughout. This raises the concluding issue in this Appendix—the organizational location of the TF function.

Given the characteristics of the innovation chain equation and in the above considerations, we see that the effective use of TF requires participation and support throughout the corporate technological base. In Chapter 4, we suggested that firms incorporate innovation as well as operations management functions into the organizational structure to expedite the innovation process. This consideration should be reflected in the location of the TF function. The high-technology company requires a dichotomous structure reflecting its dual future–present or innovation–operations orientations (conceptually illustrated in Figure A4.1.10 which is an extension of Figure 4.2). The TF function can be viewed as a "Futures Studies Function" for the organization which draws upon and interacts with individuals committed to innovation in the corporate management, R&D, and marketing functions (although not, of course, precluding interactions with the operations personnel). This TF function then performs the necessary studies to generate alternative future scenarios for the organization which can be used as a basis for technological (including R&D) planning, as described in the previous chapter (particularly Section 4.6). The relationship between TF and R&D planning will be taken up in Chapter 5.

FURTHER READING

P. Schwartz, *The Art of the Long View.* New York: Doubleday, 1991.

REFERENCES

1. H. G. Wells, "The Discovery of the Future." *Nature (London)* **65,** 326–331 (1902).
2. T. J. Gordon and O. Helmer, *Report on Long-Range Forecasting Study.* Santa Monica, CA: Rand Corporation Paper, 1964, p. 2982.
3. D. Gabor, *Inventing the Future.* Harmondsworth, Middlesex: Penguin Books, 1964.
4. H. Kahn and A. J. Wiener, *The Year 2000: A Framework for Speculation on the Next Thirty-Three Years.* New York: Macmillian, 1967.
5. E. Jantsch, *Technological Forecasting in Perspective.* Paris: OECD, 1967.
6. B. De Jouvenal, *The Art of Conjecture.* New York: Basic Books, 1967.
7. D. H. Meadows *et al., The Limits to Growth.* New York: Universe Book Publishers, 1972.
8. E. H. Carr, *What is History?* New York: Knopf, 1962.
9. P. Drucker, *Technology, Management and Society.* London: Harper & Row 1977.
10. J. R. Bright, "Evaluating the Signals of Technological Change." *Harvard Business Review***15**(4), 50–65 (1972).

11. N. S. Langowitz, "Managing a Major Technological Change." *Long Range Planning* **25**(3), 79–85 (1992).

12. J. P. Martino, *Technological Forecasting for Decision Making.* New York: American Elsevier, 1972.

13. J. Naisbitt, *Megatrends Ten New Directions Transforming Our Lives.* New York: Warner Books, 1984.

14. P. Schwartz, *The Art of the Long View.* New York: Doubleday, 1991, pp. 88–89.

15. A. Kokubo, "Japanese Competitive Intelligence for R&D." *Research Technology Management* **35**(1), 33–34 (1992).

16. J. P. Martino, *Technological Forecasting for Decision Making.* New York: American Elsevier, 1972.

17. J. P. Martino, *Technological Forecasting for Decision Making.* New York: American Elsevier, 1972, chapter 5.

18. R. N. Foster, *Innovation: The Attacker's Advantage.* New York: Summit Books, 1986.

19. G. A. Pogany, "Cautions About Using S-Curves." *Research Technology Management* **29**(4), 24–25 (1986).

20. A. W. Schmidt and D. F. Smith, "Generation and Application of Technological Forecasting for R&D Programming." In J. R. Bright (Ed.), *Technological Forecasting for Industry and Government.* Englewood Cliffs, NJ: Prentice-Hall, 1968.

21. F. J. Kovac, "Technology Forecasting—Tires." *Chem. Technol.* **1**(1), 18–23 (1971).

22. G. Wills *et al., Technological Forecasting.* Harmondsworth, Middlesex: Penguin Books, 1972.

23. E. M. Rogers and F. F. Shoemaker, *Communication of Innovations: A Cross-Cultural Approach,* 2nd ed. New York: Free Press, 1971.

24. H. A. Linstone and D. Sahal, (Eds.), *Technological Substitution Forecasting Techniques and Applications.* New York: American Elsevier, 1976.

25. J. C. Fisher and R. H. Pry., "A Simple Substitution Model of Technological Change." *Technological Forecasting and Social Change* **3,** 75–78 (1971).

26. H. Jones and B. C. Twiss, *Forecasting Technology Planning Decisions.* London: Macmillan, 1978, p. 205.

27. D. Sahal, *Patterns of Technological Innovation.* Reading, MA: Addison-Wesley, 1981.

28. W. Ascher, *Forecasting an Appraisal for Policy-markers and Planners.* Baltimore, MD: Johns Hopkins University Press, 1978.

29. J. W. Forrester, *World Dynamics.* Cambridge, MA: MIT Press, 1971.

30. H. Jones and B. C. Twiss, *Forecasting Technology Planning Decisions.* London, Macmillan, 1978, pp. 207–212.

31. B. Alenstein and D. E. Probert, "A Strategic Control Module for a Corporate Model of British Telecommunications." In O. A. Anderson (Ed.), *Proceedings of the International Public Utilities Conference.* Amsterdam: North-Holland Publ., 1980.

32. O. Helmer and N. Rescher, "On the Epistemology of the Inexact Sciences." *Management Science* **6**(1), 25–52 (1959).

33. N. C. Dalkey and O. Helmer, "An Experimental Application to the Use of Experts." *Management Science* **9**(3), (1963).

34. M. J. Cetron and C. A. Ralph, *Industrial Applications of Technological Forecasting.* New York: Wiley, 1971, p. 115.

35. G. Rowe, G. Wright, and F. Bolger, "Delphi: A Reevaluation of Research and Theory." *Technological Forecasting and Social Change* **39**, 235–251 (1991).

36. F. Woudenberg, "An Evaluation of Delphi." *Technological Forecasting and Social Change* **40,** 131–150 (1991).

37. G. Wills *et al., Technological Forecasting.* Harmondsworth, Middlesex: Penguin Books, 1972, "The Preparation and Deployment of Technological Forecasts." *Long Range Planning* **2**(3), 44–52 (1970).

38. F. Zwickey, *Monographs on Morphological Research.* Pasadena, CA: Society for Morphological Research, 1962.

39. T. Rickards, *Problem Solving Through Creative Analysis.* Epping, Essex: Gower Press, 1974.

40. J. M. Hubert, "R&D and the Company's Requirements." *R&D Management* **1**(1) (1970).

41. H. Beastall, "The Relevance Tree in Post Office R&D." *R&D Management* **1**(2) (1971).

42. T. J. Gordon and H. Haywood, "Initial Experiments with the Cross-Inpact Matrix Method of Forecasting." *Futures* **1**(2), 100–116 (1968).

43. J. P. Martino, *Technological Forecasting for Decision Making.* New York: American Elsevier, 1972, pp. 271–281.

44. H. Jones and B. C. Twiss, *Forecasting Technology Planning Decisions.* London: Macmillan, 1978, pp. 243–257.

45. J. P. Martino, *Technological Forecasting for Decision Making.* New York: American Elsevier, 1972, p. 267.

46. C. R. McNulty, "Scenario Development for Corporate Planning." *Futures* **9**(2), (1977).

47. P. Wack, "Scenarios: Uncharted Waters Ahead." *Harvard Business Review* **63**(5), 73–89 (1985).

48. P. Wack, "Scenarios: Shooting the Rapids." *Harvard Business Review* **63**(6), 139–150 (1985).

49. R. E. Linneman and H. E. Klein, "The Use of Multiple Scenarios by US Industrial Companies." *Long Range Planning* **12**(1), 83–90 (1979).

50. R. Linneman and H. E. Klein, "The Use of Scenarios in Corporate Planning—Eight Case Histories." *Long Range Planning* **14**(5), 67–77 (1981).

51. A. R. Fusfield and F. C. Spital, "Technology Forecasting and Planning in the Corporate Environment: Survey and Comment." *TIMS Special Studies in the Management Sciences* **15,** 151–162 (1980).

APPENDIX 4.2
ENVIRONMENTAL CONCERNS AND TECHNOLOGY ASSESSMENT

For the past two centuries, those who piloted the ship of industry have had a relatively free hand in managing its affairs. Now the passengers of this earthly craft are exerting their power to set the course. They believe that the quality of their lives is threatened by pollution, that the leaders of industry and government are responsible and should be held to account.

J. E. (TED) NEWELL, CHAIRMAN AND CEO, DU PONT CANADA INC.,
"MANAGING ENVIRONMENTAL RESPONSIBILITY"
BUSINESS QUARTERLY

If you talk to enough people about technology assessment, the concept begins to sound as marvelous as motherhood. Except in this case, nobody knows how to get pregnant.

NINA LASERSON, "III—TECHNOLOGY ASSESSMENT AT THE THRESHOLD" *INNOVATIONS*

A4.2.1 INTRODUCTION

The publication of Rachel Carson's *The Silent Spring* in 1962 could be said to have first aroused public concern for environmental pollution.[1] This concern grew throughout the 1960s and early 1970s, stimulated by the publication of other books such as *Limits to Growth* discussed in Appendix 4.1,[2] Barry Commoner's *The Closing Circle*,[3] Vance Packard's *The Waste Makers*,[4] and the post-1973 energy crisis.

The cumulative impacts of successive technological innovations can be viewed in the total context of technological evolution. Until the advent of the first industrial revolution in the late eighteenth century, human culture (including technology) was primarily based upon agriculture and had relatively little impact upon the "natural" environment or global ecological system or ecosphere. The cumulative effect of the growth of industrialization, human population, and the combustion of fossil fuels, etc., over the past 200 years is now believed to be modifying the ecosphere in a potentially dangerous and irreversible manner. One dramatic manifestation of such danger is the greenhouse effect—the impact that growing concentrations of carbon dioxide in the upper atmosphere, produced by the combustion of fossil fuels, may have on the amount of solar radiation reaching the earth's surface and/or the amount of heat radiated back into space. Depending upon the overall balance between these two conflicting effects, the temperature of the earth's surface may be slowly rising or falling and quite small changes in either direction could induce radical changes in the global climate patterns, sea-level, etc.

The notion of environmental regulation can be traced back some time. The burning of sea-coal was banned in English cities in the thirteenth century.[5] In the early years of the first industrial revolution, farmers in South Wales sued local copper producers for the contamination to agriculture land caused by copper extraction processes.[6] Also public health authorities have been concerned with environmental health hazards for many years, certainly since the last century when improvements in public water supply and sewage considerably reduced morbidity and mortality from infectious diseases. Furthermore, during the 1950s and early 1960s, clean air legislation in Britain, which banned the use of house-coal in domestic fireplaces and restricted the amount of smoke that could be released into the atmosphere from industrial plant chimneys, did much to reduce smog in British cities. These regulations were enacted before the "green" movement developed and serious political and legislative attention was paid to the broad assessment of the impact of technological innovations.

It should also be stressed that it has always been commercially prudent for companies to perform technology assessment, in a traditional sense, when evaluating new capabilities. However, such assessments have been based on the *primary* impact of the innovation on the company's operations and its markets. Until recently, companies have been under limited legal responsibility to assess the *secondary and higher order* impacts (particularly on third parties) of innovations. The environment was treated as a free good, like the commons in medieval agrarian society. Polluting this environment was not viewed as antisocial because it was merely exercising a citizen's right to free grazing on the commons. It has now been realized that the natural environment, like the commons, is a semirenewable resource which can be destroyed by overpollution, just as common land can be destroyed by overgrazing.

Also, it should be stressed, that (up to the present at any rate) the primary and secondary benefits of most technological innovations far outweigh their secondary disbenefits. Human social and technological evolution is inextricably linked, and it is platitudinous to observe that our present high state of material well-being is primarily based on technology. However, since a law of "diminishing psychological and social returns" may apply to technological innovations, society must critically examine further innovations to ensure that benefits outweigh disbenefits. In fact, we can adapt Maslow's hierarchy of needs to this context. This implies that a society's evaluation of the potential benefits and disbenefits of a given technological innovation is subjectively dependent upon its unsatisfied needs. Thus, a poor economy might tolerate an innovation which significantly increasès environmental pollution because these disbenefits are outweighed by the economic development benefits and the increase in employment it provides. In contrast, a rich economy might reject the same innovation on the grounds that the environmental pollution outweighs the marginal contribution of the innovation to the economy and employment opportunities. On several occasions already, we have quoted the example of the fate of the Concorde and the Super Sonic Transport (SST) to illustrate this fact. Although this is an extreme example, nowadays some evaluation of the environmental impacts as well as the commercial expectations of a prospective technological innovation must be made in many industries before substantial sums are invested in its continued development. These environmental impacts and technology assessments must typically be integrated into the technological planning and R&D project evaluation procedures of organizations, so we consider such assessments in this Appendix.

A4.2.2   THE TECHNOLOGY ASSESSMENT PROCESS AND ENVIRONMENT IMPACT STATEMENTS

Some high-technology industries, such as the pharmaceutical and aeronautical industries, have traditionally be subject to formal requirements for testing potential innovations. Moreover, the public environmental concern is only one facet of a wider concern for the safety of new technological products. The thalidomide tragedy and *Nader's Unsafe at Any Speed*[7] questioned the reliability of drug and automobile testing and safety standards, so that from the 1960s onward, governments in most developed countries enacted new laws to assess and regulate prospective technological innovations to reflect public concern for safety standards, environmental pollution, and energy conservation. Such concerns have been reinforced by more recent tragic incidents, such as those at Three Mile Island, Cherbonyl, and Bhopal. Some potential innovations may have quite diffuse impacts and unexpected side effects which affect various interest groups in society at large. Therefore, the assessment process must be a political process which admits adequate avenues of expression to all parties or constituencies which may be affected. Also, once such innovations are launched, their impacts must be monitored for any unexpected undesirable side effects, analogous to those used before and after the introduction of a new drug.

Technology assessment (TA) should be performed at two levels. First, at the societal level, where its broadest aspects should be evaluated. This exercise often requires resolutions of conflicts of interests between various constituencies, so it is becomes a political process. It should be performed by one or more public agencies with no vested interest in its outcome. Second, the technology should be evaluated at the company level, based upon technicoeconomic factors, but giving due consideration to the above broader impacts. Both levels of assessment should be performed interactively with the regulative agencies and company collaborating in joint evaluation and testing programs to assess actual and potential impacts. Thus, TA should be viewed as a technicoeconomic process embedded in a broader political process involving the key players in the environmental arena as follows:[8] *Consumers* or all citizens concerned about environmental issues. *Educators* (from the elementary school level upwards) seeking to inform their constituents about these issues. The *Media* also seeking to inform consumers about and, if appropriate, to dramatize these issues. The *Environmental Activist Groups,* which include international agencies such as *Greenpeace* and *Friends of the Earth,* as well as local *ad hoc* groups formed to dispute a local environmental issue who seek to alert and arouse consumers' concerns. In some countries, Green Parties have been formed to run candidates in national and local elections to seek political office and consequently to influence environmental decision making. *Legislators* and *Lawyers* responsible for formulating, enacting, and upholding environmental protection laws. *Producers,* including all elements in the production process leading to a consumer good or service. Several writers have treated TA as political process. For example, Gibbons[9] views it as an extension of Etzioni's "mixed-scanning" paradigm of societal decision making,[10] while Bozeman and Rossini[11] view it in the context of Allison's analysis of the resolution of the Cuban missile crisis.[12] Those who wish to examine the political process treatments of TA further may read the cited references, but many place more emphasis upon its technicoeconomic aspects. This is apparent from reading one edited text on the subject[13] where, for example, Strasser exemplifies the latter approach in his assessments of automobile emission controls, computer-communications networks, enzymes, sea-farming, and water pollution.[14] Also, all of the above citations focused on TA at the societal rather than corporate level and, since we are concerned with the latter, they are not our direct concern here.

Most Western countries have enacted laws which require individual companies to prepare an Environmental Impact Statement (EIS) before certain technological initiatives are approved. The terms "Technology Assessment" and "Environment Impact Statement," among others, have tended to be used loosely and interchangeably in the literature. For exposition purposes, here we view EIS as a product of the company-level TA process. This is broadly consistent with a distinction made by Waller.[15] The EIS constitutes the basis for the consultative process among company, regulative agency, and other parties which is integral to the assessment procedure. It takes the form of documentation which examines the full ramifications of the proposed innovation. Its scope and detailed content is obviously dependent upon the specific technological project under consideration. Typical projects in the private sector might be the building of new chemical, oil, power, or steel complexes which would require different detailed assessments from, say, a water resources project in the public sector. Assessments would also be different for the development of a new product such as a pesticide, plastic, or food additive as opposed to a new plant. Such differences are no more marked than in TF, so it seems reasonable to expect that, as with TF, generally applicable techniques for preparing EISs should have been developed. Moreover, any comprehensive technological forecast must include an impact assessment, so corporate TA is really part of corporate TF. One can also

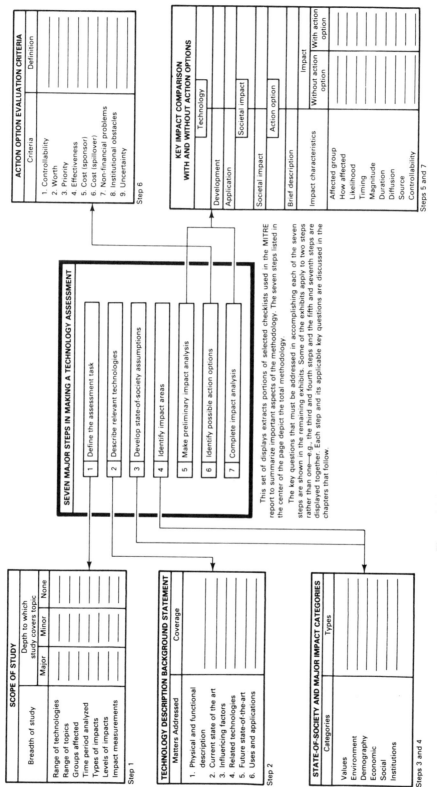

**TECHNOLOGY ASSESSMENT:
A METHODOLOGICAL OVERVIEW**

SCOPE OF STUDY

Breadth of study	Depth to which study covers topic		
	Major	Minor	None
Range of technologies			
Range of topics			
Groups affected			
Time period analyzed			
Types of impacts			
Levels of impacts			
Impact measurements			

Step 1

TECHNOLOGY DESCRIPTION BACKGROUND STATEMENT

Matters Addressed	Coverage
1. Physical and functional description	
2. Current state of the art	
3. Influencing factors	
4. Related technologies	
5. Future state-of-the-art	
6. Uses and applications	

Step 2

STATE-OF-SOCIETY AND MAJOR IMPACT CATEGORIES

Categories	Types
Values	
Environment	
Demography	
Economic	
Social	
Institutions	

Steps 3 and 4

SEVEN MAJOR STEPS IN MAKING A TECHNOLOGY ASSESSMENT

1. Define the assessment task
2. Describe relevant technologies
3. Develop state-of-society assumptions
4. Identify impact areas
5. Make preliminary impact analysis
6. Identify possible action options
7. Complete impact analysis

This set of displays extracts portions of selected checklists used in the MITRE report to summarize important aspects of the methodology. The seven steps listed in the center of the page depict the total methodology.

The key questions that must be addressed in accomplishing each of the seven steps are shown in the remaining exhibits. Some of the exhibits apply to two steps rather than one—e.g., the third and fourth steps and the fifth and seventh steps are displayed together. Each step and its applicable key questions are discussed in the chapters that follow.

ACTION OPTION EVALUATION CRITERIA

Criteria	Definition
1. Controllability	
2. Worth	
3. Priority	
4. Effectiveness	
5. Cost (sponsor)	
6. Cost (spillover)	
7. Non-financial problems	
8. Institutional obstacles	
9. Uncertainty	

Step 6

**KEY IMPACT COMPARISON
WITH AND WITHOUT ACTION OPTIONS**

	Technology	
Development		
Application	Societal impact	
Societal impact	Action option	
Brief description		

Impact characteristics	Impact	
	Without action option	With action option
Affected group		
How affected		
Likelihood		
Timing		
Magnitude		
Duration		
Diffusion		
Source		
Controllability		

Steps 5 and 7

Figure A4.2.1. The MITRE framework. Reprinted with permission.

expect TF and TA to share a common body of techniques. This is indeed so. Coates[16] and Linstone *et al.*[17] have written extensive reviews of TA modeling which include TF approaches and are published in a TF journal.

It is not surprising, therefore, that the most common TA approach reported is to generate an environmental impact matrix derived from the cross-impact matrices of TF, as suggested by Gordon and Becker.[18] A major difficulty in preparing a comprehensive and exhaustive environmental impact matrix is its herculean nature, given the innumerable possible ramifications of technology impacts from its primary, secondary, and higher-order effects. This can be seen in the approach of Leopold *et al.*,[19] which generates a matrix with 8800 cells. Their basic approach has been applied quite extensively in the United States and is reflected in California State planning requirements. It was also applied in Canada in a preliminary study of the environmental impacts of the James Bay Development Project. Both the Batelle Institute and the British government's Department of the Environment have reduced the matrix sizes to under 1000 cells, but these still remain unduly large for corporate use.[20,21] Fischer and Davies[22] suggest a TA framework which reduces the sizes of impact matrices though the initial screening out of relatively minor impacts, and the MITRE Corporation provides a broadly similar approach.[23]

The MITRE framework is a seven-step process which is illustrated in Figure A4.2.1:

1. Define the assessment task.
2. Describe the relevant technologies.
3. Develop the state of society assumptions.
4. Identify the impact areas.
5. Carry out a preliminary impact analysis.
6. Identify possible action options.
7. Carry out a complete impact analysis.

A4.2.3   ORGANIZATIONAL FRAMEWORK FOR CONDUCTING TECHNOLOGY ASSESSMENTS

We have reasoned that some procedure for conducting technology or environmental assessments should be incorporated into the innovation management process of companies developing potential hazardous and/or polluting new products or processes. We now consider how such a review might be incorporated into this process.

Unilever is a typical example of a large offensive/defensive innovator which operates a central research and engineering function and about 500 constituent companies, including some with product development functions. Its major products are foods, detergent and toilet preparations, paper, plastics, chemicals, and animal feeds, some of which generate potentially hazardous waste products. Unilever's approach to protecting the consumer and the environment from potential hazards is outlined in Philp.[24] Corporate responsibility for ensuring the safety of its products is held by the Director of Research, to whom reports an Environmental Safety Officer supported by an Environmental Safety Division (ESD). This division employed 40 scientists and 170 support staff in 1974, and also has access to the full support services of the major Unilever Research Laboratories. The corporation has an established procedure for examining proposed innovations and clearing them as environmentally safe, which is integrated into the R&D process. When a research project has proceeded to the stage of an intended new product, the intention is committed to paper and the tentative proposal with technical specifications, support data, and analyses is submitted to the clearance group of the Research Division for a *clearance forecast*. This group is an independent team of scientists, uncommitted to the project. It is vital that a proposal should be subject to an independent internal review, and this consideration could be readily incorporated in the evaluation process described in Chapter 6. They consider the extent to which the physical and chemical attributes of the proposed product could interact in biological systems relevant to consumer safety and its impact on the environment. Their scrutiny is largely theoretical at this stage and, if the outcome of this scrutiny is favorable, the *clearance forecast* will indicate the possible constraints which might affect the successful completion of product development and then specify the testing program required. This testing program then becomes an integral part of the overall project, so its costs and outcomes must be incorporated in further project evaluations. Such programs are stringent and costs and time durations may be high. It takes at least three years to test for the carcinogencity of a potential new food additive, and possible new detergents must be tested thoroughly to ensure they will not produce allergic dermatitis in a small minority of their users. Thus, test programs are not undertaken unless the probabilities of technical and commercial success for the proposed product appears to be high. Also, some projects which offered high technical and commercial promise have been aborted because they failed to pass a stringent testing program. If the project survives this scrutiny and evaluation, the original team will continue development and testing work in collaboration with the ESD.

If the project fulfills this earlier promise, it will be adopted by an operating company to develop as a new product. It will then pass through a formal clearance process made in writing. The operating company will submit the results of the testing program, together with a manufacturing and marketing plan for the product, for clearance. This plan is first reviewed at the central corporate level to compare it with past products which were similar and, assuming this review is satisfactory, the plan is then passed on to the clearance area of the Research Division. If the results of the test program are satisfactory, and it is unlikely that the project would be adopted by an operating company otherwise, this clearance process should be straightforward. The ESD will repeat its review of the proposed product, now including the results from the test program and the manufacturing and marketing plans. If formal clearance is given, the ESD will monitor outcomes during the early stages of the manufacturing and marketing of the new product. For product proposals which originate in an operating company, as against the R&D function, it would appear that the clearance forecast and formal clearance procedures are collapsed into one.

Challis[25] provides a valuable review of the problems and remedies in environ-

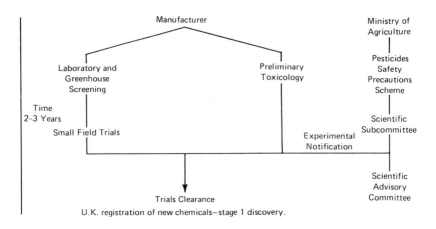

Manufacturer

Laboratory and
Greenhouse
Screening

Preliminary
Toxicology

Time
2-3 Years

Small Field Trials

Experimental
Notification

Trials Clearance

Ministry of
Agriculture

Pesticides
Safety
Precautions
Scheme

Scientific
Subcommittee

Scientific
Advisory
Committee

U.K. registration of new chemicals—stage 1 discovery.

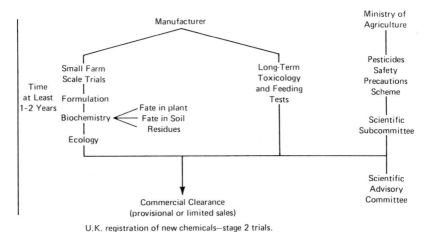

Manufacturer

Small Farm
Scale Trials

Long-Term
Toxicology
and Feeding
Tests

Time
at Least
1-2 Years

Formulation

Biochemistry

Fate in plant
Fate in Soil
Residues

Ecology

Commercial Clearance
(provisional or limited sales)

Ministry of
Agriculture

Pesticides
Safety
Precautions
Scheme

Scientific
Subcommittee

Scientific
Advisory
Committee

U.K. registration of new chemicals—stage 2 trials.

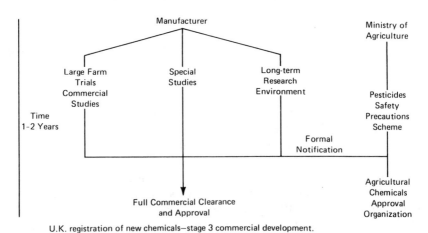

Manufacturer

Large Farm
Trials
Commercial
Studies

Special
Studies

Long-term
Research
Environment

Time
1-2 Years

Formal
Notification

Full Commercial Clearance
and Approval

Ministry of
Agriculture

Pesticides
Safety
Precautions
Scheme

Agricultural
Chemicals
Approval
Organization

U.K. registration of new chemicals—stage 3 commercial development.

Figure A4.2.2a. The clearance process. *Source:* E. J. Challis, "The Approach of Industry to the Assessment of Environmental Hazards." *Proceedings of the Royal Society,* series B, no. 1079 (1974), 183–97. Reprinted with permission.

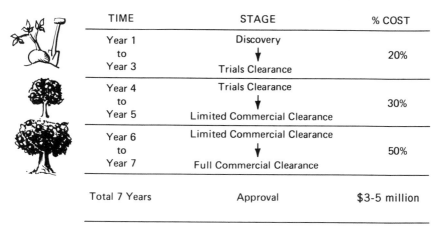

TIME	STAGE	% COST
Year 1 to Year 3	Discovery ↓ Trials Clearance	20%
Year 4 to Year 5	Trials Clearance ↓ Limited Commercial Clearance	30%
Year 6 to Year 7	Limited Commercial Clearance ↓ Full Commercial Clearance	50%
Total 7 Years	Approval	$3-5 million

Development of an agricultural chemical. It takes as long as it does to grow an apple tree to fruit and costs up to $5 million.

Figure A4.2.2.b. The time scale and costs. *Source:* E. J. Challis, "The Approach of Industry to the Assessment of Environmental Hazards," *Proceedings of the Royal Society,* series B, no. 1079 (1974), 183–97. Reprinted with permission.

mental management in the chemicals industry, written from the perspective of ICI. His review will not be discussed here (and since his and Philp's paper appear together in the same issue of the *Proceedings of the Royal Society,* it is convenient to read both together), but he provides a useful visual illustration of the registration process of new chemicals in the United Kingdom. His illustration of the process for an agrochemical product is shown in Figure A4.2.2 (adapted from his Figures 6–9). Figure A4.2.2a shows that a sequence of tests from laboratory screening to commercial studies is performed by the firm as an integral feature of the innovation process, and the regulatory body (the U.K. Ministry of Agriculture here) are continuously kept informed of progress. Figure A4.2.2b shows the time durations and cost breakdowns of the process.

These examples, which probably accurately reflect the best practices of the 1970s, could still be described as reactive. The growing evidence suggesting a manifestation of the greenhouse depletion or holes in the ozone layer at the poles as well as the deprivations of acid rain, all suggest that a proactive approach is needed. This emerged under the rubric of sustainable development.

A4.2.4 SUSTAINABLE DEVELOPMENT AND "GREEN" OR ENVIRONMENTAL MANAGEMENT

As stated at the beginning of this Appendix, the current degradations of our planetary environment are largely the cumulative products of technological developments over the past 200 years, since the beginnings of the first industrial revolution. In Chapter 2, we invoked the notion of the technosphere, an everchanging product of the Lamarckian coevolutions of technology and social needs, analogous to the natural ecosphere, itself a more slowly changing product of Darwinian biological evolution. What is becoming increasingly apparent, is that the man-made technosphere is degrading its natural counterpart to an extent that may well be reaching crisis levels. This implies that steps must be taken to halt and reverse this degradation and that future technological developments should not make matters worse. That is, in the future, we must think in terms of sustainable development.

William Ruckelhouse, the former Director of the Environment Protection Agency (EPA), argues that an attitudinal change or movement towards sustainable development is radical in its scope:

> Such a move would be a modification of society comparable in scale to only two other changes: the agricultural revolution of the late Neolithic and the Industrial Revolution of the past two centuries. Those revolutions were gradual, spontaneous and largely unconscious. This one will have to be a fully conscious operation, guided by the best foresight that science can provide—foresight pushed to its limit.[26]

This proactive rather than reactive approach to technology assessment emerged in the 1980s and is now adopted by many companies. It has two essential features:

1. Before that, the predominant approach was to clean up polluting materials at the end of production process, immediately prior to their release into the ecosystem. The sustainable development or industrial ecology approach now requires that both products and processes should be designed to be nonpolluting from the beginning of the innovation process, rather than be subject to an end-of-pipeline cleanup. Many firms, such as Proctor & Gamble, perform a life-cycle analysis (LCA) in evaluating potential innovations.[27] The dangers of the failure to perform LCAs are apparent in the nuclear power industry. By the year 2010, it is estimated that about 70,000 metric tons of nuclear waste will have to be buried in the United States.[28] Much of this waste has been generated in the Eastern United States and it will probably have to be shipped along one of the two major East–West interstate highways to disposal sites in the Southwestern United States in an end-of-pipeline cleanup. Although some of this waste, such as used laboratory coats, may have been exposed to little risk of radioactive contamination, the transportation and disposal of such a volume and weight of materials poses a challenging technical, logistical, and political problem. Many firms, like Norsk Hydro of Norway, also now perform regular *environmental audits* to ensure that they conform to environmental regulations.[29] Here a comparison may be made with Quality Management (QM), especially in manufacturing companies. The traditional QM approach was largely to delay quality inspections until the end of the process and then "pass" or "fail" products. Following the leadership of Japan and inspired by American pioneers such Deming and Juran, most companies now seek to build quality into their products from the beginning of the innovation process in order to create a quality improvement spiral as successive products are

innovated. The state-of-the-art approach to technology assessment to environmental management is essentially identical and can be described as Environmental Quality Management.

2. A characteristic of the natural ecosphere is the biological interdependencies of its parts or organisms to form locally and globally closed chemical and biological systems with no waste or extraneous products. That is, all "outputs" are "inputs" elsewhere and there is no net output or "waste." The sustainable development or industrial ecology approach seeks to replicate this characteristic by developing a range of products which are mutually interdependent or symbiotic.[30,31] This author is a member of a team applying this approach to the design of industrial parks. Stilwell *et al.*[8] expressed this concept of industrial ecology as follows.

Compatibility with the natural system.

Maximum internal reuse of materials and energy.

Selection of processes with reusable waste.

Extensive interconnection among companies and industries.

Sustainable rates of natural resource use.

Waste intensity matched to natural process cycle capacity.

That is, to create a local technosphere which enjoys a symbiotic relationship with its local ecosphere. Readers who find this notion utopian, should recognize that it has been achieved in the past. Part of the distinctive exquisite beauty of the southern English countryside in summer is based upon its hedgerows. These were originally manmade artifacts planted to enclose fields, which quickly evolved into agreeable environments for a diverse range of flora and fauna. These manmade artifacts are now, quite properly, religiously defended as part of the "natural" environment of the countryside.

Stilwell, *et al.* argues that "green" corporations are beginning to assume the mantle of environmental stewardship and one company which espouses this approach is DuPont, following the leadership of its global CEO Ed Woolard. Being an MNE with 45% of its business outside of the United States, its environmental approach must be global. Ted Newall, Chairman and CEO of DuPont Canada Inc., reports the Corporation's seven steps of environmental stewardship receive Environmental Respect Awards, analogous to those given to employees in recognition of their quality contributions. Newall cites examples of "green" process innovations in his own company and the development of new products from previously waste by-products. For example, six million pounds per year of waste nylon previously dumped at a landfill site is now recycled into engineering polymers used in office furniture, kitchen utensils, and an even better mousetrap! Also, recycling materials enables the company to reduce production costs. Its VEXAR polyethylene fencing is now made entirely from recycled resins at a lower cost and therefore a reduced price to customers. This has led to a 50% increase in sales of the product.

This proactive approach to environmental stewardship is practiced by other major corporations. As well as DuPont's approach, Stilwell *et al.* describe 3M's Pollution Prevention Pays or 3P, P&G's Environmental Quality Policy, and J&J's Worldwide Environmental Responsiveness programs. Jean Marie Hubert van Engelshoven, a senior manager at Royal Dutch Shell, describes a similar approach in his corporation.[32] He also makes the salient observation that the words economics and ecology derive from a common root, the Greek word *ekos* which means *household* or *everything in proper order*. The challenge is to reconcile these two English words with their common Greek root in order to sustain the economic development of the planetary household through the orderly management of its ecology.

REFERENCES

1. R. Carson, *Silent Spring.* Harmondsworth, Middlesex: Penguin Books, 1972.
2. D. H. Meadows *et al., The Limits to Growth.* New York: Universe Book Publishers, 1972.
3. B. Commoner, *The Closing Circle.* New York: Knopf, 1971.
4. V. Packard, *The Waste Makers.* New York: Simon & Schuster, 1960.
5. J. Gimpel, *The Medieval Machine: The Industrial Revolution of the Middle Ages.* Harmondsworth, Middlesex: Penguin Books, 1977, p. 82.
6. R. Rees, "The Great Copper Trials." *History Today* **43,** 38–44 (1993).
7. R. Nader, *Unsafe at any Speed.* New York: Grassman Publishers, 1965.
8. E. J. Silwell *et al., Packaging for the Environment.* New York: American Management Association, 1991, pp. 16–30.
9. M. Gibbons, "Technology Assessment: Information and Participation." In R. Johnston and P. Gummett (Eds.), *Directing Technology Policies & Promotion and Control.* London: Croom Helm, 1978, pp. 175–191.
10. A. Etzioni, "Mixed Scanning: A Third Approach to Decision Making." *Public Administration Review* **27**(5), 385–392 (1967).
11. B. Bozeman and F. A. Rossini, "Technology Assessment and Political Decision-Making." *Technological Forecasting and Social Change* **15**(1), 25–35 (1979).
12. G. T. Allison, *Essence of Decision: Explaining the Cuban Missile Crisis.* Boston: Little, Brown, 1971.
13. M. J. Cetron and B. Bartecha (Eds.), *Technology Assessment in a Dynamic Environment.* New York: Gordon & Breach, 1973.
14. G. Strasser, "Methodology for Technology Assessment—Case Study Experience in the United States." In M. J. Cetron and B. Bartecha (Eds.), *Technology Assessment in a Dynamic Environment.* New York: Gordon & Breach, 1973, pp. 905–940.
15. R. A. Waller, "Assessing the Impact of Technology on the Environment." *Long Range Planning* **8**(1), 43–51 (1975).
16. J. F. Coates, "The Role of Formal Models in Technology Assessment." *Technological Forecasting and Social Change,* **9**(1/2), 139–190 (1976).
17. H. A. Linstone *et al.,* "The Use of Structural Modeling for Technology Assessment." *Technological Forecasting and Social Change* **14**(4), 291–327 (1979).

18. J. J. Gordon and H. S. Becker, "The Cross-Impact Matrix Approach to Technology Assessment." *Research Management* **15**(4), 73–80 (1972).

19. L. B. Leopold *et al.*, "A Procedure for Evaluation Environmental Impact." *Geol. Surv. Cir. (U.S.)* **645** (1971).

20. I. L. Whitman *et al.*, *Final Report on Design of an Environmental Evaluation System.* Columbus, OH: Battelle Memorial Institute, 1971.

21. *Assessment of Major Industrial Applications: A Manual.* London: Department of the Environment, 1976, Res. Re. No. 13.

22. D. W. Fischer and G. S. Davies, "An Approach to Assessing Environmental Impacts." *Journal of Environmental Management* **1**, 207–227 (1973).

23. M. V. Jones, "The Methodology." *Technology Assessment* **1**(2), 143–153 (1973).

24. J. M. L. Philp, "A Multi-National Company, the Public and the Environment." *Proceedings of the Royal Society of London, Series B* **185**, 199–208 (1974).

25. E. J. Challis, "The Approach of Industry to the Assessment of Environmental Hazards." *Proceedings of the Royal Society of London, Series B* **185**, 183–197 (1974).

26. W. D. Ruckelhaus, "Towards a Sustainable World." *Scientific American Special Issue on Managing Planet Earth* **261**(3), 166–174 (September 1989).

27. S. Schmidheiny with the Business Council for Sustainable Development, *Changing Course: A Global Perspective on Development and the Environment.* Cambridge, MA: MIT Press, 1992, pp. 291–294.

28. W. R. Wells, "Management Issues of a University Nuclear Waste Transportation Center." Paper presented at the ORSA/TIMS 36th Joint National Meeting, Phoenix, AZ, November 1993.

29. S. Schmidheiny with the Business Council for Sustainable Developments, *Changing Course: A Global Perspective on Development and the Environment.* Cambridge, MA: MIT Press, 1992, pp. 197–201.

30. H. Tibbs, *Industrial Ecology: An Environmental Agenda for Industry.* Cambridge, MA: Arthur D. Little, 1991.

31. R. U. Ayres and U. Simonis *Industrial Metabolism.* New York: U.N. University Press, 1992.

32. J. M. Hubert van Engelshoven, "Corporate Environmental Policy in Shell." *Long Range Planning* **24**(6), 17–24 (1991).

PART III
THE R&D SETTING

PART III begins with some general considerations of R&D organization and planning. Effective R&D management requires the judicious balancing of technico-economic and behavioral considerations. These considerations are discussed in Chapters 6–8.

____5

R&D MANAGEMENT: SOME GENERAL CONSIDERATIONS

Management must look at the business, look at the external environment in which it is and will be operating, and initiate whatever changes are necessary to ensure the continuing profitability and survival of the business—indefinitely.

PHILIP ROUSSEL, KAMAL N. SAAD, TAMARA J. ERICKSON,
THIRD GENERATION R&D

5.1 INTRODUCTION

In Chapter 4 we saw that the scope and climate of the R&D function is dependent upon the technological plans and strategies of the firm. The most visible high-technology corporations generally follow offensive and/or defensive strategies over SBUs and product ranges so, for simplicity of exposition, we focus our discussion on such firms.

5.2 ORGANIZATIONAL AND GEOGRAPHICAL LOCATIONS OF R&D

Both the organizational and geographical locations of the R&D functions can be expected to be functions of the historical evolution and size of the individual firm. A small- to medium-sized company which exploits a comparatively narrow technological base will typically operate with a single R&D facility, often physically associated with one of its manufacturing plants, with this original location determined historically. In contrast, medium to large corporations, which may well be offensive/defensive innovators exploiting a broad technological base with several to numerous SBUs, operate between the extremes of pure *concentrated* or *distributed* R&D functions.[1]

In a concentrated function, R&D is performed in a corporate research laboratory (CRL) for the whole corporation, probably under the executive direction of a corporate VP of R&D, who also coordinates R&D efforts with divisional requirements. By concentrating its R&D capabilities in one function, the corporation creates a central pool of expertise which can be flexibly deployed on changing R&D needs and establish an outstanding reputation in its fields. In a distributed function, R&D

is performed in individual divisions with much looser coordination at corporate level. Each division exercises more autonomy over its own R&D activities, which should therefore be more responsive to divisional needs. In practice, the R&D activities of many larger corporations are performed in mixed functions, at both corporate and divisional levels. In such mixed functions, CRLs are oriented towards research and longer-term innovations and the consolidation of corporate-wide core competencies, while divisional laboratories are oriented towards development and shorter-term innovations. Lewis and Linden provide a succinct review of the comparative merits of a distributed versus a mixed functional organization.[2] They favor a mixed function provided that:

1. There are core technologies common across several SBUs, with synergies that are difficult to capture at SBU levels. That is, top management must ensure that the CRL is not merely supplementing SBU R&D efforts, but performing upstream-directed, research-seeking major breakthroughs.
2. The CRL performs some informal contract R&D for SBUs, but does not become a service department for them. At the same time, effective mechanisms for technology transfers from the CRL to the SBU R&D functions must be in place. The CRL must not become a "lab in the woods" with an ivory tower outlook aloof to downstream operations.
3. The CTO to whom the CRL reports is primarily a technical businessperson who understands the technical capabilities of the SBUs, rather than a research scientist.

Moreover, many such corporations are multinational firms which have grown through mergers and acquisitions. They may therefore duplicate both corporate and divisional R&D laboratories in different countries, especially if they grew from individual national corporations, each with its own laboratories. Starr discusses the global R&D management issues that arose from the merger of the Celanese and Hoechst Corporations in 1987.[3,4] Also, given the growing importance of global technology competition, U.S.-based corporations are placing increasing emphasis on the establishment or acquisition of R&D activities in Europe and Asia (especially Japan). Interlaboratory communication is an important issue in the effective management of such global R&D efforts, and De Meyer[5] reports on the practices of 14 large MNCs which emphasize the importance of networking, including boundary-spanning and gatekeeping roles (see Chapter 7, for a discussion of such roles). Nowadays, some MNCs rationalize their R&D activities across national boundaries by encouraging each national component to focus its capabilities within specific technology-market segments to develop world product mandates. For example, 3M conducts R&D and technical services in 36 laboratories in 21 countries, with the first overseas laboratory that it set up (in Harlow, England) having global responsibility for photographic technology development.[6] Some serve their national market needs, while others are regional laboratories serving several countries. While outside the United States laboratories enjoy considerable autonomy, there is networking among them and those in the United States to encourage the two-way exchange of information and technology. Perrino and Tipping [7] outline an approach to formulating a global R&D deployment strategy based upon a study of 16 multinational corporations. They suggest that although "pockets of innovation" develop locally, profitable deployments are based upon cross-fertilizing them with others for global commercialization. Thus, a global network model of technology management is the wave of the future.

The issue of the intranational geographical location of R&D activities also deserves a brief mention. Divisional laboratories are typically located close to a manufacturing function, as we stated at the beginning of this section. In contrast, corporate laboratories may be located so as to be geographically separate from individual divisions for two interrelated reasons. First, corporate R&D laboratories are usually competing with universities and government R&D establishments in their professional R&D staffing requirements. They therefore need to provide organizational climates comparable to academic and government institutions, free from the shorter-term pressures of divisional needs. Second, if the corporate function is physically located with a manufacturing division, there is a risk that it will become the R&D satellite of that division in the long-term and lose sight of its broader corporate mission. Given this organizational and geographical distribution of R&D efforts, longer-term innovations are typically initiated in the corporate R&D facility and then transferred to the appropriate divisional facility when the secondary development phase is reached, or later. Shorter-term innovations may be initiated in divisional facilities, often in response to specific requests from operations managers and customers. This arrangement provides response mechanism to both long- and short-term innovation opportunities.

5.3 OVERALL ISSUES

Regardless of its organizational structure, there are six interrelated factors to be considered in the management of the R&D efforts of a firm:

1. The R&D plan which should be developed to implement a technological strategy congruent with the future scenarios envisaged for the firm.
2. The overall level of expenditure or the proportion of the total budget which should be invested in R&D activities.
3. The linkage of R&D planning to the budgeting process.
4. The development of effective and realistic procedures for the evaluation of individual R&D projects and their integration into overall project selection.
5. The reconciliation of the personal development needs of professional R&D staff with the requirements for effective project management, together with the development and maintenance of a creative organizational climate in the R&D function.
6. The development of ongoing evaluation and control procedures for R&D projects once they have been approved.

The first of these requirements was examined in Chapter 4; we consider the second and third of them in this chapter, and the remainder in Chapters 6 through 8.

5.4 BUDGET SETTING—HISTORIC AND PRAGMATIC ASPECTS: FIRST AND SECOND GENERATION R&D

The scope of R&D activity is obviously dependent upon the amount of money spent on it. The process of formulation of a corporate policy and the setting of growth, profit, and market share goals is likely to evolve historically over time, during which there is a corresponding evolution of R&D activities. Although the scope and nature of these activities should (and usually does) reflect corporate policies and goals, annual budgetary allocations have traditionally often been based upon more pragmatic criteria, and Clarke[8] and Ellis,[9] respectively, provide useful reviews of Canadian and U.S. practices. Therefore, it is useful to begin our discussion by examining these traditional criteria for fixing the size of the R&D budget, as discussed by Roussel *et al.*[10]

These authors argue that the 1950s and early 1960s were an era of *first generation R&D management,* with R&D treated as an overhead cost. Some firms are still practicing first generation R&D, in which there are two alternative pragmatic criteria for setting this funding level:

Percentage of Sales Turnover or Profit

Two yardsticks are to spend a certain percentage of the previous (or a number of recent) years sales or profits. In 1987, R&D spending as a percentage of sales for nearly 900 U.S.-based firms ranged from a fraction of a percentage point to 53.7% for Centocor in the pharmaceuticals industry.[11] The two largest absolute spenders were General Motors ($4361 million) and IBM ($3998 million). These figures reflect industry and size factors. For example, 15% of sales is typical of the pharmaceuticals industry, as opposed to less than 3% in commodity chemicals.[12]

Quite possibly, one of these is the most common yardstick used, but both suffer from obvious disadvantages. If, in the absence of an industry-wide recession, sales and profits are declining, the majority of a company's products would appear to be in the declining years of their life cycles. Therefore, it would be more appropriate to spend *more* rather than *less* money on R&D to restore a flagging product range. The danger of this measure generating counterproductive budget cuts is also increased by the generally accepted methods of accounting for R&D expenditures. Many firms include R&D spending as an income statement expense in the year in which it is incurred. Therefore, if a profits downturn is expected in the following year or later, it is tempting to "improve" next year's financial results by reducing R&D expenditures. Nix and Peters[13] suggest that R&D expenditures should be at least partially capitalized and depreciated over an agreed schedule to avoid this happening. In a U.K. study, Ball *et al.*[14] found that despite the legality and advantages of partial depreciation of such expenditures, only 20% of the firms they surveyed did so, either out of fiscal prudence or fear of negative external judgments of the practice.

Conversely, if sales and profits are increasing, it does not follow R&D expenditures should be increased. First, there is no need to increase expenditures to restore flagging sales. Second, if the increased money were spent effectively, it would lead to more R&D inventions being available for potential commercial exploitation. If the company is already enjoying buoyant sales, it may lack the downstream capacity (in production, sales, and servicing) and management willpower to manufacture and promote further innovations. Admittedly, patents on surplus R&D inventions may be sold or licensed, but that may not always be feasible or congruent with corporate policy. Furthermore, increased expenditure may lead to the hiring of more R&D staff than can be usefully employed by the company in the long-term, forcing it to let go staff later.

Given these considerations, percentage of profits certainly appears to be an inappropriate criterion for R&D spending. The criterion does have some relevance in the context of tax management, however. Judicious temporary increases in R&D expenditures may be a worthy and convenient means of reducing the tax liability on the increased profits, particularly if they are spent on the purchase or construction of improved equipment or facilities, which entail no long-term employment commitments.

Interfirm Comparisons

Another commonly used yardstick is to spend the same percentage of turnover as similar companies in the same industry. A difficulty with this "keeping up with the Jones" approach is that firms resemble people in that no two are sufficiently alike to make direct comparisons valid. In companies large enough to have a substantial R&D effort, there may be a multiplicity of product lines, each with different R&D requirements, so that each firm will have a unique technological profile. Furthermore, differences in accounting systems, both among and within companies, may make the disentanglement of meaningful comparative costs from public records virtually impossible. Finally, of course, peer group firms used for comparison may have set their expenditures by other yardsticks such as those given above, so that an interfirm comparison criterion becomes a surrogate for another measure.

A more important role for interfirm comparisons is in technology monitoring, as discussed in Chapter 4, Appendix A4.1. By monitoring its competitors R&D expenditures, activities, recruiting policies, etc., a company can anticipate potential innovations coming from the opposition and recognize potential threats and opportunities. Despite reports to the contrary, U.S. companies probably continue to exercise technological leadership in many fields and maintain a fairly open attitude in their R&D activities. Consequently, companies elsewhere can still often anticipate future technological developments by observing the activities of their U.S. counterparts. In so doing, they can take due note of the expenditures of these counterparts in fixing their own.

Roussel *et al.* argue that from the mid-1960s, companies began to practice *second generation R&D management.* In the first generation era, no strategic framework existed for the R&D mission. In this era, a partial strategic framework at project level emerged; the rest of the firm was viewed as the "external customer" for the R&D function, and general management sought to balance the advocacy of the R&D function against strategic goals. In companies practicing second generation R&D, support is sought through:

An Advocacy Process Based Upon Historical Allocation

In reality, the current level of R&D expenditure, based upon whatever methods have been used historically, is likely to be the major determinant in fixing future allocations, since they are unlikely to change radically from year to year. This being so, annual appropriations are likely to be determined by a bargaining process between senior R&D and general managements. As Steele[15] describes the process, the R&D director may argue a case for an appropriation based upon the following considerations:

1. Percentages of sales or profits or industry norms as discussed previously.
2. Maintenance of a historical growth rate for R&D in real monetary terms.
3. The degree of esteem or disfavor with which general management views the R&D functions, and their satisfaction or dissatisfaction with its recent efforts.
4. The overall business and economic climate.

In practice, the exact size of the budget is likely to depend upon the senior R&D managers' skills and judgments in advocating their case at the corporate cabinet level. On the one hand, they will not wish to ask for too much money, lest they be viewed as being naively unaware of what money can be realistically made available, unsympathetic towards the problems of general management, or weak managers who are "run" by R&D staff. On the other hand, they will not wish to make too low a bid, lest they be viewed with disdain by general management as weak advocates for their functional interests. If they have the necessary grasp of corporate politics *plus* sound technological and business judgments, it is likely that they will secure a realistic financial appropriation for the R&D function.

5.5 R&D PLANNING AND BUDGET SETTING—NORMATIVE ASPECTS OF THIRD GENERATION R&D

The previous section focused on essentially pragmatic approaches to budget setting implicit in past R&D management practices, whereas a more normative approach could be deemed desirable. Roussel *et al.*[10] argue that such an approach is being adopted nowadays by leading high-technology firms who are practicing *third generation R&D management.* It largely embraces the six factors we cited at the end of Section 5.2 and approaches discussed in the rest of this and subsequent chapters. The key feature of third generation R&D management is that technology/R&D and business strategies are integrated within a holistic framework, based upon partnerships between R&D and the other functional areas, and the overall R&D budget should be set within this context.

In Chapter 4 we examined the approaches which enable this function to identify and define the alternative future scenarios in prospect for the firm and match its technological capabilities to envisaged market opportunities to constitute a technological mission or plan and strategy. Since the R&D function should provide the base for the implementation of this mission, the total R&D budget and its division between laboratories (in a distributed or mixed function) and projects should be based upon these considerations. That is, the size and allocation of this budget should be congruent with the technological mission which seeks to invent the firm's preferred scenario or future.

The criteria for evaluating individual projects within each of them should be based upon similar considerations, but vary through the succeeding phases of the innovation process. Such considerations and variations are illustrated in Okamoto's description of project assessment practices in the Japanese office automation equipment manufacturer RICOH Co. Ltd.[16] It will depend upon the phase the project has reached in the process, the resources required for its initiation or continuance, and the *immediacy* of the impact on the business of its successful completion. Given this "immediacy of impact" consideration we may divide the successive phases into two broad groups—longer-term R&D, covering research and early development, and shorter-term R&D covering advanced development onwards.

5.6 LONGER-TERM R&D

The high-technology corporation following offensive and/or defensive strategies is R-intensive and invests significant effort in nondirected research in emergent and pacing technologies supporting its technology mission which, by definition, offers payoffs which cannot yet be quantified. They are best viewed as technology platforms which may be the bases of future breakthroughs. Alternatively, investments in such research may be viewed as analogous to holding a *call option** on the stock market, since it provides a future option to exploit the technology, rather than purchase the stock.[17] Similarly, although directed applied research on key technologies may be focusing on a specific product concept, again by definition, it is questionable whether the technology will be sufficiently developed, or the product concept sufficiently delineated, to admit realistic cost projections. It should be recognized that since the technological innovation process can extend over a relatively long time scale, projects should not be subjected to the hazards of *premature economic assessment* or the ROI and marketing criteria which must be applied later in the process.[18] The ability to reconcile the paradoxical requirements of commercial stewardship and the nurturing of a climate for sustained technological innova-

---

*A call option permits its holder to purchase stock at a specified price prior to an agreed-upon expiration date.

tion is one of the difficult and vital roles for the IMF of the firm. Thus, corporate management should perhaps support these activities as acts of *faith* and *hope*, trusting in the quality of staff they have hired in these functions, rather than expecting explicitly monetary rewards from each project.

Since any commercial outcome of such research and early development is almost certainly long-term, with little immediate business impact, the detailed evaluation of individual projects cannot (and should not) be subjected to the procedures discussed in the next chapter. Rather, it should be viewed as an internal R&D laboratory responsibility, with project evaluations and selections based upon peer group evaluations and review similar to those used in R-intensive laboratories in universities and government, possibly supported by one or other of the simpler selection techniques described in the Appendix to Chapter 6. Given that high quality research staff are being recruited and retained, the quality of research output should be congruent with the company's long-term objectives and goals. If this is not so, it is likely to be due to inappropriate R&D management and staff recruitment policies, rather than inappropriate project evaluation and selection procedures.

5.7 ADVANCED DEVELOPMENT

Once one moves into the advanced development/design innovation phase of R&D activities, different considerations apply. By now, the product concept of a proposed innovation should be more clearly delineated and some reasonable estimates of costs and benefits should be possible. This means that projects can be evaluated in terms of their congruencies with SBU and corporate objectives using evaluation methods discussed in Chapter 6. It also means that function managers should participate in such evaluations under the aegis of SBU and corporate IMSCs.

Recall that Figure 4.1 illustrated the requirement of a high-technology firm to generate a continued sequence of innovations in order to survive. The ordinate in Figure 4.1 represented a measure of cash flow, but it could have readily represented another fiscal measure such as profit. Recall also that Figure 4.3 schematically represented the technology–market matrix in which such a firm might operate. The process of formulating a technology mission, described in Chapter 4, should identify which cells in present and future technology–market matrices the company wishes to occupy and specific goals (in terms of profit or other performance measures) it seeks, over time, in each of these occupied cells. Thus, the schematic representation of Figure 4.3 may be notionally disaggregated to represent the profit growth goals to be sought from each particular cell in the matrix. Given the finite life cycle of many technological products, the projected profits from currently existing company products in a particular cell might be sufficient to satisfy current profit requirements, but insufficient to satisfy future profit goals. This is illustrated in Figure 5.1. The products currently being manufactured are at varying stages of their life cycles. The sum of their financial contributions are sufficient to satisfy profit requirements until time X on the abscissa, but beyond that point, there is a growing gap between the profits which these products can contribute and the re-

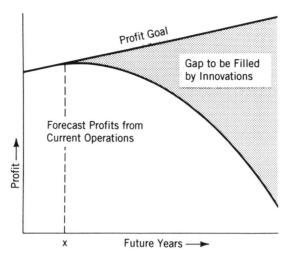

Figure 5.1. Gap analysis.

quirements of the business plan. This gap is required to be filled by new product concepts at present in the R&D phases of the innovation process.

From the viewpoint of corporate management, the role of R&D is to implement a technological strategy by generating a continuous flow of new product ideas which will fill such gaps, so once a proposed innovation has reached the secondary development/design engineering phase, it should be judged in terms of its potential to bridge the corporation's future profit gaps. Such a gap-bridging approach is described by Frohman and Bitondo.[19]

As was stated above, the detailed discussion of project evaluation is deferred until the next chapter and, for the present, it is sufficient to note three points. The first point relates to the vertical axis of Figure 5.1. Given the uncertainties of outcomes associated with potential innovations at this stage of the process, there may not be a high probability of any one project yielding a profitable product innovation. Therefore, corporate management should ensure that the R&D budget is large enough to enable a sufficient number of innovation avenues to be pursued, to provide a *high degree of insurance* that future gaps in profitability goals (in all technology–market cells) will be filled. That is, the budget should be large enough to ensure that enough "eggs" (projects) are placed in enough "baskets" (product–market segments) for a sufficient number of eggs to prove fertile and "hatch" to achieve corporate goals.

The second point relates to the horizontal axis of Figure 5.1 or the timing of new product developments. The cost associated with the R&D phase in the innovation process is typically no more than 10% of the total cost, and since (as this cost proportion implies) a new innovation places increasing demands on all company resources as it progresses further downstream, it is important to ensure that the project portfolio consists of a set of projects which are balanced in time. This protects the R&D function from harmful fluctuations in its budgets, ensures that promising projects are not aborted later through lack of (say) production capacity, and that the timing of product entry into the market is congruent with company goals. Timeliness of product introduction is also very important from a marketing viewpoint. It is by no means unknown for products to fail because they enter the market too soon rather than too late. Therefore, it is important to exercise *family planning* in project selection to ensure that the fertile eggs placed in the baskets hatch out in orderly sequence.

Third, given the need to ensure future profitability by backing more projects than would be required if all were successful, even with astute family planning it is probable that the fecundity of the R&D function will generate successful outcomes which must be aborted through lack of downstream resources and commitment. *It is most important* to recognize that such successful projects *do not* represent bad investment decisions. The best way of ensuring long-term profitability is to produce an *embarras de richesse* of new product ideas, and those which remain unexploited may still have tangible or intangible economic values. The patents filed for unexploited inventions may perhaps be either sold or licensed to other companies or used as bargaining counters to derive patent concessions from competitors.

These considerations are implicitly reflected in Hall and Nauda's interactive approach to project selection, which are briefly reviewed in the Appendix to the next chapter, after we have examined project evaluation in some detail.[20]

FURTHER READING

1. P. A. Roussel, K. N. Saad, and T. J. Erickson, *Third Generation R&D: Managing the Link to Corporate Strategy.* Boston: Harvard University Press, 1991.

REFERENCES

1. M. F. Usry and J. L. Hess, "Planning and Control of Research and Development Activities." *Journal of Accountancy* **124**(5), 43–48 (1967).
2. W. W. Lewis and L. H. Linden, "A New Mission for Corporate Technology." *Sloan Management Review* **31**(4), 57–67 (1990).
3. L. Starr, "R&D/Technology Integration Across the Atlantic." *Research Technology Management* **33**(2), 16–18 (1990).
4. L. Starr, "R&D in an International Company." *Research Technology Management* **35**(1), 29–32 (1992).
5. A. De Meyer, "Tech. Talk: How Managers Are Stimulating Global R&D Communication." *Sloan Management Review* **32**(3), 49–58 (1991).
6. L. C. Krogh and G. C. Nicholson, "3M's International Experience." *Research Technology Management* **33**(5), 23–27 (1990).
7. A. C. Perrino and J. W. Tipping, "Global Management of Technology." *Research Technology Management* **32**(3), 12–19 (1989).
8. T. E. Clarke, "R&D Budgeting—The Canadian Experience." *Research Management* **24**(3), 32–37 (1981).

9. L. W. Ellis, "What We've Learned Managing Financial Resources." *Research Technology Management* 21–39 (1988).

10. P. Roussel, K. N. Saad, and T. J. Erickson, *Third Generation R&D: Managing the Link to Corporate Strategy.* Boston: Harvard University Press, 1991.

11. W. D. Marbach with E. T. Smith, "A Perilous Cutback in Research Spending." *Business Week* June 20, pp.139–162 (1988).

12. H. A. Schneiderman, "Managing R&D: A Perspective from the Top." *Sloan Management Review* **31**(4), 53–58 (1991).

13. P. E. Nix and R. M. Peters, "Accounting for R&D Expenditures." *Research Technology Management* **31**(1), 39–41 (1988).

14. R. Ball, R. E. Thomas, and J. McGrath, "Influence of R&D Accounting Conventions on Internal Decision-Making of Companies." *R&D Management* **21**(4), 261–269 (1991).

15. L. W. Steele, *Innovation in Big Business.* New York: American Elsevier, 1975, Chapter 9.

16. A. Okamoto, "Creative and Innovative Research at RICHOH." *Long Range Planning* **24**(5), 9–16 (1991).

17. G. R. Mitchell and W. F. Hamilton, "Managing R&D as a Strategic Option." *Research Technology Management* **31**(3), 15–22 (1988).

18. M. W. Thring and E. R. Laithwaite, *How to Invent.* London: Macmillan, 1977.

19. A. L. Frohman and D. Bitondo, "Coordinating Business Strategy and Technical Planning." *Long Range Planning* **14**(6), 58–67 (1981).

20. D. L. Hall and A. Nauda, "An Interactive Approach for Selecting IR&D Projects." *IEEE Transactions on Engineering Management* **37**(2), 126–133 (1990).

___6

PROJECT–PRODUCT EVALUATION

With the aid of this model and the key to unlock it, one can then invent further plants ad **infinitum** *which, however, must be consistent; that is to say, plants which, if they do not exist, yet could exist, which far from being shadows or glasses of the poet's or painter's fancy, must possess an inherent rightness and necessity. The same law applies to all the remaining domains of the living.*

GOETHE

6.1 INTRODUCTION

We now turn to the subject of the evaluation of R&D projects as potential innovations. There is voluminous literature on this subject, and it would be beyond the scope of this text to review it in detail. Up to (and including) the primary experimental development phase, a project is probably best subject to peer group evaluations as mentioned in Chapter 5 (Section 5.41). However, prior to more advanced development, it should be subject to an increasingly comprehensive evaluation along the lines suggested in this chapter. Moreover, this initial evaluation should mark the beginning of an ongoing process, which will continue until either the project is aborted or the innovation is launched and established in its market(s). In Chapter 2 (Section 2.10), we suggested that the evaluation of prospective innovations should incorporate a conjecture–refutation process. The approaches and the ongoing evaluation process described in this chapter essentially reflect this process.

Before going further, it is useful to recall our distinction between revolutionary and incremental innovations (Chapter 2, Section 2.8). By definition, revolutionary innovations, like their scientific counterparts, often induce major changes in social as well as engineering practices and attitudes, sometimes creating new industries. Such innovations imply the application of new technology to old and often new markets; that is, the upper row cells of the technology–market matrix (Chapter 4, Figure 4.3). New technology must often displace old technology to establish itself, even when it has emerged from the same industry base. For example, the semiconductor electronics industry emerged from the vacuum tube electronics industry in the 1950s, following the invention of the transistor. Some electronics corporations found it expedient to set up separate semiconductor divisions to ensure that the growth of this embryonic technology was not stunted by persons with a vested interest in preserving the *status quo* of the old vacuum tube technology. These divisions were essentially semiautomonous new business ventures. This consideration suggests that the evaluation of a revolutionary project or innovation requires a more extensive commitment of company resources and a more extensive evaluation process, than incremental ones. We, therefore, begin by discussing the evaluation of revolutionary projects or innovations.

6.2 REVOLUTIONARY INNOVATIONS—WHITE'S APPROACH

One approach to the evaluation of the revolutionary innovations has been developed by Dr. George R. White, Corporate Vice President and Vice President, Information Technology Group of Xerox Corporation.[1] He developed the approach while he was the Carroll-Ford Foundation Visiting Professor at the Harvard Business School. As well as those from the mecca of management academia, he thus brought to it precepts and perceptions from his experience as a senior manager in a high-technology corporation which has itself been responsible for innovating one revolutionary product from which it derives its name and inventing, but failing to innovate, another (the personal computer).

White's approach evaluates the credits and debits of a proposed innovation in terms of its technology potency and business advantage. His overall conceptual framework (shown in Figure 6.1) is developed in detail through a retroactive analysis of two fairly recent revolutionary innovations—the personal transistor radio and the Boeing 707/DC 8 passenger jets (see Figure 6.2). He then applies the approach

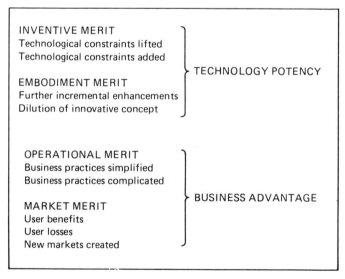

Figure 6.1. White's approach. *Source:* Reprinted with permission from *Technology Review,* Copyright 1978.

	Transistor Radios	*Boeing 707/DC 8*
INVENTIVE MERIT	Solid state vs. vacuum tube amplification. Elimination of HT power source. Increased reliability and battery life. Reduction in weight and size. Increased initial costs, but cheaper later.	Gas turbine vs. piston engine. Increased power and speed. Increased reliability. Longer run-ways required.
EMBODIMENT MERIT	Ferrile rod antennas. Smaller loudspeakers and tuning capacitors. Power signal selectivity, tuning precision. and audio quality.	Swept back wings. Dutch roll problems.
OPERATIONAL MERIT	Distribution simplified and costs reduced.	Lower costs per seat-mile. Lower maintenance costs.
MARKET MERIT	Market outlets increased Creation of personal/pocket radio Initial higher cost/price	More comfortable speedier travel plus lower fares attracted new air passengers and others from surface transportation. Decline of luxury travel (e.g. Ocean Queens)

Figure 6.2. Transistor radios and Boeing 707/DC 8. *Source:* Reprinted with permission from *Technology Review,* Copyright 1978.

to two as yet (at the time of his writing) unrealized innovations—the application of microprocessor technology to the control and operation of automobile engines and the U.S. supersonic transport (SST).

The evaluation consists of asking the series of test questions shown on the left-hand side of Figure 6.1 to determine the merits of the proposed innovation as shown in Figure 6.2. We consider each of these in turn using the first two examples cited.

Inventive Merit

At the heart of a revolutionary innovation is an R&D invention which constitutes a new combination of scientific principles and new ensemble of technical elements which has the potential to displace existing technology. That is, in terms of our Lamarckian evolution analogy, a new technological mutation. In the case of the transistor, it is amplication via positive hole and electron currents flow across semiconductor $p-n$ junctions, displacing electron currents flow generated by thermionic emission in triode vacuum tubes. In the case of the 707/DC 8, it is rotary air compressor followed by a combuster and gas turbine, displacing the reciprocating piston crankshaft engine.

The new technology may eliminate or relax previous technological constraints (basic constraints lifted), but also may add or increase others (basic constraints added). The introduction of transistors into radios eliminated the need for a high-tension (HT) voltage and reduced weight, size, power consumption, and hence battery life, and increased ruggedness and reliability. At the same time, the low manufacturing yields plus high initial start-up investments in the semiconductor industry made transistors more expensive than vacuum tubes for some years. The jet engines were superior to piston and turbo-prop engines in power and speed, but were less efficient at takeoff so their initial use had to be on routes between airports which could justify runway extensions to two miles. On balance, the credits of these inventions outweighed the debits, so they could be assigned a high inventive merit.

Embodiment Merit

R&D inventions are rarely, if ever, innovated in a technological vacuum, but are embodied in large technological products (radios or 707/DC 8 air-frames). This process of engineering embodiment may offer opportunities for further incremental enhancements which were previously technologically or economically infeasible or unattractive, albeit with some dilution of the innovative concept. The potential reduction in the size and weight of radios offered by the transistor was enhanced by the Japanese electronic industry's introduction of ferrite rod antennae and smaller loudspeakers and tuning capacitors. However, this enhancement was achieved at the cost of poorer signal selectivity, sound reproduction, and tuning precision. The increase in speed offered by jet engines was enhanced by sweeping back the wings of the planes more sharply to an angle of 35° (707) and 30° (DC8), an innovation which would be pointless on a propeller-powered airplane. This enhancement was diluted through the need to control an aerodynamic phenomenon called Dutch roll. The commercial success of both of these innovations suggests that the embodiment enhancements were overall meritorious.

High ratings on inventive and embodiment merits suggest technology potency, so attention is now turned to business advantage (the matching blade in Smookler's scissors analogy, cited in Chapter 2, Section 2.9).

Operational Merit

By definition, a revolutionary innovation can be expected to change business practice. Therefore, it is necessary to identify the extent to which the innovation will simplify or complicate such practice. The introduction of transistors reduced the size and weight and increased the reliability of radios. Therefore, distribution costs could be reduced and, more importantly, the increased reliability made a franchised dealer service distribution system redundant. This provided the opportunity for new wholesale and retail sales outlets for the product. No complications of business practice followed. In the case of the jets, their limited introduction into long-range routes plus their increased reliability simplified practice.

Market Merit

With revolutionary innovation which is displacing a prior technology in an existing market, the market merit questions in Figure 6.1 are self-evident. Nonetheless, recollection of the technology–market matrix suggests that the new market aspects of these questions invite specific attention. The size and weight reductions of Japanese transistor radios, coupled with their increased reliability, opened a previously unrealized latent product–market opportunity for the radio industry. Previously, size/weight constraints had prevented anything smaller than portable, or pick up, move, and put down, radios to be manufactured. In return for a dilution of sound quality, etc., the concepts of a reliable personal or pocket radio which could be purchased from a neighborhood chain or drugstore was introduced. This presented a new market which was steadily enlarged as the prices of personal radios dropped and the radio entertainments industry changed and increased its programming to cater for the teenage "pop" market which grew with increased personal radio sales.

The lower costs per seat-mile of the 707/DC 8s meant that they were cost competitive with turbo-prop and piston-engined airplanes over long-haul routes. Jet plane passengers quickly found that they offered a more comfortable as well as speedier journey, so airlines were forced to retire prior-technology airplanes prematurely to satisfy a final customer demand. Because costs per seat-mile were lower and occupancy rates higher, fares could be lowered, attracting new air travelers. The reduced fares coupled with reduced journey times also lured more passengers away from rail and sea travel.

The above historical analyses, which were used to illustrate the approach, are discussed in rather more detail in White's referenced paper. In the case of transistor radios, he argues that the U.S. radio industry (except for TI) failed to recognize the further engineering enhancements which could be achieved by corresponding miniaturization of other radio components to match the introduction of transistors, or the potential to develop a personal radio severed from the franchised dealer service network. Thus, the industry failed fully to exploit the innovative potential of the transistor and, at the same time, allowed its Japanese counterpart to establish a U.S. market presence. Sadly, history repeated itself in the development of the home VCR. These examples illustrate the importance of including engineering and marketing analysis and flair in the evaluation when assessing a potential new innovation.

This absence of positive thinking concerning engineering and marketing applies equally to Lockheed which developed the Electra at the same time its competitors developed the 707/DC 8. Lockheed failed to fully exploit the potential for engineering enhancement. They used turbo-prop engines and straight wings, thus maintaining some of the speed, noise, and higher maintenance cost constraints imposed by propellers, but retaining the facility for relatively short take-offs and efficient cruise which they offer. Passengers preferred the speed and comfort of the 707/DC 8 to the turbo-prop Electra, which was virtually driven off the market. Lockheed lost its number two position to Douglas in the U.S. air transport industry and did not stage a comeback until the marketing of the Tri-Star. In contrast, Boeing established a strong presence in the industry through its success with the 707, which it has reinforced through its later aircraft, as will be discussed later in this chapter.

The details of White's approach are clearly dependent upon the specific technology–market in question, but it is hoped that these illustrative examples are sufficiently explicit to enable readers to apply the approach (at least mentally) to revolutionary and possibly incremental innovations within their own organizations.

Two final comments on the approach are warranted, however. First, the distinction between the separate considerations—inventive merit, embodiment merit, etc.—is to some extent arbitrary rather than clear-cut. For example, the uncertainty concerning the impact of atmospheric pollution generated by fleets of SSTs flying at 65,000 feet is clearly a demerit. However, it is somewhat arbitrary whether it should be labeled an inventive, embodied, or operational demerit. This arbitrariness is unimportant since the key feature of White's approach is that it offers a *systematic, comprehensive, and exhaustive* identification and evaluation of *all* the technological and commercial factors which should be considered in evaluating a revolutionary innovation. Second, although comprehensive, it is obvious that the approach is qualitative rather than quantitative. *This fact is positively advantageous.* The uncertainties of both technological and commercial outcomes of a revolutionary innovation are too great to render conventional fiscal analysis meaningful. What *is* required is rigorous, logical, and systematic qualitative analysis. This point is best summed up by quoting White directly:

> In these criteria we have avoided terms such as "return on investment" and "return on assets managed." Our view is that these issues are overwhelmed whenever a new inventive concept can be placed in a beneficial embodiment which will enhance its value in a major latent market with lowered operational costs. If evaluations of an innovation must be based on assumptions of narrow differences in return on investment, they are quite possibly based on fallacy. What we proposed is a logic structure to identify a small class of innovations of greater promise whose success will transcend the cash value of any normal investment.

6.3 MICRO-RADICAL INNOVATIONS

In Chapter 2 (Section 2.8) we suggested that the term innovation embodied a continuum from very minor cost-reducing or performance-maximizing innovations to micro-radical innovations which, although technologically substantive, fell short of inducing a Kuhian paradigm shift. The scope of impacts of such radical innovations dictate that they should be evaluated by the White approach. Indeed, one example he cited—the application of microprocessor technology to the automobile —could be viewed as a micro- (as against revolutionary) radical innovation in the automobile industry. Therefore, we need not consider these evaluations further since they are implicitly considered in the above approach.

6.4 INCREMENTAL NORMAL INNOVATIONS

The reader could apply White's approach (at least mentally) to incremental as well as revolutionary and radical innovations in his own organization since it provides a useful framework for reviewing any potential new product or process. When considering an incrementally innovative project however, we should, by definition, have much more background knowledge of the relevant technology and expected market, so it should be possible to perform a more detailed evaluation. In Chapter 2 (Section 2.10) we suggested that any innovation should be evaluated in the context of its contribution to the objective and goals of the business, expressed in the notion of "gap analysis" outlined in Chapter 5 (Section 5.4 and Figure 5.1). We should therefore perform the evaluation in the context of its impact at corporate as well as functional levels, and we consider each of these in turn.

The Corporate Contest

Given that an incremental project is evaluated in the context of its contribution towards filling the gap between corporate goals and the future contributions which can be obtained from its current product lines, the evaluation should therefore provide more likely, pessimistic, and optimistic estimates of expected profits, revenues, costs, and corporate resource requirements (including a timetable for completion of successive phases of the innovation process). The time estimates are important because the organization may wish to phase in an orderly sequence of innovations both to meet corporate goals (recall Figure 5.1) and to balance the use of organizational resources. Further, as implied earlier, the timing of the market launch of an innovation may critically affect its chances of market success.

The Market Context

We have already emphasized the importance of the continued consideration of market definition and development from the secondary development phase onwards of the innovation process. Ultimately, the commercial success of an innovation is dependent upon its generation of product sales based upon the satisfaction of a perceived user need. An overall conclusion of the SAPPHO study on technological innovations was that *user need understood* was the most critical factor in determining their commercial outcome.[2] Despite the somewhat platitudinous nature of this observation, it is remarkable how frequently insufficient attention is given to this factor. The purpose of the project is to develop and produce an innovative product or products which will generate profitable sales; therefore, as Blowatt[3] suggests, one of the first questions which should be raised is: *"Who is going to buy the product, and why?"* Remarkably, the evidence suggests that innovations frequently fail because of the failure to ask or critically to answer this deceptively naive and obvious question. Three examples will be cited to support this observation.

Blowatt cites a flying saucer story from the 1970s. Two Canadian engineers developed an authentic flying saucer powered by eight propellers driven by Wankel engines located around its inner circumference. It could fly at 200 mph at 10,000 ft. carrying a 450-pound payload and required no airstrip, so that it could take off and land in, literally, one's own backyard. There were over 12,000 licensed pilots in Canada and the engineers, figuring that half would like their own flying saucer, estimated the Canadian market alone to be $60 million. Further analysis of its customer appeal proved unfortunately to be less sanguine, for the following reasons:

First, even with economics of mass production, the saucer could not be sold for under $10,000 and it was debatable whether these pilots would be willing or able to pay this price for this particular product. Second, government regulations required a two to five year testing period before an airworthiness certificate would be granted. Third, because of its aeronautical characteristics, it appeared that a person would require a helicopter pilot's license to fly it (and there were far fewer than 12,000 helicopter pilots in Canada), so that the initial market estimate was grossly optimistic.

In consequence of these further discoveries, the project was abandoned, and Blowatt reports that Sakowite's *1977 Catalogue of Unique Christmas Gifts* offered the prototype saucer for sale at a price of $600,000!

The second example, cited by Gerstenfeld,[4] was the development of polarized car headlights, which occurred on the opposite side of the Atlantic in the former West Germany. A filter was developed which polarized the light beams from the headlights of a car so that its driver could see ahead clearly without dazzling oncoming drivers. It took four years to develop a technically sound product which failed in the market, for two reasons: First, changes in legislation were required to make the device legally acceptable. Second, the product, although beneficial to oncoming motorists who were the effective "users," offered much less direct benefit to its purchaser. Therefore, the motoring public showed little interest in the device and would presumably only purchase it if polarized headlights were made mandatory. The Polaroid Corporation, which is in the polarized light business, originally planned to develop and market the same product. They abandoned it in favor of entering the polarized light-based instant photography business.

The third example was quoted to the author by a technology manager in a small company. His company test-developed a piece of laboratory equipment in light of the needs of its end users—technologists who claimed that the equipment was technically superior to anything else available and that they would use it if it were made available. The company therefore manufactured and marketed the equipment, but it proved a failure. What the company had failed to realize is that the purchasing decision for this item of equipment was made, not by laboratory personnel, but by purchasing agents. The product, as well as being technically superior, was significantly more expensive than alternative equipment offered by competitors. Purchasing agents, who were less sensitive to the technical merits of the competing products, therefore refused to sanction the purchase of a more expensive item of equipment.

All three of these products were developed to technical success after substantial fiscal expenditure and sophisticated technical thinking. If the merest fraction of the money and sophisticated thinking had been spent on the identification of the mode of end use—the end users and the purchasers—the quite naive marketing mistakes would have been avoided. In the first two innovations, only a brief discussion with government agencies would have identified the regulations which constrained their mode of end use. In the first example, the engineers would have quickly recognized their target market of end users as helicopter, not fixed wing, pilots! In the second and third innovations, cursory thought and investigation would have differentiated between end users and purchasers and discovered that the benefits accrued to the former, while the extra costs were paid by the latter!

These three management parables illustrate the vital importance of establishing a clear insight into the environmental niche or market that the innovation will fill. Blowatt[3] provides a most useful survey of the factors to be considered and data sources which may accessed in developing a market analysis and plan for a proposed new product.

The analysis should identify:

1. The market niche in terms of size, geographic distribution, and structure and the customers (individuals, companies, government agencies, etc.) who should buy the product. The use needs of these customers and (for other than individuals) purchasing procedures in the user organizations. The advantages/disadvantages of the product *in the eyes of these users* (see Section 6.5) as compared with competitive products. The dynamic behavior of the market in terms of growth/decline, seasonality, and fashion.

2. The impact of the economic and regulative environment on this market. How it might be affected by changes in the economic and political climate. The environmental, health, and safety considerations and regulations which must be satisfied.

3. The nature of the competition—its strengths/weakness, the likely competitive reactions to the introduction of the product. Whether the product has adequate patent protection and does not violate competitor's patent rights.

Data required for such an analysis should typically be available from publicly accessible (see Blowatt for suggestions) and internal company sources. By definition, an incremental innovation considered here will usually be offered through an established product promotion, distribution, sales, and servicing system, so the marketing function should be able to provide ample support for the market evaluation. Given strong downstream coupling, this support should be forthcoming, probably through the aegis of the person who enacts the role of the market gatekeeper in the project team. This role is discussed in the next chapter (Section 7.4). Sometimes an incremental innovation may be targeted onto a new market. In this case a more substantial market analysis will be needed and a new market development strategy formulated.

The market analysis should establish estimates of annual sales revenue, based upon the following interrelated considerations, possibly disaggregated over a number of market segments:

(i) Total sales estimate, including possible license income from offshore markets.

(ii) Most likely, minimum and maximum unit product price.

(iii) Most likely, minimum and maximum performance level for the product.

(iv) Most likely, latest and earliest date of product launch and the expected growth and decline pattern over its life cycle.

 (v) Channels for promoting, distributing, and selling the product. Any customer education and after-sales servicing programs required.

(vi) The launch and overall strategy for marketing the product.

In making these analyses, it is vital to consider potential competitive developments or reactions (3 above). Competing companies may be developing new products to compete in the envisaged market and, insofar as possible, this contingency must be allowed for. Gerstenfeld quotes another example to illustrate this point and the need to monitor competitive developments regularly while the project continues.[5] Over two years, a company developed a method for embedding aluminum oxide whiskers in a plastic base to produce a reinforced plastic material of far greater tensile strength for little increased weight compared with those currently available. At the beginning of the project, it had identified a ready market for such a product. Unfortunately, by the time development was completed and the innovative product was ready for launching, the market had disappeared! A competing company had developed a similar but superior product at lower cost using boron carbide instead of aluminum oxide. The moral is try and keep tabs on the competition and be prepared to terminate a project or reorient its direction in light of competitive developments.

Another marketing consideration was illustrated earlier in the Polaroid Corporation's decision to enter the instant photography market rather than the headlights market. Sometimes during the process of technical and marketing development, a company recognizes that the original target market envisaged for the innovation will prove unacceptable and a new market is identified. ABCO, a small company which pioneered the development of fiberglass-reinforced plastic piping systems, originally intended to market in its already familiar pulp and paper industry market. The "new" technology proved to be unacceptable in that market, but the company found a profitable "new" market for it in the utilities industry.[6]

The Technical Development Context

The development phases of the innovation process are primarily concerned with reducing technical uncertainty and establishing the technical feasibility of the product at varying performance levels. In consequence, the estimates of technical outcomes, development duration times, and costs can clearly be expected to be subject to large errors. The level of uncertainty, and these consequent errors, may vary with both technology, project, and the technological strategy of the firm. For an offensive innovator seeking to exercise technological leadership, working close to the state of the art with scarcity of precedence, the level of uncertainty may be relatively large. For a firm following an applications-engineering strategy, it may be relatively small. Whatever the situation, estimates of the times and costs of achieving milestone events will be needed to complete the overall evaluation exercise. They will also be needed to monitor project progress as an input to overall R&D management and planning, and the subject is discussed further in Chapters 7 and 8.

The Manufacturing Context

Apart from the limits to accuracy imposed by the uncertainties of the development process (including pilot/prototype development), in general manufacturing parameters should be the least difficult to estimate. An incremental innovation is unlikely to require radical changes from current manufacturing methods so, when compared with marketing and development estimates, relatively accurate estimates of capital and operating cost versus output levels over the estimated sales range should be readily derivable, taking into account possible economies of scale and learning curve savings (see Chapter 9), as experience is gained with product manufacture. The costs of distributing, selling, and providing an after-sales service for the product plus a spare parts inventory must also be estimated.

Regulative and Environmental Impact Context

For some industries, notably the pharmaceuticals and aviation industries, the development process includes extensive government-regulated clinical trials or prototype testing procedures. Since the late 1960s, however, widespread concern for environmental pollution has led to the enactment of environmental control legislation in most Western developed countries (see Chapter 4, Appendix 4.2). Clearly, nowadays, any thorough project evaluation procedure should include an appropriate assessment of the environmental impacts of the proposed innovation, and the effects of these impacts on estimates of the costs and benefits to be derived from the project. Wisemma describes a useful "checklist" approach for doing this.[7]

6.5   EVALUATION AS A CONCOMITANT PROCESS

In Chapter 2 (Sections 2.10 and 2.11) we viewed an invention or potential innovation as a technological mutation in the evolutionary process which should be evaluated by identifying the niche (or niches) which it might competitively occupy in the evolving technosphere. The frameworks considered so far in this chapter are consistent with that approach, but at first sight appear to emphasize a static or instanteous evaluation followed by a go/no-go decision. This impression is misleading. Trial-and-error learning selection is really a *trial-and-elimination-of-error* process. In our context, this learning or elimination-of-error process must be conducted at both the technological and *marketing (or commercial) levels*. The successive phases of Bright's treatment of the technological innovation process represent error-elimination or uncertainty reduction at the technological level, but successful innovation management requires that error-elimination or uncertainty reduction be simultaneously or concomitantly conducted at the marketing level (as reflected in

Figure 4.2). That is, the technico–commercial evaluation of proposed innovation should be a dynamic or ongoing process. The arguments and framework for conducting this process are outlined in this section.

Schmidt-Tiedemann, who is a member of the Executive Board of Philips GmbH and Director of the Philips Research Laboratory, Hamburg, Germany, outlines one such concomitance model which is of interest.[8] It is an extension of the Bright type of linear and the cybernetic frameworks we introduced in Chapter 2, and that author claims its application increased the acceptance rate of R&D projects by downstream functions in one Philips laboratory to 53%.

Cooper has also proposed a framework for the ongoing evaluation of industrial new products based upon extensive studies of successful and unsuccessful innovations. In Project NEWPROD, he studied the outcomes of 195 innovations (102 successes and 93 failures) introduced by firms in Ontario and Quebec, Canada.[9] From his analysis of the results from this project plus those from about a dozen similar ones in the United States, Western Europe, Hungary, and Japan, he identified six lessons for innovation managers and project team members.

A Much Stronger Market Orientation is Needed. This is essentially the point stressed in Section 6.4. Cooper identifies the lack of market research as the most frequent cause of failure. He suggests that market research results should be integrated with technological outcomes from early in the process and, most importantly, as inputs to the product design, engineering, and product development activities.

New Product Success is Largely Amenable to Management Action. Most studies suggest that the actions of team members (especially the product or project champion, see Chapter 7, Section 7.4) and the impact of key (especially market research) activities determine innovation outcomes. That is, the design and implementation of the innovation process critically determines its success.

There is No Easy Explanation for What Makes a New Product a Success. Most studies identified a large number of factors influencing success and, quoting Cooper, *"success depends on doing many things well, while failure can result from a single error."*[10] Innovation management requires the management of numerous multidisciplinary tasks into an uncertain venture in which the failure of any one task may lead to overall failure. Thus, an overall planning framework is required.

The Product Itself—A Unique Product with Real Customer Advantage— is Central to Success. At first sight, this observation appears to be platitudinous, but more considered thought suggests otherwise. The product must be unique and superior *in the eyes of the customer,* which implies that clear understanding of the customers' needs, performance criteria, use practices, etc., is required before advanced product development begins. To obtain such understanding requires significant market analysis and research effort and the subsequent marriage of technological design and development to customer needs.

A Well-Conceived, Properly Executed Launch is Vital to Success. A well-conceived marketing strategy and plan leading to a properly targeted and executed launch effort must be integrated into the overall innovation process.

Internal Communication and Coordination Between Internal Groups Greatly Foster Successful Innovation. Since successful technological innovation requires the multidisciplinary and multifunctional participation of scientists, engineers, marketing, and production staff, etc., the creation and maintenance of good internal communication within the project team and with other organizational groups is vital. This issue is discussed further in Chapters 7 and 8.

Based upon his research analysis and these six lessons, Cooper proposes an innovation process model which constitutes a seven stage action guide for innovation managers.[11] The overall process is illustrated in Figure 6.3 in the context of corporation with a central, as well as divisional, R&D facilities. As the project progresses through the successive phases of the innovation process, it is viewed as moving from corporate to divisional R&D and then into production. The corporate-divisional transfer would clearly not apply to a company operating a single R&D facility. We consider each of the seven steps in turn.

I. *Product Idea.* The product idea is likely to emerge out of nondirected research, coupled with technology and marketing monitoring activities—that is, the matching of a technological possibility with a potential market niche. There should be an informal peer group screening of the idea at this early stage. This should identify whether the idea is of sufficient technological interest, feasibility, and challenge, congruent with the company's technological strategy and "doable" within the company's resources. This screening and the successive evaluations may be made in competition with other projects or product ideas, so may be subject to the sampler of the screening methods described in the Appendix to this chapter. If the project survives this screening it will pass into the *primary development* phase of the process.

II. *Preliminary Assessment.* Primary development should clarify the technical feasibility of the idea and give a rough idea of the resources and money required to yield a commercial product, so a preliminary market assessment should be made. This can be a fairly "quick and dirty" desk marketing study eliciting information from the company salesforce, readily accessible statistical data and research reports, industry experts, and perhaps a few potential customers. If it survives a more rigorous screening process after this assessment, the project will continue to the *secondary development* phase, possibly being transferred from the corporate to the divisional R&D facility.

III. *Product–Market Concept Definition.* Secondary development should further clarify and delineate the performance characteristics of the proposed product, thus facilitating the identification of its target market and how it will be positioned vis-à-vis market segments and competitive products. Cooper comments that this

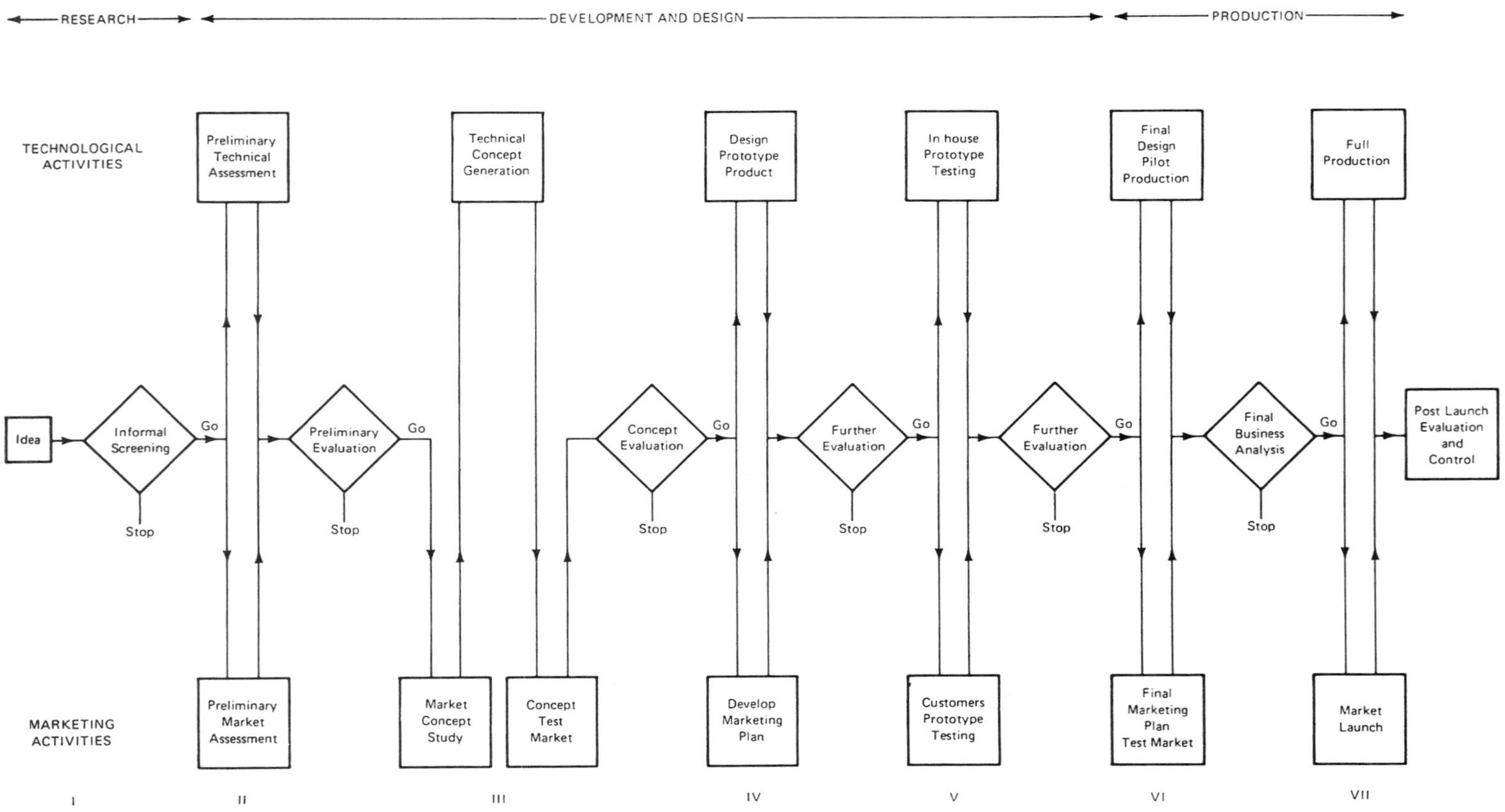

Figure 6.3. Cooper's evaluation process.

concept definition stage is frequently omitted by industrial product firms, often with disastrous results. Concept definition should include several elements. First, it should identify a market niche for the product. That is, a customer segment dissatisfied with competitive products currently available or a niche where new technology can gain a competitive advantage. Second, it should identify what improved benefits or features are desired by these customers and the product concept and design (and its technical feasibility) required to satisfy them. Third, the concept should be informally market tested. That is, potential customers should be shown diagrams, models, or descriptions of the proposed product and their reactions noted. Product concept changes may be required in light of these reactions, but they should enable an assessment of market acceptance and tentative sales forecast to be made. It will also initiate the market planning process and enable reasonable estimation of production costs to be made. Thus, a reasonable financial and commercial analysis can be made by this stage.

To illustrate concept definition, Cooper cites the example of a high-quality truck manufacturer who wished to enter the construction dump truck market. The initial product concept was to build a high-quality dump truck which appeared to be technically feasible and for which there existed a large growing market, but a detailed definition of the product concept and market positioning was lacking. A market study identified the likes, dislikes, and preferences of dump truck fleet operators. This showed that truck downtime due to breakdowns was a major and costly problem for operators. The revised product concept was defined as "*a truck that can be repaired and back on the road overnight . . . no matter how serious the breakdown.*" To implement this concept, a modular vehicle was developed in which every major unreliable component could be removed and replaced quickly, using standard parts available from local suppliers. The minimum downtime product concept proved to be technically feasible at a competitive price. Paul Cook (founder and CEO of Raychem Corporation) uses almost identical words in describing the series of questions his firm asks potential buyers when developing a new product concept.[12]

The concept can now be subjected to a more detailed evaluation and, if continued, proceed to tertiary design and development leading to prototype or small-scale pilot production. By now, such work will definitely be performed in the divisional R&D facility.

IV. *Tertiary Development.* Design and Market Planning Detailed technical design and development and market planning now continue simultaneously. The market planning will determine the pricing, selling, and advertising strategies, and the distribution and after-sales servicing networks required. It may require further buyer behavior studies. The technical output from this stage will be prototype products which may be subject to customer testing.

V. *Prototype Trials.* One or more prototypes (depending upon the product) will now be subject to detailed test trials both within the company and with selected users. These trials, conducted in a possibly more rugged and demanding user environment than those in a development laboratory, may identify technical and design flaws before larger scale production is begun. It may also identify product modifications which should be incorporated to improve user acceptance.

Cooper cites the example of a design fault in a dial-in-hand wall telephone which was being developed by a major telephone manufacturer. Fifty prototypes were subjected to in-house durability and reliability tests, and a further 50 were installed in nontechnical employers' houses for customer testing. This latter test identified a potential disastrous design fault which had remained undetected in laboratory testing—the receiver fell off the hook if a nearby door was slammed in certain types of houses! This fault was eliminated by a quite minor and inexpensive design change. After these trials, the project is subjected to a further evaluation and, if continued, passes onto:

VI. *Larger Scale Pilot Production and Test Marketing*. The project incorporating any product design changes identified above will probably now be transferred to the production facility and a larger scale pilot production line set up there. This stage may identify any changes in manufacturing methods which may be required in scaling up to full-scale production. It will scale up output by at least one order of magnitude and provide sufficient units to conduct a formal test market—that is, to sell the product using the proposed marketing plan to a limited sample of customers or in a limited marketing area. This should identify any remaining design faults to be eliminated or changes to be made to increase customer acceptance. Most importantly, it should also identify any changes which may be required in the marketing plan.

When this stage is completed and any desired product design, manufacturing methods, and marketing plan changes have been made, the final evaluation is performed. This will constitute a comprehensive business analysis and plan for the product, incorporating the considerations listed earlier in the chapter. If accepted the innovation now proceeds to:

VII. *Full Production and Product Launch*. The final stage incorporates the adoption of the innovation by the operations management functions of the firm, the scale up to full production output, and the launching of the product through the implementation of the marketing plan in the full market area. Although further technical and manufacturing problems may need to be debugged during this final scaling up process, if the previous stages have been thoroughly and effectively managed, barring any major unexpected changes in market circumstances, this launch should go ahead in a straightforward manner. Post-launch monitoring will be conducted to identify customer acceptance, market share, sales, unit production costs, etc.

Cooper's process model provides a useful conceptual framework for managing the innovation process, which implicitly implements the error-elimination process. Obviously, the detailed stages may vary between differing product/process technologies and it is presented in the context of an offensive–defensive innovator with a comprehensive technological base (as defined in Chapter 3) here. However, a company following an applications-engineering strategy from a truncated technological base could readily apply the later stages of the process.

6.6 THE STANFORD INNOVATION PROJECT AND THE NEW PRODUCT LEARNING CYCLE

The SAPPHO and NEW PROD studies were performed mainly in Great Britain and Canada, respectively, so the question arises as to whether similar results would be obtained in the United States. Maidique and Zirger studied 158 new product successes and failures in mainly West Coast located companies in the U.S. electronics industry.[13–15] Their results largely agreed with the Canadian and U.K. studies. Maidique and Zirger also emphasized new product development as a trial-and-error-elimination learning process in three ways:

1. Learning by trying and doing as exemplified in the learning or experience curve concepts pioneered by the Boston Consulting Group (considered in Chapter 9).
2. Learning by using on the part of the customers, particularly the lead customers as discussed in the next section, which leads to subsequent improvements in product designs and operations.
3. Learning by failing or, more accurately, by trial-and-error elimination.

Like Cooper, they take an implicitly cyclical learning view of new product development. Successful organizations learn from their mistakes made in developing one product when they develop the next one. They cite the experiences of Apple Computer with the unsuccessful Apple III. Knowledge learned from this failure was embodied in the design of the very successful Apple IIE. In fact, if "failure" is a precondition for "success," "success" is often a precondition for "failure." Firms that enjoy one or two product successes often become complacent, losing their in-depth understanding of the customer and the marketplace, so that their next product fails. Suitably chastened, they learn from this unexpected failure to produce a successful product the next time around. Thus, successive new product development efforts frequently generate failures as well as successes. This is reflected in the still short, but somewhat checkered, history of Apple Computer and IBM's own efforts in the microcomputer market, as well as the spectacular failures of other large mature corporations. The PCjr, Corfam, and the Edsel are all examples of failures sandwiched between successes.

Another important lesson to be learned from new product development learning cycles is that organizations should think in terms of product *families* rather than products. This important lesson will be discussed further in Section 6.8, after we have considered the important roles played by lead users in new product development and user-dominated innovation.

6.7 LEAD USERS AND USER-DOMINATED INNOVATION

As Mowery and Rosenberg argue, it is often difficult to identify user needs.[16] This is obviously true for innovations toward the revolutionary pole of the continuum where the White approach is needed, but can also apply to situations where more incremental innovations are being developed for a rapidly evolving or turbulent market. Also, as Burgelman and Sayles point out, there is always the danger of choosing a "lunatic fringe" customer who has very atypical, if not eccentric, needs.[17] In this situation, a key task is to identify lead users who wish to maintain themselves at the leading edge of technology–market trends, but are not part of a lunatic fringe. Hamel and Prahalad[18] describe this as *expeditionary marketing,* and stress the need to develop *marketers with technological imagination and technologists with marketing imagination.* In Chapter 4, Appendix 4.1 (Section A4.1.8) we discuss the adoption and diffusion of innovations, including the seminal work of Rogers who categorized the first two groups of users to adopt a new product as "innovators" and "early adopters" (Figure A4.1.5*b*). Typically, in any marketplace, there are such individuals who perceive an advantage in being among the first to adopt an innovation. They can also be labeled *lead users* and are usually the immediate target customers for a new product. SAPPHO, NEW PROD, and the Stanford Innovation Project, as well as others, all stress the vital importance of an intimate understanding of user needs which, in practice, means lead users needs. To achieve this intimacy, Maidique and Zirger repeat the thesis of Chester Barnard[19] that customers should be seen as part of the organization. The well-informed innovating firm can usually identify its lead users, but if it cannot, von Hippel suggests an approach for so doing.[20]

The concept of the lead user being part of the innovator's organization can be extended to the lead user *becoming* the innovating organization. Von Hippel describes this quite widespread situation as *user-dominated innovation.* It is particularly common in the development of scientific instruments and semiconductor manufacturing equipment. A (say) university laboratory may develop a new prototype piece of equipment for its own experimental use. It may then recognize that the equipment could be used elsewhere and so, possibly after further development, may grant a license to a scientific instrument company to manufacturer and market it. For example, von Hippel studied four types of instruments: gas chromatographs, nuclear magnetic resonance spectrometers, ultraviolet absorption spectrophotometers, and transmission electron microscopes. He found that all four of the first-of type of these instruments were developed by users. Also, in his sample of 44 major and 63 minor improvements to the original instruments, 36 (82%) and 32 (70%) were developed by users, respectively.[21]

6.8   RE-INNOVATION AND ROBUST DESIGNS

In Chapter 2 we saw that, following the introduction of a revolutionary technology, user needs are often clarified with the introduction of one or more dominant designs (such as the Boeing 707) that define a framework or paradigm for the subsequently evolutionary development of an emerging technology (such as jet airliners). Although some dominant designs, such as the QWERTY typeface, may be poor and only survive because of the putty-clay features of technological evolution, others are good in that they inherently allow for further evolutionary developments within the overall design paradigms. Once a dominant design has emerged and the technological path or trajectory evolves towards maturity, market needs become more clearly delineated or segmented and technological evolution continues through the design of products targeted towards specific market segments. Therefore, a design that is inherently capable of flexible evolution to produce a family of products to satisfy a range of market segments or customer needs will be proven to be inherently superior to one lacking this flexibility. This implies that product design should be viewed as a strategic issue in the technology oriented firm, as outlined by Eschenbach and Geistauts.[22]

Rothwell and Gardiner[23] articulate this notion through their concept of re-innovation and robust designs. They reason that innovation should not merely embrace the development of a single product, but rather be viewed as generating a family of products as illustrated in Figure 6.4. The first product will be conceived, developed, manufactured, and marketed ideally through the interactive dialogs discussed in the previous sections of this chapter. After market launch, the company will enjoy access to continual user feedback, so design and development can continue to generate improved versions of the product more accurately targeted onto specific market segments. The authors define this activity as *re-innovation*. Some product designs are more open to improvement than others, and they describe these

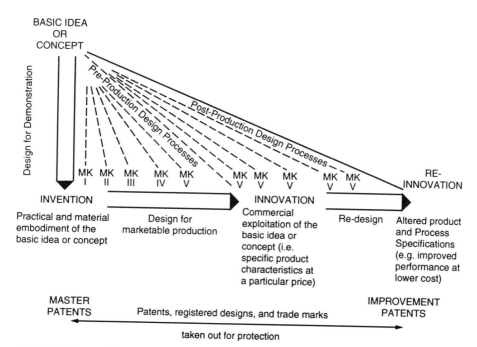

Figure 6.4. Innovation and re-innovation. *Source:* Reprinted with permission from Rothwell and Gardiner, 1983.

as *robust designs*. Here, the adjective "robust" is not being used as a synonym for strength and ruggedness in use, but rather as a synonym for inherent flexibility or the "technological slack" available for further development. They cite the original Boeing 707s, 727s, and 747s as robust designs because they generated a family of different versions of these aircraft. In contrast, they cite the original Lockheed Tri-Star or L-1011-1 as a *lean* design since it generated only two further versions, the L-1011-100/200 and L-1011-500, as compared to the family of 747s shown in Figure 6.5*a*. It is perhaps ironic to note that although the L-1011-1 proved to be a lean design, its first power unit, the RB 211 engine whose development bankrupted the original Rolls Royce Company, proved to be a robust one. Once it had been bailed out by the British government, the newly structured Rolls Royce was able to market a family of derivatives from the original design as shown in Figure 6.5*b*. Other examples of robust designs cited by the authors are the Ford Cortina car and the IBM PC.

What might be called the design of the robust designing process is illustrated in Figure 6.6 which bears some resemblance to the convergent-divergent cycles of the creativity generation process of Chapter 7, Appendix (Figure A7.6). It begins with basic invention which generates a divergent set of design ideas. These can be evaluated against a range of often conflicting cost/performance criteria during the composite design phase when some ideas are filtered out. Those remaining provide the workable compromise that is proven in the consolidated design phase. This may be stretched later in various ways such as upsizing, downsizing, and/or incorporating cost reducing or performance enhancing improvements.

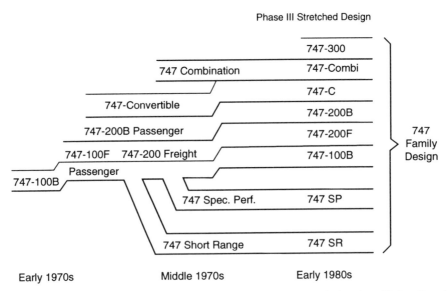

Figure 6.5*a*. The Boeing 747 family. *Source:* Reprinted with permission from Rothwell and Gardiner, 1983.

RB211 FAMILY

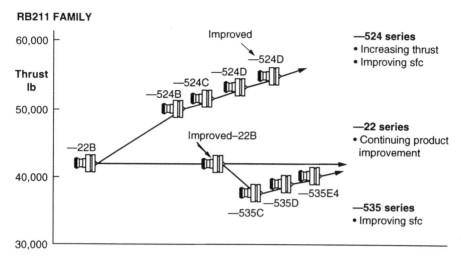

Figure 6.5b. The Rolls-Royce RB211 family. *Source:* RB211-22B Technology & Description, Rolls-Royce Limited, TS2100, Issue 20, July 1980. Updated and revised by authors in 1987 figures. Reprinted with permission from Rothwell and Gardiner.

Rothwell and Gardiner point out that user-led innovation coupled with robust design can generate good designs, as illustrated by the following verbal equation:

Tough Customers + Robust Designs = Good Designs

where a "good" design is one which satisfies user, technical, and commercial needs and yields an acceptable market share and profit to the producer. They illustrate their equation with a brief history of the development of the Boeing 747.[24]

In December 1965 Boeing signed a contract with Pan Am (its tough customer) to produce an airliner capable of satisfying the transportation needs of the 1970s. The airliner would be capable of carrying 350 passengers or 200,000 lbs. of freight and cover 5100 nautical miles at a cruising speed of 547 knots. The development program was expected to cost Boeing $2.0 billion. Pan Am agreed to staged progress payments to Boeing of $500 million for 25 airplanes, with 50% due six months before delivery of the first one. These fiscal arrangements meant that both parties effectively bet their companies on the success of the project.

This contract was based upon the design composition phase of the process. Boeing first considered and proved the design feasibility of double-decking the 707 to provide the required capacity, but had to reject this design for safety reasons. The FAA stipulated that its 350 passengers could be evacuated from the airplane within 90 seconds in an emergency. This was considered impossible with two decks, so an essentially single-deck design was required. The single-deck design was determined by the freight rather than passenger carrying requirements. Pan Am required that the airplanes should be capable of convertible usages as passenger or cargo carriers and, in the later capacity, should be capable of carrying two 8 ft × 8 ft wide containers

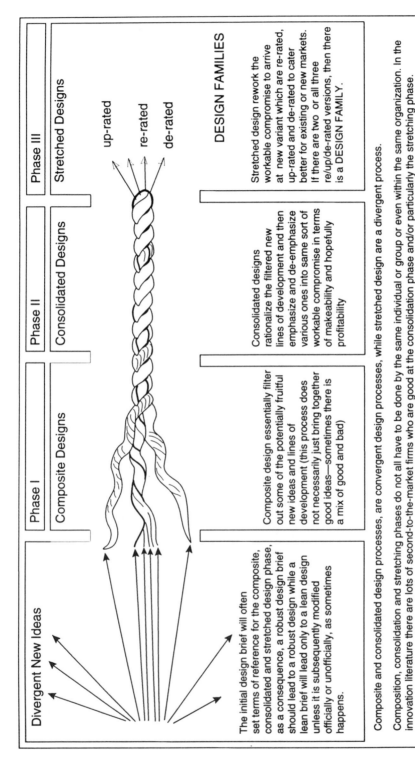

Phase I | **Phase II** | **Phase III**

Divergent New Ideas | **Composite Designs** | **Consolidated Designs** | **Stretched Designs**

up-rated

re-rated

de-rated

DESIGN FAMILIES

The initial design brief will often set terms of reference for the composite, consolidated and stretched design phase, as a consequence, a robust design brief should lead to a robust design while a lean brief will lead only to a lean design unless it is subsequently modified officially or unofficially, as sometimes happens.

Composite design essentially filter out some of the potentially fruitful new ideas and lines of development (this process does not necessarily just bring together good ideas—sometimes there is a mix of good and bad)

Consolidated designs rationalize the filtered new lines of development and then emphasize and de-emphasize various ones into same sort of workable compromise in terms of makeability and hopefully profitability

Stretched design rework the workable compromise to arrive at new variant which are re-rated, up-rated and de-rated to cater better for existing or new markets. If there are two or all three re/up/de-rated versions, then there is a DESIGN FAMILY.

Composite and consolidated design processes, are convergent design processes, while stretched design are a divergent process.

Composition, consolidation and stretching phases do not all have to be done by the same individual or group or even within the same organization. In the innovation literature there are lots of second-to-the-market firms who are good at the consolidation phase and/or particularly the stretching phase. Big aviation and automobile firms normally proceed through all three phases, but if competitors come up with something new which allows the whole process to be shortened, most firms will take it up and incorporate it into their own designs.

Figure 6.6. The robust designing process. *Source:* Reprinted with permission from Rothwell and Gardiner.

side by side. This requirement meant that the fuselage had to be 19 ft wide. It also meant that a small cockpit had to be mounted above the main fuselage, to avoid the 200,000 lb. cargo crashing through into it in an emergency. The aerodynamics of a second-story cockpit dictated a bubble design, and provided further second story space which was initially used as a first class cocktail lounge in some passenger configurations and promoted as a selling point to first-class travelers.

The design composition imposed tough demands on Boeing's major subcontractors. First, Pratt and Whitney had to design engines with the required thrusts. Second, it was decided to replace a flight crewman navigator with computer-based navigational aids that could be used by the pilots; AC Electronics had to design a system that would be accurate and reliable enough to do this. The design consolidation phase which followed contract signature has been described as a four-year-long nightmare which involved, at times, acrimonious disputes among Pan Am, Boeing, and Pratt and Whitney.[25] Once detailed work began, it appeared that the airplane's takeoff weight would have to be increased. This, in turn, reduced its cruising speed and altitude as well as its range, so that it became dearer to fly than the existing 707s. Boeing considered seven alternative design configurations and ways of optimizing the aerodynamic efficiencies of these designs, but was, in turn, forced to impose tough customer requirements on Pratt and Whitney. The original composite design envisaged that the airplane would be powered with Pratt and Whitney JT9D engines with a thrust of 41,000 lb. The JT9D design had to be modified to produce a thrust of 43,000 lb. to achieve the desired takeoff weight, and the first 747 test flight took place on February 9, 1969. Pan Am's inaugural operational 747 flight took off from New York bound for London only a day late on January 22, 1970. By the late 1980s, over 800 747s had been produced in stretched design variations.

Re-innovations can exploit economies of technology, analogous to economies of scale and of scope in manufacturing, as shown in Boeing's subsequent development of the narrow-bodied 757 and wide-bodied 767 families.[26] The cockpit and avionics designs were common to both airplanes, reducing development and production costs, and have also been applied to 747-400 aircraft. It also reduced flight crew training costs since pilots could be trained on common simulators and readily retrained for switching from one aircraft to the another. The incorporation of the 757-767 cockpit and avionics designs in the 747-400 represents economic re-innovation by applying "new" technology to an "older" robust design and reverse economies may also be obtained. Like Boeing, Airbus Industries has developed their A330 and A340 aircraft as a single program with many common parts, but has also incorporated parts from their earlier A300, A310, and A320 aircraft, to apply "old" to "new" technology. Black and Decker applied both philosophies in developing their heatgun paint stripper. When they first launched it, they designed it around the "old" technology of power drills to reduce the cost of developing a commercially unproven new product. Two years later, once it had gained market acceptance, it was totally re-designed or re-innovated with "new" technology to provide a lower cost, higher-performance product which has subsequently evolved into a design family of heatguns customized for individual market segments.

6.9   PROCESS, PRODUCT, AND SERVICE INNOVATIONS

The approaches discussed in this chapter have been primarily directed at product innovations, but that does not imply that process and service innovations are of any less importance. The three constitute an interdependent triad or trinity. Process innovations facilitate the efficient manufacture of product innovations (particularly when a technology–industry has reached the mature plateau of its S-curve) which, in turn, facilitate the provision of new or improved services. Furthermore, as Chase and Garvin argue, the Japanese-led developments in manufacturing management, philosophy, and technology, mean that a manufacturing plant must increasingly behave as a service factory if it is to retain its competitive edge.[27] The complementary roles of process and product innovation are apparent from the results of one study, which examined successes and failures in a mixed sample of 211 process and product R&D projects in six mature technology–industry segments.[28] The researchers found that those projects with process improvement objectives were more likely to succeed and make bigger contributions to a sponsor firm's profits by reducing costs, while product-oriented projects contribute to profits through increased sales and the enhancing of the firm's image. De Brentani and Cooper have also extended the latter's approach into services.[29,30]

The process–product–service triad is apparent in the evolution of the information/telecommunication industry. It is manufacturing process innovations in photolithographic and related techniques in the microelectronics industry that are facilitating product innovations in the forms of successive generations of DRAMs and microprocessors. These product innovations are, in turn, facilitating service innovations. Similar triads can be identified in the evolution of other industries. Mitchell discusses the interdependence of process, product, and service innovations in the United States, and argues that R&D efforts must be increasingly directed towards providing new and improved services.[31] He argues that when presented with a new technological capability, rather than concentrating on how it can be used to upgrade existing products, emphasis should concentrate on the imaginative consideration of new user service needs that it might satisfy. Presented with the capability providing plentiful inexpensive bandwidth in the future, he argues that the critical question for the telecommunications company is not "How can we incorporate this capability into our equipment?" Rather, it is "What will customers seek from this new bandwidth? What new video, data or voice should we providing to satisfy these needs?" It was, of course, this creative thinking that led to the emergence of the personal transistor radio, the Walkman and the PC. Mitchell argues that R&D thinking must become increasingly oriented towards satisfying user service needs, with service industry organizations enacting lead-user roles for their manufacturing product suppliers, as was illustrated in Pan Am's role in the development of the 747.

6.10   FINAL COMMENTS

Finally, it is of interest to note that the genesis of new products in general, as well as robust designs in particular, reflects a transition from the divergent to convergent thinking discussed in the Chapter 7 Appendix. The early stages reflect a divergent exploration of alternative potential approaches, while the latter stages converge on a specific focused product concept. Schmidt-Tiedemann, in describing his process model, argues that as a R&D project proceeds, the largely divergent search for new ideas or problem solutions is replaced by the convergent pursuit of a certain goal.[8] As he puts it, switching at the right time from divergent to convergent thinking is largely a matter of technical and commercial judgment or gut feel. To converge too late means spending too much effort on "losers." To converge too early may mean throwing away "winners." Some project selection approaches to help differentiate between potential "winners" and "losers" are discussed in the Appendix to this chapter and we then turn to some of the behavioral aspects of project management.

FURTHER READING

1. R. G. Cooper, *Winning at New Products.* Agincourt, Ont.: Gage Educational Publishing Company, 1987.

REFERENCES

1. G.R. White, "Management Criteria for Effective Innovation." *Technology Review* **80,** 15–23 (1978).

2. *Success and Failure in Industrial Innovation.* London: Centre for the Study of Industrial Innovation, 1972.

3. K. R. Blowatt, "Marketing for the Technical Entrepreneur." In D. S. Scott and R. M. Blair (Eds.), *The Technical Entrepreneur,* Ontario: Press Porcepic, 1979, pp.135–183.

4. A. Gerstenfeld, *Innovation: A Study of Technological Policy.* Washington: University Press of America, 1977, Chapter 5.

5. A. Gerstenfeld, *Effective Management of Research And Development.* Reading, MA: Addison-Wesley, 1970, pp. 29–30.

6. M. J. C. Martin and P. J. Rosson, "R&D Philosophy at ABCO." In *Four Cases on the Management of Technological Innovation and Entrepreneurship,* Technological Innovation Studies Program. Ottawa, Ont.: Department of Industry, Trade and Commerce, 1984.

7. J. G. Wissema, "Putting Technology Assessment to Work." *Research Management* **24**(5), 11–177 (1981).

8. K. J. Schmidt-Tiedemann, "A New Model of the Innovation Process." *Research Management* **25**(2), 18–21 (1982).

9. R. G. Cooper, *Project Newprod: What Makes a New Product a Winner?* Montreal: Quebec Industrial Innovation Centre, 1980.

10. R. G. Cooper, "A Process Model for Industrial New Product Development." *IEEE Transactions on Engineering Management* **EM-30,** 2–11 (1983).

11. R. G. Cooper, *Winning at New Products.* Agincourt, Ont.: Gage Educational Publishing Company, 1987.

12. W. Taylor, "The Business of Innovation: An Interview with Paul Cook." *Harvard Business Review* **68**(2), 97–106 (1990).

13. "Why Products Fail." *Inc. Magazine* May(1984).

14. M. A. Maidique and B. J. Zirger, "A Study of Success and Failure in Product Innovation: The Case of the U.S. Electronics Industry." *IEEE Transactions on Engineering Management* **EM-31**(4), 192–203 (1984).

15. M. A. Maidique and B. J. Zirger, "The New Product Learning Cycle." *Research Policy* **14**(6), 299–313 (1985).

16. D. C. Mowery and N. Rosenberg, "The Influence of Market Demand upon Innovation: A Critical Review of Some Recent Empirical Studies." *Research Policy* **8**, 102–153 (1979); also reprinted in N. Rosenberg, *Inside the Black Box: Technology and Economics.* New York: Cambridge University Press, 1982.

17. R. A. Burgelman and L. R. Sayles, *Inside Corporate Innovation Strategy, Structure and Managerial Skills.* New York: Free Press, 1986.

18. G. Hamel and C. K. Prahalad, "Corporate Imagination and Expeditionary Marketing." *Harvard Business Review,* 81-92 (1991).

19. C. Barnard, *The Functions of the Executive.* Cambridge, MA: Harvard University Press, 1968.

20. E. von Hippel, "Lead Users: A Source of Novel Product Concepts." *Management Science* **32**(7), 791–805 (1986).

21. E. von Hippel, *The Sources of Innovation.* New York: Oxford University Press, 1988.

22. T. G. Eschenbach and G. A. Geistauts, "Strategically Focused Engineering: Design and Management." *IEEE Transactions on Engineering Management* **EM-40**(2), 62–70 (1987).

23. R. Rothwell and P. Gardiner, "Re-Innovation and Robust Designs: Producer and User Benefits." *Journal of Marketing Management* **3**(3), 372–386 (1988).

24. P. Gardiner and R. Rothwell, "Tough Customers: Good Designs." *Design Studies* **6**(1), 7–17 (1985).

25. L. S. Kuter, *The Great Gamble: The Boeing 747.* Birmingham: University of Alabama Press, 1973.

26. R. Rothwell and P. Gardiner, "The Strategic Management of Innovation," Unpublished presentation. Manchester: R&D Conference, 1988.

27. R. B. Chase and D. B. Garvin, "The Service Factory." *Harvard Business Review* **67**(4), 61–69 (1989).

28. N. R. Baker, S. G. Green, and A. S. Bean, "The Need for Strategic Balance in R&D Project Porfolios." *Research Management* **29**(2), 38–43 (1986).

29. U. de Brentani, "Success and Failure in New Industrial Services." *Journal of Product Innovation Management* **6**(4), 239–258 (1989).

30. R. G. Cooper and U. de Brentani, "New Industrial Financial Services: What Distinguishes the Winners." *Journal of Product Innovation Management* **8**(2), 75–90 (1991).

31. G. R. Mitchell, "Research and Development for Services." *Research Technology Management* **32**(6), 37–44 (1989).

APPENDIX 6
PROJECT SELECTION APPROACHES

If you can look into the seeds of time,
And say which grain will grow and which will not.
MACBETH, ACT I, SCENE III

A6.1 INTRODUCTION

Given the fecundity of a good R&D function in generating potential innovations, it is virtually certain that a firm's R&D budget will be insufficient to support all the projects which are competing for funding. Therefore, any R&D-based firm will be required to operate some explicit or implicit procedure for R&D project ranking and selection. The purpose of this Appendix is to examine some of the considerations and difficulties which are inherent in this issue and outline some approaches which may be of value in some situations. It does not suggest any one methodology which can be applied to all situations. R&D project selection, like most areas of business decision making, is best performed by informed management judgment at the appropriate level in the organization, using formal techniques as guides and disciplines to promote effective decision making.

A6.2 SELECTION OF PROJECT SELECTORS

Whether projects should be selected by R&D or general management depends upon the phase reached by each project in the innovation process, the resources it requires and the *immediacy of impact on the business of successful completion*. In Chapter 5 (Section 5.4) we argued that projects in the basic research and primary development phases are best selected by internal R&D function peer review. Once they completed these phases, typically reaching Stage III in Figure 6.3 of Chapter 6, corporate resource and impact considerations dictate that selections should now be made by general management. The selectors should be (and in many companies are) either a review committee drawn from the Innovation Management Function (IMF) of the firm or qualified individuals who form a panel which reports to such a committee. The acronym IMF is perhaps appropriate since, like the International Monetary Fund, it is responsible for disbursing funds to worthy causes, though some readers may question whether its more august international level namesake always does so!

There are four major considerations that should define the mission and membership of such a selection panel:

1. As already suggested, the IMF should operate at general management level. In a technologically diversified corporation, a committee will probably be needed in each SBU, with their recommendations reviewed by a corporate level committee, to avoid any duplication of efforts and to encourage the exchange of knowledge and the exploitation of potential synergies between units.

2. Project selections should be dictated by strategic considerations and gap analysis (Figure 5.2), as discussed in Chapters 4 and 5.

3. In making selections, it should be recognized that projects are typically interdependent rather than independent. That is, the knowledge and outputs obtained during the course of performing one project may impact upon others. Probably more importantly, projects compete for scarce personnel, laboratory space, and experimental equipment as well as finance, so that the extent that a project requires these scarce resources must be considered. Since projects are beset by technological, manufacturing, marketing, and environmental impact uncertainties, the probability of *any one* project yielding a commercially successful outcome may not be high. Such factors dictate that the intent of project selection should be to produce a balanced project portfolio, which will deploy resources to maximize probability of closing the future gap between business aspirations and current achievements.

4. A good R&D facility and climate is prolific in generating projects with new product potential through the encouragement of proposals from its creative staff. Since such a climate generates interpersonnel competition for R&D resources, it is important to employ a selection panel team which, while incorporating expertise in the relevant technological and functional areas, also enjoys the respect and esteem of the R&D staff. In this way, the organization can seek to maximize the probability of achieving its goals *and* convince R&D staff of the equity of its project selection decisions. Abetti and Stuart suggest that, as well as people with varied experience of new product development, the panel should include people who enjoy peer group esteem, regardless of their technical or managerial standings.[1] They recommend that a panel should include at least one project champion and, reflecting the importance of learning by failing, people with *unsuccessful* as well as successful experiences in developing new products. 3M claims to use information from nine failed projects to achieve one success.[2] Abetti and Stuart also recommend that membership should be rotated to avoid the fossilization of the evaluation criteria.

A6.3 PROJECT SELECTION TECHNIQUES

Project evaluation and selection have both subjective and objective aspects, and in order to ensure an equitable selection process, it is desirable to use objective selection methods whenever possible. Over the last 25 years or so, massive efforts have been made to develop R&D project evaluation and selection approaches to satisfy this need. As long ago as 1964, Baker and Pound[3] found that there were then over 80 different formal models available in the literature to help management make project selection decisions.

For this reason, and because of the critical importance of the subject to R&D management, it is useful to overview the approaches to project selection and evaluation which have been developed. The approaches are described in order of increasing mathematical sophistication and vary in applicability to the different stages of the R&D process, some being more applicable to basic research as opposed to advanced development. Some evaluate each project independently using subjective or numerical measures of merit, while others allow for the interactions between projects and the selection of project portfolios. The overview is mainly based upon Souder,[4] Fusfield,[5] and Vertinsky and Schwartz.[6]

Intuitive Individual or Group Evaluations and Selections

An individual or committee selects projects based on a number of subjectively evaluated criteria.

1. The halo effect—that is, the past track record or halo of each project team.
2. The "squeaky wheel method (A)"—project teams which complain the loudest get their projects funded.
3. The "squeaky wheel method (B)"—projects are selected so as to minimize the disruption from the selection of the previous period.
4. The "white charger technique"—selection based upon direct individual influence or power of the project team, its power of advocacy, or possibly it being the last team to present its case to the selectors.

Intuitive judgment is a traditional approach which, in the absence of any alternatives, was also probably used exclusively up to the 1950s. In 1965 Seiler[7] found that half of the R&D managers he surveyed used this method, and Cetron reported that 56% of a sample which he surveyed used individual subjective assessments.[8]

Check Lists

The next stage in analytical sophistication from purely intuitive judgment is project evaluation against a checklist of factors which determine the likelihood of an innovation becoming commercially successful. Dean studied 34 firms and identified 32 R&D, manufacturing, and marketing factors which were most frequently considered in evaluation projects.[9] These considerations are listed in Table A6.1 which, although incomplete, does incorporate many of the criteria that are relevant to project evaluation. Unfortunately, Dean found that the average company used only about a quarter of these considerations in project evaluation, so that they could hardly be said to be operating comprehensive checklists! However, R&D or corpo-

TABLE A6.1   Dean's Checklist

 1. Compatibility with company objectives;
 2. Compatibility with other long-term plans;
 3. Availability of scientific skills in R & D;
 4. Critical technical problems likely to arise;
 5. Balance of R & D program;
 6. Interaction with other R & D projects;
 7. Competitors R & D programs;
 8. Size of potential market;
 9. Factors affecting expansion of the market;
10. Influence of government regulations and control;
11. Export potential;
12. Probable reaction of competitors;
13. Possibility of licensing and know-how agreements;
14. Possibility of R & D cooperation with consultants or other organization;
15. Effect on sales of other products;
16. Availability and price of materials needed;
17. Possibilities of 'spin-off' exploitation of innovation;
18. Availability of production skills and equipment;
19. Availability of marketing skills and experience;
20. Advertising requirements;
21. Technical sales and service provision;
22. Effects on company 'image';
23. Risks to health or life;
24. Probable development, production and marketing costs;
25. Possibility of patent protection;
26. Scale and timing of necessary investment;
27. Location of new or extended plant(s);
28. Attitude of key R & D personnel;
29. Attitude of principal executives;
30. Attitude of production and marketing departments;
31. Attitude of trade unions;
32. Overall effect on company growth.

Source: B. V. Dean, *Evaluating, Selecting and Controlling R & D Projects* (New York: AMACOM, a division of American Management: Association, 1968), p. 49. Reprinted with permission. All rights reserved.

rate management should find little difficulty in developing a checklist similar to Dean's and congruent with the technological strategy of the firm. Dessauer[10] reports that at Xerox, R&D projects are evaluated by a corporate committee composed of R&D engineering, finance, and marketing functional representatives using a comparable checklist.

Project Profiles

The disadvantages of a simple checklist, even if it exhaustively lists all evaluation criteria, is that a project is checked against each criterion purely for its presence or absence without consideration for the varying degrees to which that criterion is satisfied. Bright[11] suggests extending the checklist to a profile in which the relevance of a project against each criterion is rated on a five-point Likert scale, as shown in Figure A6.1. This literally produces a project profile by which the strengths and weaknesses of a project can be readily perceived.

Merit Numbers

The scoring of each criterion on a Likert scale represents only a first step towards quantification of project features. Clearly, some project features or criteria will be more important than others, and it would be desirable to reflect this fact in the evaluation scheme. If each criterion is given a weighting (perhaps derived from a

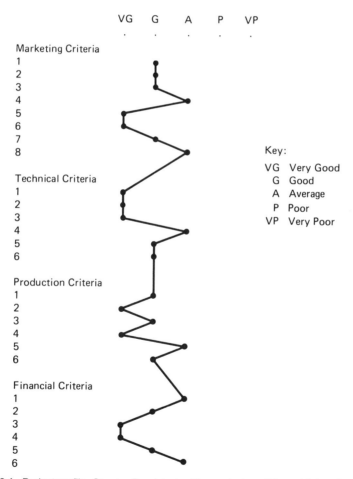

Figure A6.1. Project profile. *Source:* Reprinted, with permission of the publisher, from *Evaluating, Selecting, and Controlling R&D Projects* (New York: AMACOM, a division of the American Management Association, 1968), p. 40. All rights reserved.

relevance tree analysis), then we may extend the above method to calculate an overall merit number for the project. Wells is reported to have incorporated an interactive weighting system and relevance trees in his evaluation model.[12] If the criteria listed are comprehensive and include considerations of costs, profits, and probabilities of outcomes, a comprehensive ranking should be possible. The projects can then be selected for funding based upon the ranking subject to the constraint of the overall budget available.

Steele suggests that scoring techniques such as checklists and merit numbers are popular in industry because they lend themselves to "home-grown" development, do not make excessive demands upon scarce management time, and can elicit the participation of the line functions (see later in this Appendix).[13] As he puts it:

> To the methodological purist, scoring techniques seem primitive, but to the harried manager making estimates in the face of great uncertainty, they may seem the most that he or she would subscribe to.

Watts and Higgins, in a survey of current U.K. practice, supported this observation since they found that 35 of their 47 respondents used mostly only simple techniques such as checklists and project profiles.[14] Despite the above observations, it is worthwhile reviewing the more sophisticated methods because they do enjoy a limited use, particularly in public agencies selecting competing projects proposed by outside organizations for sponsorship by government.

Benefit–Cost Index Methods

Benefit–cost analysis is a well-established technique for selection among competing public sector investment proposals, so it is hardly surprising that it has been applied to the R&D project selection. Both expected benefits and costs are evaluated, separately summed, and then expressed as an overall index of total benefits divided by total costs or *benefit–cost ratio* as a measure of return on financial investment. Projects can then be ranked in order of the decreasing benefit–cost ratios, and the projects can be funded ordinally until the available budget is exhausted. The most popular model used for R&D project evaluation was developed by Ansoff and is shown in Table A6.2. As Souder points out, index models have a seductive appeal because of their apparent simplicity, which is, in reality, a mirage. All the terms listed in Table A6.2 must be estimated subjectively, so the realistic implementation of the method requires the application of consistent estimation procedures across the board to all projects. The problems of parameter estimation in R&D projects were discussed in Chapter 6. The method is likely to be most useful in selecting among alternative incremental innovations when the terms may be estimated with reasonable accuracy and consistency. A more sophisticated indexing method which is used in the Central Research Laboratory, Hitachi, Ltd., Tokyo is described by Kuwahara and Takeda.[15] The authors have developed an evaluation algorithm called the research–contribution-to-profit method which may be extended across projects and laboratories. The annual profit contributions of projects are identified and plotted either individually, or in an aggregation, over five plus years. The cost-effectiveness

TABLE A6.2 Ansoff's Index Formula

$$\text{Figure of Merit (profit)} = FM_p = \frac{(M_t + M_b) \times E \times P_s \times P_p \times S}{C_d \times J}$$

$$\text{Figure of Merit (risk)} = FM_r = \frac{C_{ar}}{FM_p}$$

Where M_t = Technological merit
  M_b = Business merit
  E = Estimated total earnings over lifetime
  P_s = Probability of success
  P_p = Probability of successful market penetration
  S = Strategic fit
  C_d = Total development cost
  J = Savings factor from shared resources
  C_{ar} = Total cost of applied research
  F = Total cost of extra facilities, staff, etc.
  M_p = Figure of merit (profit)

of both individual and aggregations of projects can thereby be tracked to make comparative performance evaluations.

Risk Analysis Models

Risk analysis, or the computer simulation of expected future positive and negative cash flows generated by a capital investment project to derive an overall risk profile or statistical distribution of ROI, has been used extensively in other areas of project appraisal. The principles of the approach are well described by Hertz.[16,17] Furthermore, in principle anyway, it should be possible to readily extend index models into computer simulation programs which will generate risk profiles such as in Figure A6.2, providing more effective measures than simple benefit–cost indices. Unfortunately, the uncertainties of parameter estimation discussed above limit the credibility which might be attached to such profiles, so the approach is of limited applicability in the early phases of the innovation process. It is of more value in later phases of the process, say from advanced development onwards, when many R&D uncertainties have been resolved. Equally, it offers a useful approach to evaluating a new product development with minor technological changes or new combinations of existing technological knowhow so that parameters may be more readily estimated. Bobis et al.[18] describe a risk analysis type model in the Organic Chemicals Division of American Cyanamid Division to redistribute its research budget. Another application of the approaches, this time at Xerox, is described by Zoppoth.[19]

Risk–Return Profiles

If, by the use of an index method, risk analysis, or other methods, an overall estimate of success/failure and expected return for each project can be derived, a

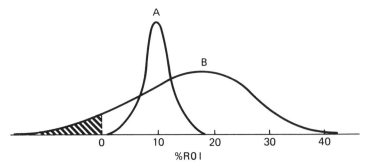

Project A offers a modest expected ROI of about 10%
Project B offers an higher expected ROI of about 20%,
but runs the risk of incurring a loss (shaded).

Figure A6.2. Risk profiles.

risk–return profile may be constructed. This is illustrated in Figure A6.3 for nine hypothetical projects: A, B, C, D, E, F, G, H, and I. The axes of the graph are return (expected profitability) against risk (probability of failure), with boundaries marking the minimum acceptable return and maximum acceptable risk levels respectively. Each project may be located on the graph through its return/risk coordinates. A offers an unacceptable low return, B offers an unacceptable high risk, and C is unacceptable in terms of both risk and return. The remaining projects offer acceptable risk–returns, with those further to the right offering the best risk–return combinations. A risk–return profile or frontier can be drawn through the best projects and, if sufficient funds are available, these projects should be funded. Projects located to the left of this profile may be funded if sufficient money is available (provided they also lie within the acceptable risk–return boundaries). Subject to the unreliability of estimates discussed above, this approach provides a useful visual aid for ranking projects and perceiving the risk characteristics of a project portfolio. In the (admittedly artifical) example shown, projects D, E, H, and I lie on the frontier, and so, in principle, should be funded. However, projects D and E offer low returns and low risks, while in contrast, projects H and I offer high returns and high risks so that, collectively, D, E, H, and I constitute a questionably balanced portfolio. Management may therefore seek out one or two medium risk–return projects to obtain an improved portfolio balance.

In evaluating risk–return profiles to achieve a balanced portfolio, Morris *et al.*[20] point out that the multiphase risk–investment profiles of projects must be compared. A riskier project X may be preferable to a less risky one Y with the same overall expected return because of the differences in risk and investment requirements for R&D and subsequent commercialization between the two. Although Project X may be risker, if successful, it promises a higher return than Project Y, and the major investment required for its commercialization will only be made if R&D results are promising. Thus, if the R&D fails, only the modest R&D investment is lost, whereas if it is successful, an exceptionally high return can be expected

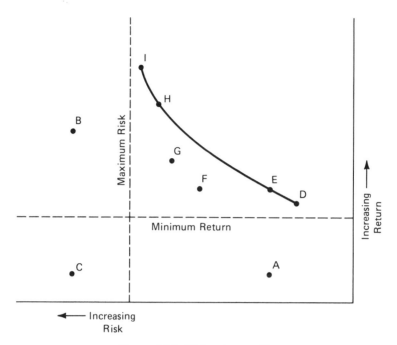

Figure A6.3. Risk–return profile.

from its commercialization. In contrast, because Project Y is less risky, it is more likely to require the investment for its commercialization *and then lose money*. Therefore, the riskier Project X is preferable because it has a higher expected payoff *if its R&D is successful*.

Business–Technical Profiles

Hall and Nauda[21] describe a variation of this approach in which the coordinates are Business Evaluation and Technical Evaluation. Projects are evaluated by business and technical merits on 10-point scales which reflect the criteria suggested in Chapter 2.

Technical evaluative criteria include:

1. Is the customer's technical problem clearly identified? Are its objectives clear, measurable, and attainable?
2. Does the R&D approach include novel techniques or replicate known work? Will it advance the state of the art? Can stronger technical approaches be suggested? Will the work boost the firm's competitive technical position?
3. Is the expected project cost appropriate for the level of the effort required? Does the firm have the capacity to develop and manufacture the resulting product?

Business evaluative criteria include:

1. Will the performance of the project lead to the satisfaction of a customer need, and is the customer convinced of this?
2. Is there a "window of opportunity" which could be missed if the project were postponed? Could its completion yield a sustainable competitive advantage and significant long-term commercial benefits to the firm?
3. Would the manufacture and marketing of the resulting product be congruent with the overall corporate or business strategy of the firm?

Statistical Decision Analysis Models

Given the uncertain nature of outcomes in the R&D, some workers have formulated R&D project selection as a problem in statistical decision theory. Decision pay-off matrices of projects against alternative outcomes are constructed and projects which offer the highest expected value are selected.

A related decision analysis approach, particularly when used in conjunction with Bayesian subjective probability estimation, is decision tree analysis, which has been particularly associated with new product development and is described in that context by MacGee.[22] Since decision tree analysis is an approach to the analysis of sequential decisions under uncertainty, it can be used as a tool both to evaluate and manage R&D projects. The expected benefits from a project can be estimated as an input to other evaluation and selection methods. Once a project is funded, a decision tree model, possibly in conjunction with a PERT/CPM network model, may be maintained and updated by the project manager to monitor the progress of the project. In the event of disappointing project outcomes, it can provide a useful aid in deciding whether to continue or abort a project.

Statistical decision analysis offers valuable tools and insights for project evaluation, selection, and control to the R&D manager who has the statistical understanding to use it intelligently, particularly if used in conjunction with the improved cost and outcome estimation procedures discussed in Chapter 8 (Section 8.4). Interesting applications of risk and decision analyses approaches are described by Hudson, Chambers, and Johnston (again at Xerox)[23] and Rzasa, Faulkner, and Sousa (at Eastman Kodak).[24]

Mathematical Programming Techniques

The most sophisticated methods of project evaluation and selection are based upon mathematical programming models. These approaches may be based upon the selection of a portfolio of projects as divisible entities simultaneously (using linear programming), or sequentionally (using dynamic programming), or as indivisible entities (using integer programming). They all offer, as Thompson and Vincent[25] express it, the *lure of optimality*—that is, they select projects and allocate resources to those selected so as to maximize the benefits or returns from the overall portfolio.

Bell and Read report that the Central Electricity Generating Board of the United Kingdom used linear programming as an aid to project selection,[26] while Lockett and Gear[27] and Madey and Dean[28] describe selection models incorporating decision analysis and mathematical programming approaches. Souder describes a dynamic programming approach applied at Monsanto.[29] Integer programming models have been applied in pharmaceuticals at Smith, Kline, and French[30] and in the steel industry.[31] Another pharmaceutical company, Johnson and Johnson, has used a mixed integer model (where some projects are indivisible and the remainder divisible) for R&D portfolio selection.[32]

A6.4 LIMITATIONS OF SELECTION TECHNIQUES

The absence of the widespread use of other than the simplest of the above selection techniques can be attributed to a number of quite valid reasons.

1. In Chapter 5 we argued that the overall funding of long-term research should be set by corporate management and individual projects selected by peer group evaluation. Research seeks to create new knowledge, and its outcomes are uncertain. This means that the numerical estimates in any formal mathematical evaluation must be based upon subjective, and often, unavoidably crude judgments or "guestimates." The research staff most closely involved with the project may be the only people capable of making these judgmental estimates and, since they may well have a vested interest in its funding, they may (either consciously or subconsciously) bias their estimates to favor its support. Furthermore, the expected amount and quality of new knowledge generated from a research project depend upon the professional quality of the participating researchers, which can only be subjectively assessed by the researchers' peers. Therefore, subjective peer group evaluation appears to be the best selection method at the research end of the innovation process.

2. In Chapter 5 we also argued that both the overall funding and individual selection of projects from advanced development onwards should be determined by the IMF within the context of corporate gap analysis. The considerations discussed in Chapter 6, as well as others discussed in later chapters, emphasize the importance of the active participation of line management if an innovation is to be successful. When a project enters into advanced development, estimates of production and marketing costs as well as revenues involved have to be made. Just as research staff can only estimate research outcomes, these costs and revenue estimates must be made by production and marketing management. The more sophisticated selection techniques are typically demanding in their estimation requirements, and so impose a substantial added burden on the line management's time. The imposition of this added burden will probably be resented by the line managers, particularly as it is required of all projects entering into advanced development. Thus, it may well harm relations between the R&D and other functional areas, reduce the potential to develop inventions into commercially successful new products, and harm the overall climate of the organization.

3. The selection techniques mostly take a snapshot (single point in time) view of either an individual or a portfolio of projects. The reality in many R&D laboratories is that most projects, once started and provided they progress acceptably, continue for some years, so that the year-to-year project turnover is 20% or less. This means that selection techniques can only be effectively used on projects as they enter the advanced development phase, and the small number of these candidates in any annual (or shorter) review cycle may not justify the effort required to apply sophisticated techniques.

4. Just as the professional reputations of the researchers need to be considered in selecting a research proposal, there is ample evidence to suggest that research inventions cannot be continued through advanced development to successful commercialization without the presence of a project champion and other skills in the project team. This important issue is discussed in Chapter 7, and is a crucial consideration which is not readily incorporated into selection techniques.

A6.5 SELECTION COMMITTEE PROCEDURE— THE Q-SORT METHOD

Whatever method of project evaluation is used, once a project has reached the advanced development stage, the selection decision should ideally be made by a committee which includes line managers who will assume responsibilities for the ultimate commercialization of the product. Such managers may use one or more of the above methods of appraisal in their deliberations, but must ultimately achieve a group consensus in project portfolio selection.

The dangers of bias developing in committee or group decision making were discussed in Chapter 4, Appendix 4.1, and the DELPHI method of expert opinion forecasting was described as a method of reducing this bias. It almost goes without saying that similar bias may develop in committees responsible for project selection. Souder, therefore, has proposed a group project selection procedure, which he calls the Q-Sort/Nominal Interacting (QS/NI) process, which bears some similarity to the Delphi method. This procedure is as follows.[2]

Each member of the group or committee is given a deck of cards, one for each project to be considered, and outlining the characteristics of the project. He begins by privately and anonymously classifying each project into one of five categories ranging from a very high to a very low level of priority, using the sorting process illustrated in Figure A6.4. The sorting criteria will be based on an agreed-upon set of project selection criteria (see Figure A6.5). Once this nominal period is completed, interaction begins. The results of the classifications of the whole group are tabulated on a tally sheet anonymously and displayed to the group on an overhead projector, as shown in Figure A6.6. Agreements and disagreements can be identified without violating individual anonymity, and discussion can now take place. Individuals may challenge others' classifications and the latter may choose to defend their positions. Alternatively, they may choose to remain silent and preserve their anonymity, particularly if they wish to retain a minority position. During this

RESULT AT EACH STEP

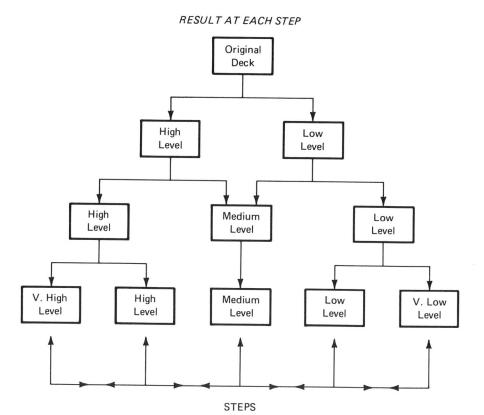

STEPS

1. An individual is given a stack of cards, each card bearing the name, title or number of one project. The individual is asked to perform the following sorting operations. A specified criterion (e.g. "priority") is the basis for sorting.

2. Divide the deck into two piles, one representing a high level of the specified criterion, the other a low level. (The piles need not be equal.)

3. Select cards from each pile to form a third pile representing the medium level of the criterion.

4. Select cards from the high level pile to yield another pile representing the very high level of the criterion; select cards from the low level pile to yield another pile representing the very low level of the criterion.

5. Finally, survey the selections and shift any cards that seem out of place until the classifications are satisfactory.

Figure A6.4. Q-Sort mechanics. *Source:* W. E. Souder, "A System for Using R&D Project Evaluation Methods," *Research Management* 20, No. 5 (Sept. 1978), 29–37. Reprinted with permission of Industrial Research Institute.

interaction, however, the arguments in favor of the various classifications are likely to be canvassed, so individuals may wish to modify their choices in light of the diversity of opinions which has been expressed. Group members are then allowed a second anonymous nominal period during which they modify their choices. The revised tally sheet is then projected and a second interaction period begins. As in the

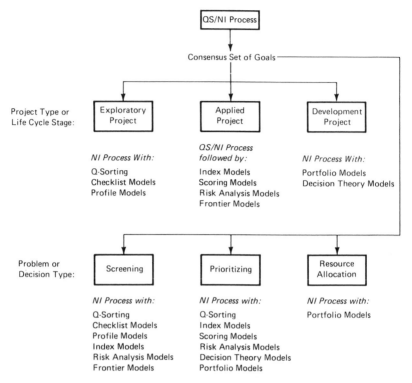

Figure A6.5. Project selection and evaluation system. *Source:* W. E. Souder, "A System for Using R&D Project Evaluation Methods." *Research Management* 20, No. 5 (Sept. 1978), 29–37. Reprinted with permission of Industrial Research Institute.

Delphi method, nominating–interaction "rounds" may be repeated until, as nearly possible, a consensus is reached. Souder reports that this consensus is usually reached after two or three rounds.

Souder further proposed a multistage application of the QS/NI process which would be applicable across the range of R&D projects performed in the offensive–defensive innovative firm, and which reflects the technological and budgetary considerations in the R&D resource allocation process. This is illustrated in Figure A6.6. Projects are first classified as exploratory, applied, and developmental within the framework of corporate goals, which are themselves ranked by a QS/NI process. Projects in each classification may then be evaluated and selected by second level QS/NI processes, which also employ appropriate criteria or models. Thus, for exploratory projects, where information is meager and uncertainties are large, simple checklist or profile methods may be used. On the other hand, for applied and developmental projects, where more information is available and more resources are required, more sophisticated models may be incorporated as decision aids in the QS/NI processes. Souder's multistage QS/NI scheme appears to provide a framework for the intelligent use of project evaluation and selection models. More impor-

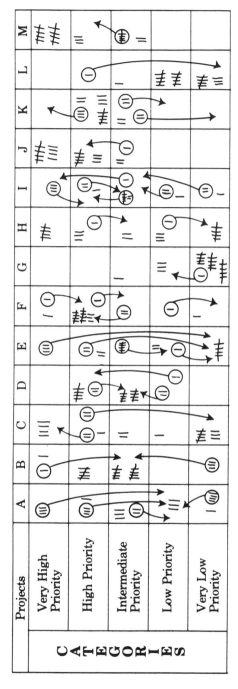

Tally chart for 20 subjects and 13 projects at the end of the following sequence of periods: Nominal—Interfacing—Nominal. Projects evidencing consensus at the end of the second nominal period = A, B, D, E, F, G, I, J, L, M. All projects evidenced consensus at the end of an additional subsequent interacting—nominal sequence. Consensus is measured by standard statistical tests and the subject's feelings.

Figure A6.6. Example of tally sheet. *Source:* W. E. Souder, "A System for Using R&D Project Evaluation Methods," *Research Management* 20, No. 5 (Sept. 1978), 29–37. Reprinted with permission of Industrial Research Institute.

tantly it provides a potential framework for integrating at both corporate and functional management levels the considerations of technological strategies, goals, forecasts, and R&D resource allocations we have discussed earlier.

A6.6   DECISION SUPPORT AND EXPERT SYSTEMS

Fahrni and Spatig suggest a set of guidelines that managers might use in evaluating and selecting R&D projects.[33] Expert systems can be used to improve manufacturing design, and these authors suggest that developments in Decision Support Systems (DSS) and Expert Systems (ES) should provide managers with improved evaluation and selection measures, so we conclude this appendix with an illustration of these approaches.

Probably the best example of the sustained exploration of many of the alternative project selection techniques discussed above with consequent experiential learning that led to the development of such a DSS is described by Islei et al.[34,35] Still evolving, it represents the outcome of over a decade of collaboration between the R&D staff of ICI Pharmaceuticals, consultants in ICI Corporate Management Services, and faculty researchers at the Manchester Business School, England. One senior R&D manager in ICI Pharmaceuticals made the following comments on their work:

> The techniques (they) developed have been particularly helpful in monitoring progress in our large cardiovascular research effort, and they are also being used with effect in other Divisions of ICI. . . . we are now extending its use even further so that it will be applied across the whole of our research department.
>
> —(Islei et al.,[35] p. 22)

The project selection features of their approach incorporate behavioral (notably the personality attributes of the project champion) as well as technical and risk facets, and have been extended to the monitoring and control of a project after its selection. Because it has been so successful, the approach has also been used in marketing applications.

In a related study, Wilkinson is developing an expert system (ES), using Lotus 1-2-3 for project evaluation, which is being tested in six companies.[36] His original intent was to design an ES to assist managers in project selection, but he quickly found that there was no agreement on how projects were selected among the "expert" R&D managers he consulted. He therefore concentrated on developing an ES to aid ongoing project evaluation where there was some consensus among his subjects. Based upon this consensus, he found that project evaluation could be defined in terms of three mutually orthogonal axes:

1. Time to completion.
2. The need of innovation (roughly corresponding to the technological uncertainty reduction dimension of Figure 8.1b).
3. Corporate urgency for its completion.

Each was measured on a three-point scale: "low," "medium," and "high" to create a 27-cell cube based upon the $3 \times 3 \times 3 \times 3$ possible combinations spanned by the three axes. When a project begins, it typically has a high time to completion and need for innovation and a low corporate urgency for its completion. As it nears completion, it should have a low time to completion and need for innovation and a high corporate urgency for its completion. Progress should therefore ideally follow the leading diagonal cells of the cube.

Still (at the time of writing) under development, the ES is designed to identify the cell location of a project and, if it is located off the leading diagonal of the cube, the appropriate remedial action that should be taken to return it to a leading diagonal location. Also, although directly designed for evaluating projects, by comparing projects which are similar in technical specifications and phases in the innovation process, the ES can offer advice on project selection and termination decisions.

REFERENCES

1. P. A. Abetti and R. A. Stuart , "Evaluating New Product Risk." *Research Technology Management* **31**(3), 40–43 (1988).

2. P. D. Klimstra and J. Potts, "Managing R&D Projects." *Research Technology Management* **31**(3), 23–39 (1988).

3. N. P. Baker and W. H. Pound, "R&D Project Selection: Where We Stand." *IEEE Transactions on Engineering Management* **EM-11**(4), 124–134 (1964).

4. W. E. Souder, "A System for Using R&D Project Evaluation Methods." *Research Management* **21**(5), 29–37 (1978).

5. A. R. Fusfield, " A Review of the Major Types of Project Selection Techniques and Suggestions for New Approaches." Unpublished paper. Pugh-Roberts Assoc., 1976.

6. I. Vertinsky and S. L. Schwartz, *Assessment of R&D Project Evaluation and Selection Procedures*, Report No. 49, Technological Innovation Studies Program. Ottawa, Ont.: Department of Industry, Trade and Commerce, December 1977.

7. R. E. Seiler, *Improving the Effectiveness of Research and Development.* New York: McGraw-Hill, 1965.

8. M. Cetron, *Innovation Group Newsletter*, No. II. New York: Technology Communications Inc., October 1970.

9. B. V. Dean, *Evaluating Selecting and Controlling R&D Projects.* New York: American Management Association, 1968.

10. J. Dessauer, "Some Thoughts on the Allocation of Resources to Research and Development Activities." *Research Management* **10**(2), 77–89 (1967).

11. J. R. Bright, *Research, Development, and Technological Innovation.* Homewood, IL: Richard D. Irwin, 1964, P. 424.

12. H. A. Wells, "Weapons Systems Planner's Guide." *IEEE Transactions on Engineering Management* **EM-14**(1), 14–16 (1967).

13. L. W. Steele, "Selecting R&D Programs and Objectives." *Research Technology Management* **31**(2), 17–36 (1988).

14. K. M. Watts and J. C. Higgins, "The Use of Advanced Management Techniques in R&D." *Omega* **15**(1), 21–29 (1987).

15. Y. Kuwahara and Y. Takeda, "A Managerial Approach to Research and Development Cost-Effectiveness Evaluation." *IEEE Transactions on Engineering Management* **37**(2), 134–138 (1990).

16. D. B. Hertz, "Risk Analysis in Capital Investment." *Harvard Business Review* **42**(1), 95–106 (1964).

17. D. B. Hertz, "Investment Policies that Pay Off." *Harvard Business Review* **49**(1), 96–108 (1968).

18. A. H. Bobis, T. F. Cooke, and J. H. Paden, "A Funds Allocation Method to Improve the Odds for Research Success." *Research Management* **14**(2), 34–39 (1971).

19. R. C. Zoppoth, "The Use of System Analyisi in New Product Development." *Long Range Planning* **5**(1), 23–26 (1972).

20. P. A. Morris, E. O. Teisberg, and A. L. Kolbe, "When Choosing R&D Projects, Go With Long Shots." *Research Technology Management* 35–40 (1991).

21. D. L. Hall and A. Nauda, "An Interactive Approach for Selecting IR&D Projects." *IEEE Transactions on Engineering Management* **37**(2), 126–133 (1990).

22. J. F. MacGee, "Decision Trees for Decision Making." Harvard Business Review **42**(3), 126–138 (1964).

23. R. G. Hudson, J. C. Chambers, and R. G. Johnston, "New Product Planning Decisions Under Uncertainity." *Interfaces* **8**(1), part 2, 82–96 (1977).

24. P. V. Rzasa, T. W. Faulkner, and N. L. Sousa, "Analyzing R&D Portfolios at Eastman Kodak." *Research Technology Management* **33**(1), 27–32 (1990).

25. P. N. Thompson and D. Vincent, "OR in Research and Development—A Critical Review." *R&D Management* **6**(3), 109–113 (1976).

26. D. C. Bell and A. W. Read, "The Application of a Research Project Selection Method." *R&D Management* **1**(1), 35–42 (1970).

27. A. G. Lockett and A. E. Gear, "Representation and Analysis of Multi-stage Problems in R&D." *Management Science* **19**(8), 947–960 (1973).

28. G. R. Madey and B. V. Dean, "Strategic Planning for Investment in R&D Using Decision Analysis and Mathematical Programming." *IEEE Transactions on Engineering Management* **32**(2), 84–90 (1985).

29. W. E. Souder, "A Scoring Model Methodology for Rating Management Science Models." *Management Science* **18**(10), 526–543 (1972).

30. M. A. Cochran *et al.*, "Investment Model for R&D Project Evaluation and Selection." *IEEE Transactions on Engineering Management* **EM-18**(3), 89–100, (1971).

31. C. J. Beattie, "Allocating Resources to Research in Practice." In E. M. L. Beale (Ed.), *Applications of Mathematical Programming Techniques.* New York: American Elsevier, 1970, pp. 281–292.

32. D. Grossman and S. N. Gupta, "Dynamic Time-Staged Model for R&D Portfolio Planning—A Real World Case." *IEEE Transactions on Engineering Management* **EM-21**(4), 141–147 (1974).

33. P. Fahrni and M. Spatig, "An Application-Oriented Guide to R&D Project Selection and Evaluation Methods." *R&D Management* **20**(2), 155–171 (1990).

34. G. Islei, A. G. Lockett, B. Cox, and M. Stratford, "A Decision Support System Using Judgmental Modeling: A Case of R&D in the Pharmaceuticals Industry." *IEEE Transaction on Engineering Management* **38**(3), 202–209 (1991).

35. G. Islei, A. G. Lockett, B. Cox, S. Gisbourne, and M. Stratford, "Modeling Strategic

Decision Making and Performance Measurements at ICI Pharmaceuticals." *Interfaces* **21**(6), 4–22 (1991).

36. A. Wilkinson, "Developing an Expert System on Project Evaluation. Part I: Structuring the Expertise." *R&D Management* **21**(1), 19–29 (1991); "Part II: Structuring the System." *Ibid.* **21**(3), 207–213 (1991); "Part III: The Managerial Questions Raised by the Work." *Ibid.* **21**(4), 309–318 (1991).

———7
MATCHING ORGANIZATIONAL AND INDIVIDUAL NEEDS IN R&D

The ideas, talents and skills of scientists, engineers and other technical professionals are an R&D laboratory's greatest asset. In organizations whose most valued product is essentially ideas, the importance of effective utilization of human resources cannot be overemphasized.

M. BADAWAY, "MANAGING HUMAN RESOURCES"
RESEARCH TECHNOLOGY MANAGEMENT

There was, it appeared, a mysterious rite of initiation through which, in one way or another, almost every member of the team passed. The term that the old hands used for this rite . . . was "signing up." By signing up for the project you agreed to do whatever was necessary for success. . . . From a manager's point of view, the practical virtues of the ritual were manifold. Labor was no longer coerced. Labor volunteered.

TRACY KIDDER, "BUILDING A TEAM"
IN THE SOUL OF A NEW MACHINE

7.1 INTRODUCTION

In this chapter we explore these skill requirements in some detail and in the context of the personal developments needs of R&D staff. Arguably, the most important skill requirement of creativity is examined in some detail in the Appendix to this chapter. In Chapters 8 and 9 we shall examine alternative organizational frameworks for successfully carrying innovation beyond the R&D phase.

7.2 CONFLICT BETWEEN INDIVIDUAL AND ORGANIZATIONAL VALUES IN R&D

It is generally accepted that the dissonance between individual and organizational values and goals is greater in the R&D function than anywhere else in the organization. Furthermore, this dissonance is likely to be particularly marked if the firm follows an offensive/defensive strategy, pursuing nondirected research in a professionally prestigious central laboratory which is both geographically and culturally distant from its operating divisions. Such a company may have a continued requirement to recruit creative higher degree graduates in science and engineering from universities to maintain its proximity to the state of the art in research, graduates who, for their parts, will be attracted by the prestige of the laboratory. The personal development of such graduates is a matter requiring the sincere and concerned attentions of both individuals and the organization.

The specialist educational process of a higher-degree graduate in science or engineering is essentially a Kuhnian apprenticeship in "puzzle solving" skills (recall Chapter 2, Section 2.4), which makes him *au fait* with the currently dominant paradigm (or paradigms) in his field and provides him with his initial research training. By virtue of his personality and education, his internal commitment is to his discipline and the goal of achieving significant professional recognition and esteem by making significant research contributions in his field. He is thus interested in publications rather than innovations. The existence of this attitude among government (as opposed to industrial) R&D staff was illustrated in a study of the problems faced by small companies attempting to innovate inventions made in government R&D laboratories.[1]

The investigators interviewed the presidents or vice presidents of some 50 companies who had taken government R&D inventions and sought to develop them to commercially viable products, and asked them to comment on the problems they faced in transferring the invention from the government's to their own laboratories. Among these interviewees, there was a widespread feeling that some government "bench-level" R&D staff were concerned only with the generation of new knowledge, and were disinterested in its commercial exploitation and the values and pressures of business. Two comments made which illustrate this feeling were:

> The objective of government and university research is publish or perish while in industry it is publish **and** perish. There is a major gap here in objectives between the two.

and

> There is a stigma on profit. Government scientists' assistance could be valuable but unfortunately there is a boy scout concept of purity in Government R&D.

The commitment of research staff, at least early in their careers, to their discipline and their development as researchers rather than to the corporate goals must be accepted by R&D management. It must be recognized that research staff identify more closely with the informal network and status system of their professional peers rather than that of the organization. This, of course, is in contrast to most managers who identify with the formal and informal networks and status system of the organization. Nowadays, it is accepted that the implicit employment contract does not dictate that the employee should be totally committed to the values and goals of the employer. One role of a manager is to reconcile the goals of the organization and his subordinates, as far as possible, in the best interest of both. That is, he should adopt a [9,9] orientation in the terminology of Blake and Mouton's managerial grid.[2] The adoption of this [9,9] orientation is probably more important in R&D management than in any other functional area. It implies that staff should be encouraged to participate in the formulation of R&D goals and to match them with their own. Such participation is implicit in approaches to technology planning, R&D budget setting, and project selection discussed in Chapter 4 onwards. By so doing, the problems created by differences of outlook between R&D staff and managers will be ameliorated, and hopefully the former will become more sympathetic to the total innovation process and business culture. Furthermore, as we argue in the remainder of this chapter, a systematic examination of R&D staff development needs and innovation needs does suggest that the two are broadly congruent. Badaway provides an excellent broad overview of human resource management issues in R&D organizations,[3] and here we focus on individual career development alternatives for R&D staff and project team needs.

7.3 DEVELOPMENT ALTERNATIVES FOR R&D STAFF

A technological graduate whose first appointment is as a researcher in an industrial R&D function has three alternative personal development paths available to him within the organization. Sooner or later, either consciously or unconsciously, he probably chooses to follow one of these. They are:

1. *Research.* An individual can remain as a bench-level researcher seeking self-actualization through continued research output, gaining increasing insight into his field, and earning increasing professional esteem. Following this path implies seeking the minimum management and administrative responsibility, congruent with the initiation and supervision of perhaps an expanding portfolio or research projects. Such a role may imply modest formal status within the management structure since it may be effectively executed at (say) section leader level. Such an individual may enjoy considerable esteem among his professional peer group both within and outside the organization. Thus, an individual may achieve maximum *personal* self-actualization by adopting this role, particularly as it may offer the maximum scope for personal autonomy, something creative people value highly. One possible dis-benefit of his role is that it may not offer what we might label maximum *family* self-actualization. First of all, salary distributions may be generally lower for researchers than managers, so the individual may forgo substantial long-term income by selecting this path. Therefore, he may have to forgo the opportunity to offer his family a higher material quality of life—perhaps a higher quality private education for his children, higher quality family healthcare, etc. Secondly, although he may enjoy the high esteem of his peers, this may not extend to society at large. Apart from the high-status professions, society at large tends to evaluate a family's status through its manifest material standard of living. Therefore, unless a person is a

really outstanding researcher, both he and his family may forgo some social esteem through his decision to remain a researcher, rather than becoming a manager. This is a factor which can affect a person's choice of career development.

2. *Research Management.* A person may choose to remain in the R&D function, but pursue increasing managerial responsibilities. Particularly if he moves from the research to the development phases of the innovation process, a person may be able to build up his management experience by taking increasing responsibility in administering increasingly complex projects. He may thereby progress to senior R&D management responsibility.

As we stated in Chapter 5, one role of senior R&D management is to mediate between goals and needs of corporate management (and other functional areas) and those of the R&D function. To be effective in this role, a person must possess both management and R&D credibility. That is, he must enjoy the esteem and professional respect of both managers and R&D staff in the firm, and given that he does, R&D management may offer a viable option.

3. *Other Management Positions.* Although R&D management is an attractive personal development avenue, except possibly in high-growth situations, it is unlikely that the demand for R&D managers will exceed the supply—that is, R&D staff with the personal resources and motivation to develop their career in that direction. Thus, after some project management experience, if they are to remain with the firm, some R&D staff may wish to seek further growth opportunities through management positions elsewhere in the organization. Some high-technology companies do use the R&D function as a nursery for future managers.

The frameworks for developing R&D staff into innovation managers is discussed later in Section 7.4. We now briefly discuss the development needs of people who wish to remain researchers, within the framework of a dual-ladder system.

7.4 THE DUAL-LADDER STRUCTURE

First of all, it is reasonable to assume that a creative individual will only wish to remain in research as long as he can maintain sufficient productivity to satisfy his internally set standards. This raises the issue of the aging process in creativity. There is a body of lay and, for that matter, management opinion which holds that creativity is a youthful quality, and that an individual will have produced his best work before the age of 30 or 40. There is some, but by no means conclusive, evidence to support this belief. Outstanding mathematicians do have a habit of producing their best work before the age of 30, but other outstanding scientists as well as artists have maintained highly creative outputs over a relatively extensive lifespan. Newton's massive contributions to physics and mathematics were spread over several decades while Edison (an inventory–entrepreneur) maintained his inventive output over a similar period. In literature, Milton produced his masterpiece (*Paradise Lost*) when comparatively old. Similarly, at a more modest level of creativity, some researchers are able to maintain a satisfying level of output (to themselves and their employers) throughout a lengthy professional career. Although the larger industrial laboratories will wish to maintain a steady recruitment of young researchers from leading university research schools to maintain their proximity to the state of the art, a cadre of mature experienced researchers *who have maintained their personal proximity to that state* can provide research coaching and leadership that would otherwise be lacking. Therefore, the offensive/defensive innovative company has a definite long-term incentive for encouraging its most able researchers to make a permanent career in research. The issue then becomes one of providing such people with adequate remuneration and status. One method of dealing with this issue has been for laboratories to create *dual hierarchies*—one managerial and one technical—through which people can progress as managers or researchers.

One traditional method of rewarding R&D staff for good technical performance is to promote them into management positions—a method whereby the firm may lose the services of a good researcher and gain the services of a poor manager—clearly an undesirable outcome! In order to develop a satisfying career plan for R&D staff which motivates and rewards their good performances as technologists rather than managers, a number of major companies employ dual-ladder systems of personnel recognition and promotion in their R&D functions. Although successes and failures have been reported with such systems, detailed examinations of individual corporate experiences suggest that the dual-ladder system provides an intrinsically worthy and effective means of R&D staff development, and it only fails when it is misapplied or corrupted in some way or other. Two U.S. oil companies—Mobil and Amoco—report* having used such systems since the 1940s and, despite having had some periods of frustration with them, both companies speak highly of the value of dual-ladder systems and are pledged to their continued use.[4,5]

As its name implies, the central feature of the dual-ladder system is to create two parallel hierarchical sets of positions or titles in the organization—one for researchers and one for managers. Most organizations which use it appear to try exactly to match "rungs" in the research and management ladders and, as far as is possible, to provide similar fringe benefits and status connotations as well as a common salary scale. Military officers of identical rank may perform widely different duties or tasks, dependent upon the specialist branch of the service to which they belong, but will enjoy a broadly identical status which is primarily dependent upon their identical rank. Similarly, researchers and managers occupying equivalent rungs in the dual ladder enjoy broadly similar rewards and outward recognition in the organization—success which readily communicates itself to others in the organization, his family, and his friends.

Needless to say, the detailed formats of dual-ladder systems vary from company to company, but Table 7.1 illustrates the concept by showing the dual-ladder structure for two large companies in different high-technology industries—Westinghouse

---

Research Management **20**(4) (1977) contains a series of short articles on individual corporate experiences with the dual-ladder system.

TABLE 7.1   Two Ladders

ICI		
Scientific Ladder		Administrative Ladder
	1st rung	
Senior Research Officers		Some Section Leaders
	2nd rung	
		Section Heads
		Section Managers
	3rd rung	
ICI Research Associate		Group or Assistant Research Managers
	4th rung	
Senior ICI Research Associate		Associate Research Managers Research Managers etc.
	Terminal	
		Research Director
Westinghouse		
Management		Individual Contributor
Director		Director
Department Manager		Consultant
Section Manager		Advisory Scientist or Advisory Engineer
Supervisor		Fellow Scientist or Fellow Engineer

Source: S. L. Meisel, "The Dual Ladder—The Rungs and Promotion Criteria." *Research Management* **20**(4), 24–26 (1977). Reprinted with permission of Industrial Research Institute.

Research Laboratories in the United States and ICI Ltd. in the United Kingdom.[6,7] In both firms there are four matching rungs, although ICI does have a terminal position which is purely administrative. Table 7.2 shows the criteria that Westinghouse reported using in evaluating R&D staff for promotion in their technical ladder. The consensus of opinion of firms which have *successfully* instituted a dual-ladder system is that its success is dependent upon maintaining high standards of undoubted integrity in its application. That is, ensure that the system recognizes and rewards professional excellence. Firms which have *unsuccessfully* instituted such a system appear to have failed because (possibly, for quite legitimate reasons) they have been unable to maintain such standards. These considerations suggest the following "dos" and "don'ts" for installing dual-ladder systems.

Do ensure that the ladder structure is integrated into the overall human resource planning procedures in the corporation. In particular, ensure that the system is incorporated into the job evaluation and career planning procedures used in the R&D function and elsewhere in the corporation. The critical functions approach to the analysis of R&D project needs, which is described shortly in Sections 7.4 and

7.5, provides a framework for job evaluation and career planning which could be readily linked with a dual-ladder system.

Do ensure that common technological and managerial rungs have at least roughly equivalent salaries, responsibilities, statuses, and fringe benefits. Such equivalence would be reflected in personal locations and titles in organizational charts, committee memberships, reporting procedures, responsibilities (including project selection), laboratory, and office space, etc.

Do ensure that promotions up the technological ladder are based upon reasonably defined criteria (as illustrated in Table 7.2), and professionally agreed-upon and recognized standards of excellence appropriate to the remuneration and status offered by the position. The establishment and maintenance of these standards appear fundamental to the acceptance of the system by the R&D staff and to its continued success. Promotion to the more senior positions on the technological ladder should be made at the corporate level by a committee consisting of senior and distinguished technologists (not all of whom need necessarily belong to the corporation).

TABLE 7.2 Promotion Criteria

Basic Criteria for Technical Appointments

Advisory Scientist or Engineer

1. Mastery of scientific and technical field. Guide: Ph.D. or equivalent in research accomplishment.
2. Accomplishments as demonstrated by:
   a. Papers (external and internal), reports, memos, and other publications—number and quality.
   b. Discovery of invention leading to company benefit.
   c. Recognition by peers.
   d. Strong consulting activity.
   e. Invited participation in professional societies.
3. Demonstrated ability to plan independently and follow through on significant programs.
4. Demonstrated ability to:
   a. Exert influence and exercise judgment in matters that affect the laboratories or company (mandatory).
   b. Assume responsibility for special tasks.
   c. Guide work of others.

The supervision of others is neither a necessary nor a sufficient condition for the position of Advisory Scientist or Engineer.

Fellow Scientist or Engineer

The Fellow Scientist and Engineer classification is based on same criteria as for the Advisory Scientist and Engineer except that, while high, they are applied less stringently.

Source: S. L. Meisel, "The Dual Ladder—The Rungs and Promotion Criteria." *Research Management* **20**(4), 24–26 (1977). Reprinted with permission of Industrial Research Institute.

The composition and membership of such a committee must obviously be dependent upon the local situation, but the success of the system will be largely dependent upon the integrity and credibility of its decisions. Given the seniority of the more senior positions and the caliber of suitable candidates for promotion to this level, decisions ideally need to be made by one central (that is, corporate level) committee with an arm's-length relationship to the candidates. If possible, a candidate's local supervisors and superiors should *not* sit on this committee, but rather act as advocates on her or his behalf. Obviously, the strength of their advocacy would be dependent upon the extent to which these superiors thought the candidate merited promotion, but such an arrangement also protects the integrity of the procedure from corruption by local pressure of political patronage. Promotion to the less senior rungs may have to be made at more local levels in a large organization, if for no other reason than the relatively large numbers of R&D staff involved, but promotion criteria and procedures should be designed to maintain the integrity of the system.

Whatever detailed criteria for promotion are used, a mandatory criterion should be an appropriate past and future level of contribution to corporate goals (see 4a, Table 7.2). At first sight, it might appear that this mandatory criterion is in conflict with the arguments presented earlier. However, we can expect candidates for the more senior rungs to be distinguished technologists in their fields who consequently enjoy continued job mobility. In these circumstances, it is unlikely that such individuals would have stayed with the corporation unless they could reconcile their personal research goals with those of their employers. Conflicts between senior R&D staff and R&D and corporate management can and do arise, but they are more typically concerned with which new innovations or products should be supported in order to achieve corporate goals, rather than the goals themselves. Furthermore, persons seeking to promote the adoption of an innovation by the firm are likely to be entrepreneurs or project champions (see next section) and managers rather than being solely concerned with a research career. In other words, their appropriate career path may be the management rungs on the ladder.

The pitfalls to be avoided mostly concern the protection of the integrity of the rank structure.

Never promote staff to a rank that their R&D track record does not justify—that is, ensure that promotion recognizes legitimate R&D performance and is not used in some honorary capacity to reward non-R&D contributions.

Never, ever use R&D rungs as "dumping grounds" for failed managers. If a manager proves ineffective in a line R&D management position, there is a humane and pragmatic temptation to move him sideways into the corresponding technological rung. If the dual system is working well otherwise, it might be argued that by moving him to that rung he will retain his salary and status but occupy a position in which apparently he can do no harm. *Nothing could be further from the truth.* The use of the R&D rungs in this way immediately discredits the dual system in the eyes of R&D staff and rapidly destroys its credibility. Smith and Szabo of Union Carbide also indicate that the credibility of their firm's system was compromised when some managers were appointed to technological positions (often one or two rungs lower) to protect their job security during a period of corporate cutbacks.[8] Although the desirability of the retention of such personnel was not disputed in this case, the use of the ladder in this way eroded its credibility and viability.

Whatever the "dos" and "don'ts" may be, it appears certain that the maintenance of a credible system requires a continuous commitment on management's part. As Dr. Meisel (Vice President, Research for Mobil R&D Corp. at the time he presented his paper) put it, "the dual-ladder system is no perpetual motion machine."[4] It needs to be subject to changes in light of changing disciplinary needs and corporate objectives and staff cutbacks or recruitment increases due to economic fluctuations. Typically, there may be several technological ladders required for different scientific and engineering specialties, and staffing requirements in each specialty may change markedly over a few years due to shifts in the state of the art or corporate goals. For example, after reviewing a long-established dual-ladder system in the early 1980s, Dow Corning extended it into a multiple-ladder (or lattice?) system, with three technical categories: Research, Technical Service and Development, and Process Engineering.[9]

Because a specialist researcher's professional life may last 40 years, this can create difficult human resource management problems—as any contemporary university president will confirm! As far as possible, promotion criteria and standards need to be maintained equitably across different disciplines and (if a multinational corporation is attempting to use the system in its R&D laboratories in different countries) across national geographic boundaries. These considerations require a significant ongoing administrative effort which must be initiated and maintained without undue bureaucracy. Furthermore, the existence of a dual-ladder significantly impacts upon management as well as R&D staff. R&D managers may find that they have researchers reporting to them who are their seniors in the system. This means that their managerial authority is less well-defined and the researchers may be able readily to bypass them in the formal and informal communications networks of the organization. Managers may also envy the relative freedom from corporate disciplines enjoyed by researchers—indeed, overall, members of the technical ladder may exist as a kind of counterculture to management in the organization. According to the proponents of the dual systems, these inevitable tensions can be made productive rather than counterproductive. As Moore and Davies of ICI express it: *"Like the effect of grit in oysters, pearls have been produced but at some discomfort to the host—and at some discomfort also to the Scientific Ladder people."*[7] Finally, perhaps the best overall argument for using a ladder system is summed up by Dr. Meisel, who quotes the words of one of Mobil's young research chemists: *"I don't think I would have come here if Mobil didn't have a professional ladder. But most other places have a dual ladder, too."*[4] In other words, corporate R&D laboratories may not find it easy to attract good R&D staff if they have a dual-ladder system, *but they will find it even more difficult to do so if they don't!*

7.5   STAFFING NEEDS OF THE R&D FUNCTION

Earlier, we argued that R&D managers should adopt [9,9] orientation in terms of Blake and Mouton's managerial grid. Readers familiar with that grid recall that it postulates two orthogonal dimensions of concern—"concern for production" and "concern for people," with terms "production" and "people" really being used as synonyms for "organizational needs" and "individual needs." In Section 7.2 we explored the individual needs of R&D staff, and in this section we explore the other dimension of that grid—the organizational needs of the R&D function. In Section 7.5 we shall outline one approach to marrying these two sets of needs, that is, establishing an R&D function which has a [9,9] orientation.

Blake and Mouton's choice of the designation "concern for production" as a synonym for organizational needs is quite apposite in the present discussion. Readers conversant with the history of management thought will recall that the development of a discipline of production management as a systematic body of knowledge could be said to have started with the pioneering work of F. W. Taylor. He argued that efficient production imposed a requirement of task specialization and the obligation to match the right person to the right task in production as a line-team process. Although his ideas were carried too far in subsequent developments in mass production methods and may be superceded by developments in computer integrated manufacturing, the broad concept of task specialization remains valid. Furthermore, production-line thinking, rooted historically in his ideas, has been applied quite recently, both effectively, to service as well as manufacturing industries. For example, in Chapter 4 we referred to the McDonald "technological hamburger" as an example of the innovative use of production line methods in the fast-food industry. Since the authority and effectiveness of task specialization has been recognized in other functional management areas, it is reasonable to suppose that it is also valid in R&D and innovation management. However, despite this comparison with Taylor's classic work on Scientific Management, it must be stressed that the management of science in a business (or indeed any other) setting cannot be based upon purely rational cosiderations. The potential for conflict between individual and organizational needs in R&D plus the competition for support between R&D projects precludes this. Thus, management at both the R&D organizational and separate project levels is best viewed in an organizational politics model framework as suggested by Dill and Pearson.[10] Much work on communication networks and role needs in R&D has been performed since the late 1960s.[11-14] We shall not examine such work here, but rather focus on one approach which reflects both rational and political considerations and can be readily developed by expanding the considerations implicit in the innovation chain equation analogy introduced in Chapter 2 (Section 2.2). It is also consistent with Cooper's ongoing evaluation process described in Chapter 6 (Section 6.5).

The nature of task specializations and roles in R&D management has been explored extensively, and Chakrabarti and Hauschildt[15] identify numerous titles that have been used for them. These roles can be delineated according to the division of labor through the successive phases of the innovation process and the power sources of their incumbents. One such well-regarded delineation was developed by Pugh-Roberts Associates, Inc.[16,17] The R&D functions in a number of organizations were studied to determine the tasks required to expedite the innovation process. The researchers identified six distinct tasks (idea generation, technological gatekeeping, market gatekeeping, project championing, project managing, and coaching) which must be performed effectively if innovation is to be successful. They describe them as the *critical functions* in R&D management. The tasks relate directly to the terms or links in the innovation-chain reaction, as is shown in Figure 7.1. Although any one person may be able to perform, say, up to three of these tasks effectively, no one person possesses either the aptitude or desire to effectively perform them all. Therefore, to ensure that an innovation is potentially successful, it is imperative that R&D management selects a project team which is *collectively capable* of performing *all* of these tasks. This implies that the project team is built upon task and role specialization, in which one member of the team is recognized as having prime responsibility for the effective performance of each of the tasks. The activities and personal aptitudes required by these critical functions are now discussed.

Idea Generating

Idea generators are inventors of creative individuals. Therefore, they are likely to exhibit the characteristics discussed in the chapter appendix. They will possess specialist expertise and be good at problem-solving and at generating or testing the feasibility of new ideas. They will enjoy conceptualization, working with abstractions, and seeing new and different ways of viewing things. They may be individual contributors to the team and prefer to work alone.

Technological Gatekeeping

Allen first identified information gatekeepers in R&D as people who provide human interfaces between the R&D staff and the outside environment, particularly the external R&D community.[18] They thus provide information channels whereby R&D staff can keep abreast with the state of the art both in terms of new technological ideas and new commercial exploitations of the technology. Rhoades *et al.*[16] have separated these two aspects of the role into two critical functions—the technological and market gatekeeping. We discuss each in turn.

Technological gatekeepers possess a high level of technical competence and esteem and are very widely read and known. They keep informed of related developments outside the R&D function through reading a wide range of journals, attending conferences, and maintaining a wide network of outside professional contacts. They are personable and enjoy helping people, so they serve as information resources for colleagues in the organization. They can provide invaluable support for the idea generator since they can expand the latter's repertoire of ideas, concepts, and facts which can be incorporated into the latter's conscious and non-conscious creative mentation. They can also provide informal coordination between project personnel.

Marketing Gatekeeping

In the framework of the technology–market matrix, the technological gatekeepers may be viewed as providing the information resources for generating new technological realizations and capabilities. Likewise, the market gatekeepers may be viewed as providing the information resources for identifying potentially exploitable market opportunities. Their personality characteristics are similar to their technological counterparts, but their expertise and interests will be in technological marketing. Again, they will possess technical competence and esteem, but in the area of product application rather than realization. Again, they will read journals, attend conferences, and maintain a network of professional contacts concerned with the markets in which the firm exploits its generic technological skills. They will discuss customer problems and needs with (say) the customer's researchers and design engineers, thereby monitoring the market and identifying with the technological gatekeepers of customer firms. They can play an important role in ensuring that the innovation is meeting an identified market (avoiding the dangers of technology-push) which may change during project development and help coordinate the required marketing effort.

Project Championing

Most case histories of innovations stress the importance of the project or product champion role. Technological innovation, which is a synonym for technological change, requires not only that the uncertainties and hazards associated with a novel technical ensemble must be resolved and surmounted, but also that internal and external resistances to change must be overcome. Unless there is at least one person, often in a position of some influence, who is totally committed to the innovation and has the dogged determination to overcome all the barriers, it is unlikely to succeed. Expressed more succinctly, this means that: *"Behind every successful technological innovation there growls a bulldog!"* The project champion is the individual who provides for the continuance of the project, despite resistance and setbacks, until it succeeds. Champions have wide technological interests which are applications-oriented, but they are without strong propensities to contribute to the basic knowledge in their fields. They are energetic and determined and aggressive in championing their own careers and are willing to take risks. They are skilled in selling ideas to others and securing organizational resources. Pinto and Slevin[19] discuss several types of project champions, and suggest that they all share the following attributes:

1. They have some personal or positional power in the organization and are willing to deploy it in support of the project.
2. They use their power somewhat nontraditionally and above and beyond the call of duty, as defined by their formal job responsibilities.

Don Frey (a former CEO of Bell & Howell) describes his personal experiences with the Ford Motor Company, where he championed several projects and products, including the Ford Mustang.[20] His *Coaching Tips for Aspiring Champions* are particularly shrewd. One type of project champion is the typical technological entrepreneur whose personality characteristics are discussed further in Chapter 10.

Project Managing and Leading

Although champions or entrepreneurs provide the drive or momentum for the project, they may be unsuited to the task of managing the project—that is, the attention to detail required to coordinate the various facets of the project and ensure that they are completed on time. This requires a person with management and administrative interests who is recognized as the focus for questions, information, and decision making. This is the project manager or leader. Project managers provide the team leadership and motivation and are sensitive to accommodating others' needs. They are interested in all the management functions, know how to use the corporate structure to get things done, and seek to match the project goals with organizational needs. They plan, organize, and coordinate the project, ensuring that administrative requirements are met and that the project moves forward smoothly.

Coaching

The first five critical functions can be directly related to the constituents of the innovation-chain reaction as shown in Figure 7.1, but the final function requires more explanation. Not infrequently, an innovative project is conceived and promoted by an informal coalition of bright, young, ambitious, but relatively junior individuals in the corporate hierachy. In that case, although the project team may incorporate all the above critical functions, it may still be ineffective because its youthfulness undermines its organizational credibility. This is particularly so if the project championship and leadership roles are performed by relative juniors in the organization hierachy. The project may consequently be aborted through inadequate corporate support. In some cases, this may lead to the resignation of the project team from the firm and the formation of a spin-off firm dedicated to the commercial exploitation of the proposed innovation, possibly funded by private and venture

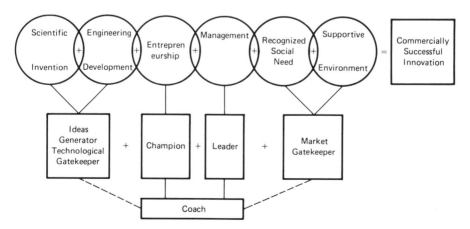

Figure 7.1. Critical functions and the innovation chain equation.

capital. This spin-off phenomenon is, of course, well documented in the innovation literature and is discussed in Chapter 10. If the project looks viable, its abortion is clearly undesirable from the firm's viewpoint, whether or not it is resurrected elsewhere. To avoid such outcomes, a project coaching role is needed.

Project coaches act as patrons and advisors to the project team. They are often more senior people in the organization whose patronage invests the project with credibility and confidence. They are themselves experienced in developing new ideas, and they provide objective guidance and information or coaching to the less experienced team members, helping to develop the latters' talents. They provide access to the power base of the organization, helping the team obtain organizational resources and buffering it from unnecessary constraints. They thus ensure that the project does not fail through lack of organizational support. Don Frey, in describing his experiences championing the Ford Mustang, implies that Lee Iacocca (then Vice President and General Manager of his Ford Division) enacted the coaching role.

The critical functions provide a useful typology of innovative project team needs and so we now discuss how it may be used to develop a [9,9] orientation in R&D management.

7.6 MATCHING INDIVIDUAL AND ORGANIZATIONAL NEEDS IN R&D

In Section 7.2 we suggested three possible personal development paths which young R&D staff might follow. Although these may be viewed as the extremes of a triangular continuum, they do suggest the critical function roles which would appeal to staff following each of the three paths, as is shown in Figure 7.2. This figure is essentially self-explanatory, but some brief comments may be made. Clearly, researchers are going to be the idea generators (because in the present context, the terms are almost synonomous), and they may also play technological gatekeeping roles. Anyone might become a project champion since the role implies emotional commitment and determination, qualities which any researcher or manager may possess. It is of interest, however, to note that the term inventor–entrepreneur (applied to those such as Edison) is essentially a synonym for idea generator–project champion. Championing their own inventions through the successive innovation phases can provide a valuable and chastening experience for researchers since they may thereby experience for the first time, and in an intensely personal way, the real difficulties of the innovation process. Research managers are likely to develop through project leading roles and, once they gain the experience and seniority, are well-equipped to enact the coaching role. Other managers may also develop through project leader and/or market gatekeeper roles, and again, later they may be very well-placed in the firm to enact the coaching role.

To match individual staff and organizational development needs, a *critical function assessment* is performed. This assessment identifies:

1. The personnel required in each critical function, based upon the R&D goals and strategy and the portfolio of projects.

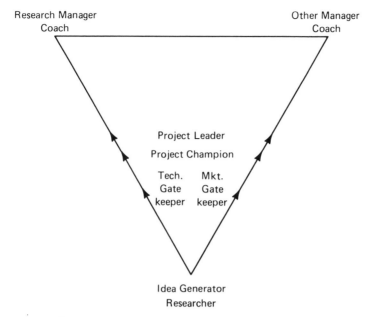

Figure 7.2. Alternative personal development paths.

2. The existing strengths in each function.
3. The factors supporting and inhibiting the performance of each function.

The above information is obtained through questionnaires, discussions, and workshops with R&D staff. Individuals complete questionnaires to identify their critical function strengths and those required in their current R&D roles. From the assessment, mismatches between ideal requirements and the strengths of staff can be identified to highlight weaknesses and suggest areas for improvement.

Fusfield quotes the results of such a survey of a unit with 90 R&D staff.[21] This unit appeared to have an unbalanced allocation of skills to needs and suffered from a shortfall of project leaders and market gatekeepers. It appeared to be an organization which will be good at generating ideas or inventions, but relatively weak at carrying them through to innovations or perceiving their true market relevance. Despite a declared market-pull orientation, the distribution of critical strengths was biased towards technology-push.

It was obvious that senior management was obligated to remedy the weakness exposed, and the following actions were taken:

1. Every individual discussed her or his critical function score with management to reach agreement as to the individual's skills as a basis for future career planning. The jobs of staff who were mismatched to their innovation roles were modified accordingly.

2. Training in project leading skills was provided, which involved the participa-

tion of personnel from outside units, to enhance the recognition of project leadership in the total organization. This exercise clarified the project leading role and ensured that upper management was sensitive and supportive towards it.

3. Recruiting practices were changed to ensure that applicants were evaluated in terms of their critical functions, as well as technical strengths and interests. Employment offers were based upon critical function as well as disciplinary needs. Interviewers were chosen to reflect a balance of critical function to ensure no unconscious bias in recruiting. Historically, it appeared that idea generators had unconsciously favored and attracted more idea generators.

4. All future jobs were defined in terms of the critical functions as well as technical factors. The management-by-objectives procedure being used was expanded to incorporate critical functions and innovation teams were selected to ensure a balanced distribution of these functions.

The critical functions treatment provides a pragmatic typology for developing [9,9] innovation management, and one other facet needs to be discussed. Both the discretionary scopes and the measuring and reward structures vary across the critical functions, but it should be recognized that they must honestly adhere to the performance criteria of the critical function if the approach is to succeed. If idea generators perceive that the firm awards most of the credit for a successful idea to the project champion, they are either gong to be their own project champions (even if ill-suited to it), or they may well suppress the idea rather than let someone else take the credit for it. Similarly, people will not undertake gatekeeping roles which produce intangible outputs, unless the organization recognizes the worth of these activities. Therefore, it is imperative that individuals' rewards and recognition are based upon the criteria set by the critical function roles they are enacting.

7.7 MATRIX STRUCTURE

The technical demands of R&D (particularly in an offensive–defensive innovator) impose conflicting requirements on the laboratory organizational structure as well as its staff.

The fundamental and applied research phases of innovation are primarily discipline oriented. Furthermore, young scientists and (to a lesser extent) engineers entering an industrial research laboratory direct from a university identify with their respective disciplines. These two factors dictate that a strong research capability should be based upon research groups in the appropriate disciplines (chemistry, physics, etc.). Even when innovations have progressed into development projects, they place varying demands upon expensive laboratory services and facilities such as engineering, machine shop, computing services, etc., so that it makes economic sense to manage these centrally to provide a common service to all project users. To take an extreme example, projects in an oceanographic R&D establishment may require the performance of sea-going trials. Oceanographic research vessels are expensive items, and it is obviously economically absurd to lease or purchase a ship for *one* voyage for *one* project. In reality, any one voyage of such a vessel would be

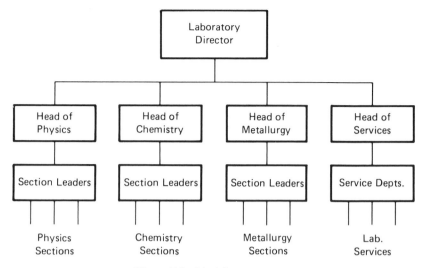

Figure 7.3. Discipline structure.

designed to accommodate the needs of a number of projects, including, say, ocean-ographic prospecting devices, diving-gear, etc. Such considerations dictate a labora-tory management structure based upon disciplines and laboratory service functions, as shown schematically in Figure 7.3.

However, as an innovation progresses from research through to development, it becomes more multifunctional in character and places an increasing demand on a widening range of laboratory and other firm services. These demands must be made in the face of competition from other projects. Furthermore, if a project team is cohesive, as it should be, team members may increasingly identify with the project, as against their disciplinary or service function backgrounds. This cohesion and identification will be weakened if members still see their disciplinary or service function divisional heads as their only bosses—who will largely determine their future career prospects. Effective project management must therefore cut across the structure shown in Figure 7.4, with team members identifying an allegiance to the

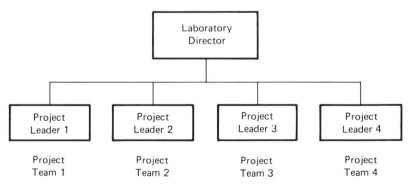

Figure 7.4. Project structure.

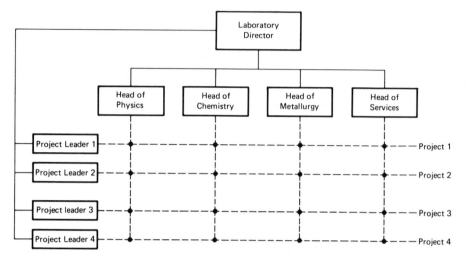

Figure 7.5. R&D matrix structure.

project and its leader, as much as to their divisional head. Many laboratories have accommodated this requirement by adopting a matrix organizational structure as shown in Figure 7.5. The team membership is seconded from a number of divisions, depending upon the technical requirements of the project and the manpower available, and reports to a project leader. Given that project teams may have a transient existence and/or include part-time members, individual members will usually retain their divisional affiliations. They thus become "two-boss" persons, with reporting responsibilities to divisional heads *and* project leaders. Matrix structures therefore violate the classical management dictum of a single line of command, and they have been criticized for this reason. However, despite doubts concerning the efficiency of matrix structures, many firms claim to have adopted them to manage projects successfully, both in R&D laboratories and manufacturing operations. They are discussed further in Chapter 9, when we consider the transfer of projects from R&D to operations.

FURTHER READING

1. T. Kidder, *The Soul of the New Machine.* New York: Little, Brown, 1981.

REFERENCES

1. M. J. C. Martin, J. H. Scheibelhut, and R. Clements, *Transfer of Technology from Government Laboratories to Industry,* Report No. 53, Technological Innovation Studies Program. Ottawa, Ont.: Department of Industry, Trade and Commerce, November 1978.

2. R. R. Blake and J. S. Mouton, *The Managerial Grid.* Houston, TX: Gulf Publishing Co., 1964.

3. M. Badaway, "Managing Human Resources." *Research Technology Management* **32**(4), 19–35 (1989).

4. S. L. Meisel, "The Dual Ladder—The Rungs and Promotion Criteria." *Research Management* **20**(4), 24–26 (1977).

5. E. W. Cantrell *et al.*, "The Dual Ladder Success and Failures." *Research Management* **20**(4), 30–33 (1977).

6. E. X. Hallenberg, "The Dual Advancement Ladder Provides Unique Recognition for the Scientist." *Research Management* **13**(3), 22–227 (1970).

7. D. C. Moore and D. S. Davies, "The Dual Ladder Establishing and Operating It." *Research Management* **20**(4), 14–19 (1977).

8. J. J. Smith and T. T. Szabo, "The Dual Ladder—Importance of Flexibility, Job Content and Individual Temperament." *Research Management* **20**(4), 20–23 (1977).

9. C. W. Lentz, "Dual Ladders Become Multiple Ladders at Dow Corning." *Research Technology Management* **33**(3), 28–34 (1990).

10. D. D. Dill and A. W. Pearson, "Managing the Effectiveness Project Managers: Implications of a Political Model of Influence," Unpublished paper. Lausanne, Switzerland: EURO V-TIMS XXV International Meeting, July 1982.

11. T. J. Allen, "Roles in Technical Communication Networks." In C. E. Nelson and D. K. Pollock (Eds.), *Communication Among Scientists and Engineers.* Lexington, MA: Heath Lexington Books, 1970, pp. 191–208.

12. W. A. Fischer, "Scientific and Technical Information and the Performance of R&D Groups." *TIMS Special Studies in the Management Sciences* **15** (1980).

13. W. A. Jernakowicz, "Organizational Structures in the R&D Sphere." *R&D Management* **8**, 107–113 (Special Issue) (1978).

14. R. Katz and M. Tushman, "An Investigation into the Managerial Roles and Paths of Gatekeepers and Project Supervisors in a Major R&D Facility." *R&D Management* **11**(3), 103–110 (1981).

15. A. K. Chakrabarti and J. Hauschildt, "The Division of Labour in Innovation Management." *R&D Management* **19**(2), 161–171 (1989).

16. R. G. Rhoades, E. B. Roberts, and A. R. Fusfield, "A Correlation of R&D Laboratory Performance with Critical Functions Analysis." *R&D Management* **9**(1), 13–17 (1978).

17. E. B. Roberts and A. R. Fusfield, "Staffing the Innovative Technology-Based Organization." *Sloan Management Review* **22**(3) (1981).

18. T. J. Allen, *Managing the Flow of Technology.* Cambridge, MA: MIT Press, 1977.

19. J. K. Pinto and D. P. Slevin, "The Project Champion: Key to Implementation Success." *Project Management Journal* **20**(4), 15–20 (1989).

20. D. Frey, "Learning the Ropes: My Life As a Product Champion." *Harvard Business Review* **69**(5), 46–56 (1991).

21. A. R. Fusfield, Unpublished Executive Seminar presentation.

APPENDIX 7
CREATIVE THINKING

Creative activity could be described as a type of learning process where teacher and pupil are located in the same individual. Creative people like to ascribe the role of the teacher to an entity they call the unconscious, which they regard as a kind of Socratic daemon—while others deny its existence, and still others are prepared to admit it but depore the ambiguity of the concept.
ARTHUR KOESTLER, *THE DRINKERS OF INFINITY; ESSAYS 1955–1976*

Machines and computers cannot be creative in themselves, because creativity requires something more than the processing of existing information. It requires human thought, spontaneous intuition, and a lot of courage . . .
AKIO MORITA WITH EDWIN M. REINGOLD
AND MITSUKO SHIMOMURA, *MADE IN JAPAN*

A7.1 THE NEED FOR CREATIVE THINKING

At the beginning of Chapter 1 we stressed that innovation is much more than invention. But since the innovation process is based upon invention, innovation cannot occur without it. That is, the innovative organization must be an INVENTIVE organization. Although on several occasions already, we have indicated that a good R&D function is prolific in generating such inventions, the innovative company will still be concerned to enhance both their quality and quantity. Furthermore, *inventiveness* is a quality usually required, and always desirable, *in all phases* of the innovation process. Recall that Sony and Texas Instruments invented the personal radio, following the American R&D invention of the transistor. Later, in Chapter 10, we shall also argue that entrepreneurship is another manifestation of the human creative urge.

It is hardly surprising, therefore, that many firms seek to foster creative behavior among their employees and that it has achieved cover story status in *Business Week*.[1] The *Concise Oxford Dictionary* defines "to invent" as to *create by thought*. In this Appendix, we explore the nature of this creative thought process, how it may be stimulated, and how organizations may identify people and foster a climate which stimulates their creativity.

A7.2   THE CREATIVE PROCESS—THE SOCRATIC DAEMON

As the quotations at the beginning of this Appendix imply, there is no irrefutable means of defining the creative process, and although psychiatrists and psychologists have developed numerous approaches to explaining the phenomenon, they appear to be of limited practical value. Perhaps the most important inference to be drawn from these approaches is that creativity is not markedly dependent upon intelligence (as measured by IQ tests), but is correlated with a number of other personality attributes. These inferences will be discussed later in Section A7.4. Therefore, managers may most fruitfully view the process as a black box which has had its input–output characteristics fairly well documented, even though the nature of its Socratic daemon is shrouded in obscurity.

Arthur Koestler is probably the most eloquent and perceptive writer on this subject. In the first two works of a trilogy concerned with the relation between reason and imagination—*The Sleepwalkers* and *The Act of Creation*—he explores in some detail the nature of scientific discovery and the creative act.[2,3] The first includes biographical and psychological sketches describing how some of the major figures in modern science, ranging from Copernicus to Kekule, discovered their ideas. The second examines the process of creative thinking in humor, science, and art. To illustrate the creative process, we quote two examples from the latter book. Koestler is almost exclusively concerned with human creativity, but also postulates its capability in animals too. Other primates, as well as man, have developed tools or technological artifacts, so we first quote an example of technological invention in the animal kingdom which Koestler extracts from *The Mentality of Apes* by Wolfgang Kohler.[4]

> Nueve, a young female chimpanzee, was tested 3 days after arrival (11th March, 1914). She had not yet made the acquaintance of the other animals but remained isolated in a cage. A little stick is introduced into her cage; she scrapes the ground with it, pushes the banana skins together in a heap, and then carelessly drops the stick at a distance of about three-quarters of a metre from the bars. Ten minutes later, fruit is placed outside the cage beyond her reach. She grasps at it, vainly of course, and then begins the characteristic complaint of the chimpanzee: she thrusts both lips— especially the lower—forward, for a couple of inches, gazes imploringly at the observer, utters whimpering sounds, and finally flings herself on to the ground on her back—a gesture most eloquent of despair, which may be observed on other occasions as well. Thus, between lamentations and entreaties, some time passes, until—about seven minutes after the fruit has been exhibited to her—she suddenly casts a look at the stick, ceases her moaning, seizes the stick, stretches it out of the cage, and succeeds, though somewhat clumsily, in drawing the bananas within arm's length. Moreover, Nueve at once puts the end of her stick behind and beyond her objective.
> —(Koestler,[3] p. 100)

Our second example is the possible apocryphal but usefully illustrative example which most readers know—Archimedes in his bath.

Hiero, tyrant of Syracuse and protector of Archimedes, had been given a beautiful crown, allegedly of pure gold, but he suspected that it was adulterated with silver. He asked Archimedes' opinion. Archimedes knew, of course, the specific weight of gold—that is to by its weight per volume unit. If he could measure the volume of the crown he would know immediately whether it was pure gold or not; but how on earth is one to determine the volume of a complicated ornament with all its filigree work? Ah, if only he could melt it down and measure the liquid gold by the pint, or hammer it into a brick of honest rectangular shape, or . . . and so on. At this stage he must have felt rather like Nueva, flinging herself on her back and uttering whimpering sounds because the banana was out of her grasp and the road to it blocked. . . One day, while getting into his bath, Archimedes watched absentmindedly the familiar sight of the water-level rising from one smudge on the basin to the next as a result of the immersion of his body and it occurred to him in a flash that the volume of water displaced was equal to the volume of the immersed parts of his own body—which, therefore, could simply be measured by the pint. He had melted his body down, as it were, without harming it, and he could do the same with the crown.

—(Koestler,[3] p. 105)

Both of these examples (and numerous others which Koestler quotes) have four stages in common:

1. Perception of the problem.
2. Frustration at the inability to solve it.
3. Relaxation.
4. Perception of solution—EUREKA!

Green,[5] who wrote from the perspective of an R&D management background (he was Vice President, Bell Telephone Laboratories when he wrote the article) essentially expands on the above stages in his description of the creative process. First of all, he proposes that creative thinking in science can be typified as a continuum of styles of thinking from systematic thinking to intuitive thinking, as he puts it:

The one is a deliberate act of the conscious mind, the other the gracious gift of the subconscious in return for the previous labors of the conscious mind.

Green suggests that systematic thinking is a combination of empiricism and *omphaloskepsis*. The first is self-explanatory, and the second was devised by a British officer in India to describe oriental meditation. Green uses it to describe the contemplative and mediative process of the rational and analytical formulation of a theory, tested by guided empiricism, sometimes supplemented by serendipitious outcomes. Intuitive thinking yields the sudden flash of insight of genius or hunch as is an output of nonconscious thought processes. This flash of insight or genius will be defined further shortly.

Green formulates the creative process as:

1. The problem faced by the individual arises by the nature of the R&D process. The individual develops at least one preliminary conception of the problem and moves to:

2. *Accumulation* of data, ideas, and concepts through literature reading, discussion, investigation, and experiment.

3. *Incubation* when the conscious and nonconscious mind assimilates and digests the information gathered.

4. *Intensive thinking* next occurs whereby the individual seeks to solve the problem by weaving ideas in new combinations, but despite the intense effort, without success. This leads to:

5. *Frustration,* dissatisfaction, and fatigue, so in the face of this psychological blockage, the individual abandons conscious consideration of the problem, takes some:

6. *Relaxation* and "sleeps on it," leading later to:

7. *Illumination* or sudden inspiration, the so-called flash of genius or EUREKA, and (in science and engineering anyway) finally:

8. *Solution,* verification, and embodiment.

Green's treatment provides a useful outline of the creative process, and later we examine the techniques which may be used to enhance the effectiveness of stages 1, 3, and 4 in his process. Before doing so, however, we elaborate upon the progression through the process, returning to Koestler's writings on the subject.

Although the explicit mental mechanics of the creative process are, as yet, largely inexplicable, it is important to avoid viewing it in purely passive terms. The Socratic demon, like the poet's muse, may provide inspiration, but *not in an autonomous, arbitrary or capricious manner.* R&D inventions like good poetry are not generated spontaneously in the inventor's or poet's conscious mind. *Inspiration* and illumination only occur after the inventor has undergone the *perspiration* of extensive and frustrating conscious deliberation and mediatation. Unfortunately, although laying on a beach in Fort Lauderdale, Florida or the Cote d'Azur of France may lead to a flash of genius, it will only do so *after* you have exhausted yourself in preparation. Therefore, the Socratic daemon hypothesis cannot be used as an excuse for plain idleness!

Medieval thinkers described the flash of genius as a flash of lightning or *fulguratio,* implying a sudden thunderbolt from Zeus or intervention from God above. Contemporary intellectual fashions make us look for mystical explanations from depth psychology rather than Hellenic or Judeo–Christian theology, and Koestler offers such a description. He points out that mental activity or mentation occurs with or without personal awareness. He postulates that there is a continuum or gradient of states from begin totally unaware or unconscious (say, from being knocked senseless by a hard blow on the head) to (as he puts it) the "laser beam of focal consciousness." Some physiological control activities are performed entirely unconsciously, while others such as breathing can be performed unconsciously or consciously. Similarly, some complex consciously learned skills such as tying one's shoelaces or driving a car may be performed either unconsciously (absent-mindedly) or consciously. Equally, some activities (both unconscious and to lesser degree conscious) are nonverbal and nonrational and cannot be articulated in the framework of language and logic. Furthermore, these activities or thoughts are characteristically exposed to one level of awareness in dreams or dream-like states of mind, which are rejected by the linguistic and logical articulations of the focal consciousness. Indeed, the preservation of human sanity requires the permanent repression of some of these archaic and prelinguistic thought forms to nonconscious levels of mentation.

Now, the concrete realization of an R&D invention must be verified and embodied in the linguistic and logical framework of the scientific body of knowledge; therefore, it must be finally *articulated* in this form. However, Koestler provides substantial anecdotal evidence to support the postulate that the mental act of creation invention *does not occur in this way*. The creative process appears to occur at a prelinguistic level of mentation. He suggests that creative people soak themselves thoroughly in the subject matter of the problem and then cogitate. The verb to cogitate is derived from *coagiture* meaning to shake together two or more previously separate entities. Now, this process of cogitation can occur at two levels of mentation. First, at the linguistic and logical level of systematic thought in Green's treatment. Second, at the prelinguistic and dream-like or fantasy level, which provides the flash of genius in Green's treatment. At the conscious level, this cogitation involves exploring new associations, combinations of data, etc., within the existing linguistic and logical framework of the state of the art, that is, within an existing framework of assumptions and rules. This framework constitutes a dominant perceptual pattern within which the problem is viewed. A characteristic of many creative acts in science is their formulation of new associations or patterns which overturn the existing framework through a gestalt switch of perception. Systematic and rational mentation does not readily make these novel associations because its perception is dominated by the existing pattern. Thus, systematic and rational thinking leads to failure and frustration. The individual is frustrated by his failure to solve the problem so he abandons it and relaxes or does other work.

Now, what happens when he abandons the problem? Up to this moment, the process of mentation may have been occurring at both a conscious linguistic and rational level *and at a preconscious, prelinguistic, and nonrational level*. From now on, further cogitation is performed at the latter preconscious level, untrammeled by the rules and regulations of rationality. Further associations of ideas, concepts, data, etc., however irrational and fantastic, may be combined and associated, but never reach the threshold of awareness. Finally, a novel association or new perceptual pattern is found which has credibility at the rational level. It is, therefore, permitted to emerge into conscious awareness through a sudden thought in the bath or in a dream or dream-like state.

Koestler postulates that the act of creation consists of the novel association of two previously unrelated concepts or ideas by *BISOCIATION* to form a new perceptual pattern. He represents this act of bisociation as shown in Figures A7.1 and A7.2. Both component concepts of frames of reference are represented by two-

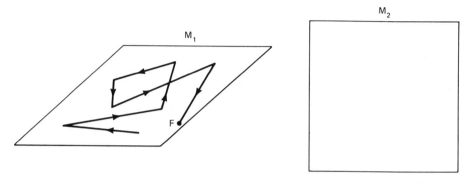

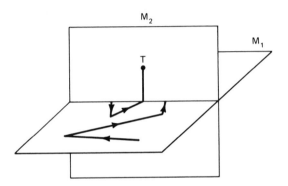

Figure A7.1. Koestler's bisociation.

dimensional matrices of association M_1 and M_2—one in a horizontal and the other in a vertical plane. The individual, prior to the act of creation, does not associate them with each other, and may reject their association as irrational, fantastic, and outrageous. Thus, prior to this act, he searches for a problem solution within one matrix of association (the arrowed line in M_1), experiencing increasing frustration (point F) because the solution cannot be found there. He relaxes, but preconscious cogitative mentation continues exploring novel matrix associations which his conscious mind would reject. The preconscious mind discovers a novel association or bisociation between M_1 and M_2 which holds promise of solution. Because of its promise, the bisociation is allowed to pass through the threshold of awareness and become a flash of genius or EUREKA in the conscious mind of its creator, leading to the problem-solution T (Figure A7.1). This intuitive solution can then be proved using systematic, logical, and rational thinking.

This treatment may be readily applied to the two creative acts described earlier. The chimpanzee Nueva's problem was to obtain the bananas outside her cage. Her previous method of doing this—matrix M_1—had been to squeeze an arm or a leg through the bars of her cage and rake the fruit towards her until she could grasp it.

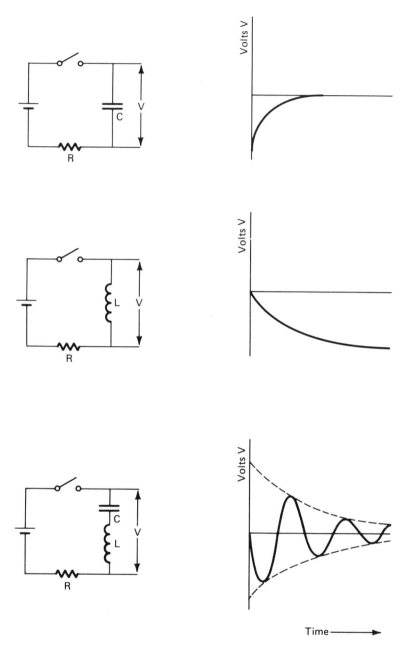

Figure A7.2. LCR circuit.

The method is unsuccessful in this instance as the fruit is too far away. She reaches point F in Figure A7.1, expressing despair. The other matrix M_2 is the stick which she associates with pushing and scraping objects on the floor of her cage (including banana skins). There is no recognition of the potential commonalities between the two matrices. A few minutes after her lamentations, she makes the bisociation, grasps the stick, and uses it to reach the bananas, reaching the solution point T.

Archimedes' problem was to assay the King's crown, which he defined as the problem of determining its volume. The methods available for doing this based upon his current scientific knowledge required its destruction—matrix M_1. He was unable to devise a nondestructive method within this framework, so he experienced frustration (point F). He probably enjoyed the relaxation and sensual pleasure and relief offered by a daily bath, and doubtlessly frequently observed the rise in water level as he sat down in it. Unfortunately, this observation was perceived within the pattern of the taking of a bath frame of reference—matrix M_2—and assigned no significance. Again, during relaxation, Archimedes' mind made the bisociation between the two matrices, reached the solution point T, and (if legend is true) jumped out of the bath and ran naked down the street crying Eureka. His verbalization of this act of bisociation was bequeathed to us as the Principle of Archimedes.

Koestler provides numerous other examples of bisociation in the references cited earlier, including Gutenburg's invention of the printing press by the bisociation of the coin or seal stamp and the wine press. Perhaps his most charming example is that in which Kepler bisociates *Gravity and the Holy Ghost.* To justify his laws of planetary motion, Kepler (who was educated initially as a theologian) postulated that a force emanates from the sun which influences the planets in an analogous manner to God the Father influencing the Apostles through the Holy Ghost. This postulate provided a basis for Newtonian gravitation and the integration of astronomy and physics. One might conjecture that classical physics could have developed differently had there been a different outcome to the conflict concerning the doctrine of the Trinity in the Early Christian Church!

It is important to note that the creative bisociation of two frames of reference is not simply an additive association. Rather, it exhibits an ontological discontinuity whereby the bisociated whole is greater than the associated frames of reference sum of the parts. In Chapter 2 we suggested that an R&D invention might be viewed as a technological mutation in technological evolution. Konrad Lorenz, the eminent ethologist, points out that in biological evolution, through etymological accident, new organic systems or capabilities are created with properties which cannot be described in terms of the properties of the associated subsystems.[6] He draws an analogy between the creative process and this manifestation of creativity in biological evolution, using an example from elementary circuit physics for illustration. This illustration is shown in Figure A7.2. In the first circuit (a), a battery is connected to a resistance and capacitor in series, and as the capacitor is charged with electricity, the voltage across it rises exponentially. In (b), the battery is connected to a resistance and inductor in series, and voltage across the latter falls exponentially. In (c), the battery is connected across all three component in series, and the voltage across the capacitor and inductance exhibits oscillatory behavior with exponentially falling amplitude. Circuit (c) represents an association of circuits

(a) and (b), but its phenomenological behavior is not readily expected from the phenomenological behaviors of circuits (a) and (b). It is in this sense that creative bisociation is more than simple association and the bisociative whole is more than the sum of the constituent parts. The unexpected emergent properties of systems synthesised from component subsystems is a well-known feature of cybernetics and general systems theory.

Recapitulating and summarizing this section, we can say that it provides a description of the creative process which displays the following salient features:

1. The creative act typically requires the synthesis or bisociation of two frames of reference or conceptual patterns, previously not associated.

2. Although after the event such an act may appear obvious, because of its revolutionary novelty it often constitutes a feat of some psychological courage on the part of its creator.

3. The mentation leading to bisociation involves cogitation or the shaking together of nonassociated conceptual patterns. Because these perceptions frequently occur at a preverbal and irrational level, the process requires a regression to preverbal, prerational, and preconscious levels of mentation and to child-like fantasy, without losing ultimate rational control and evaluation.

4. Since cogitation involves combining nonassociated matrices, perhaps on a random trial and error basis, the more matrices there are to combine the more likely a creative combination is to be found. Therefore, creative individuals are likely to be people with wide as well as specialist scientific interests, people with child-like eclectic intellectual curiosities who are persistently asking the why of everything and delighting in the answers which they obtain. They are indefagitable idea and fact collectors. They steep themselves in the literature of their own fields, and are curious about everything else too. This way, they store a large repertoire of matrices in their minds from which a bisociative combination is more likely to be found.

5. Finally, they must be very industrious to generate the enthusiasm, persistence, and tenacity that is to generate the requisite perspiration which precedes inspiration.

A7.3   CREATIVITY AND ARTIFICIAL INTELLIGENCE

In contrast to the views of Akio Morita, quoted at the beginning of this Appendix, the fourth point above poses an intriguing question. If the cogitation process possibly occurs on a random trial and error basis, could a computer be programmed to replicate it? Boden in *The Creative Mind* argues that the answer to that question might, in a limited sense, be "yes."[7] In its opening chapter, she states:

> This book takes up the question of creativity from where Koestler left it. . . The central theme of the book is that these matters can be better understood with the help of ideas from artificial intelligence (AI).
>
> —(Boden,[7], p. 5)

Her book is similar to Koestler's writings in that much of it explores examples of human creativity in diverse fields, and is written in an equally engaging style. She discusses it in terms of the exploration of conceptual spaces in the mind, rather similar to Koestler's conceptual patterns, but extends such ideas by describing some of the AI programs that have pursued analogous explorations to yield creative results. She cites the example of meta-Dendral which generated new ideas in biochemistry and was essentially cited as a "coauthor" in a paper published in a chemistry journal. As she puts it, it has a refereed publication on its *Curriculum Non-Vitae!* Another example she describes is BACON (and other programs named after creative scientists of the past), developed by a team of AI researchers led by Nobel laureate Herbert Simon, to explore the nature of scientific creativity. BACON has made numerous "discoveries" including Boyle's law and Kepler's third law of planetary motion.

AI programs like meta-DENDRAL offer promise as aids for researchers in certain fields such as biochemistry, while those like BACON offer improved insights into the nature of human creativity. However, as Hofstadter comments in his review of *The Creative Mind,* their achievements need to be viewed critically.[8] Such programs work within a narrow domain of data relevant to the problem under study and defined by their designers. In contrast, creative human individuals often work with a fog of data, where the distinction between the relevant and irrelevant is unknown. This was the situation that confronted Kepler. However, they do enjoy the advantage of being able to draw on a diverse repertoire of prior experience and knowledge. Much of this material may be irrelevant, but some may be made relevant by the insightful use of analogies and metaphors. Thus, according to Koestler, Kepler was able to draw upon his prior familiarity with the Christian doctrine of the Holy Trinity in discovering his laws of planetary motion. It is debatable whether computational psychology will ever be able to replicate the essentially poetic imagination and diversity of the experience and knowledge of the prepared and courageous human mind, so Akio Morita's contention should remain unchallenged. Therefore, let us turn to more pragmatic managerial concerns.

A7.4   IDENTIFYING CREATIVE PEOPLE

Although creativity, like intelligence and physical strength, may be a trait possessed by everyone, just as some people are manifestly more intelligent or physically stronger than others, it can be expected that some people are more creative than others. High-technology companies, particularly those following offensive/defensive strategies, are in the competitive business of exploiting technological creativity for profit. Consequently, it is axiomatic that such organizations must recruit and retain a cadre of creative individuals if they are to maintain their business competitiveness. In this section we identify the personality attributes which, according to the current state of the art in the field, are indicators of a person's creative ability.

The previous fairly detailed discussion of the creative process broadly identifies these indicators. Creativity requires the ability to think at the nonverbal level without a decline in the capacity for abstract thought and logical analysis. This viewpoint strongly suggests that it requires a mixture of Hudson's convergent and divergent thinking.[9] Indeed, Kuhn argues that the interplay between convergent and divergent thinking provides the essential tension for the creative talent of creative scientists.[10] Good academic performance in science subjects at the high school and undergraduate levels mainly requires convergent thinking. Therefore, it is important to look for other attributes as well as academic performance when seeking creative R&D staff.

This observation is possibly reflected in the results of Getzel and Jackson[11] who, in a study of U.S. high school students, found that a threshold IQ score of 120 was required for creativity, but that there was no correlation between IQ and creativity beyond that score. These findings were supported in research on teachers by Torrance and Hansen.[12] Getzel and Jackson also found that creative students did as well as their more intelligent classmates, suggesting that they were overachievers. This observation—which can be expressed as commitment to internal standards of performance, rather than those imposed by an external peer group—is supported by other research. MacKinnon conducted extensive research on architects.[13] He chose architects because they must combine elements of the artist and the scientist and also require business skills. From his research, he concluded that highly creative, as opposed to less creative, architects stressed the importance of performing to an inner standard of excellence, rather than that required by the profession. More importantly here, Roe[14] and Chambers[15] obtained similar results in their studies of eminent scientists versus other scientists, and scholars versus administrators. Again, the creative individuals set their own internal standards which they adhered to religiously, being capable of considerable self-discipline and perseverence. Barron also obtained similar results with writers and mathematicians.[16]

Their implacable commitment to high standards and goals coupled with a breadth of interest and curiosity is reflected in the interpersonal behavior of creative individuals. They seek personal autonomy and are indifferent to group standards and control. But although independent, they are also dependable. They have good communication skills and are eager to discuss ideas with others. However, the purpose of such discussions is to promote a two-way traffic in the ideas themselves, rather than human fellowship, so they are unconcerned with policing their own or other people's feelings. On the other hand, although primarily interested in communicating ideas, they are sensitive to interpersonal aggression and prefer to repress or avoid interpersonal conflict. The commitment to internal, rather than group standards and control does present a difficulty to the creative child. The educational system imposes continued external evaluative criteria upon a person from the age of six onwards, and generally disapproves of nonconformist personality traits. Getzel and Jackson, in their study, found that the teachers preferred the more intelligent to the more creative pupils. A person seeking an R&D career is likely to have survived the external evaluative criteria of the educational system to master's or doctorate degree level. Some people may have had their creative potential irrecoverably stifled in reaching this level, but persons who manage to identify and preserve this creative talent throughout this 20-year process are worthy of recognition and encouragement. It may also relate to a point made by Boden in her book cited earlier. She suggests that creative people need the challenge of legitimate rules and constraints to enrich their perceptions and insights, so it may be that the requirements of the formal education system provide them with a similar challenge.

Before concluding this section, one other attribute can be reported. Because of the nature of the creative process Koestler suggests that the creative person has multiple potentials—that is, the latent capacity to be creative in a number of fields (Leonardo da Vinci is an outstanding example). Koestler quotes Dr. Johnson as saying that true genius *is a mind of large general powers, accidentally determined to some particular direction, ready for all things, but chosen by circumstances for one.* In other words, chance and necessity determine the outlet for a person's creative potential. Kepler was going to be a theologian, but was offered a job as a mathematician. Darwin was expecting to be come a curate when invited to join the voyage of the Beagle. Therefore, probably the best indicator of an individual's creative potential is evidence of past creative achievement, not necessarily in the field of immediate interest to the prospective employer.

A7.5 CREATING A CREATIVE CLIMATE

A company will only derive the inventive benefits from potentially creative R&D staff if it provides an environment to support their potential. This implies "creativity-squared," that is, the creation of an organization climate in which personal creativity can thrive. As Badaway expresses it, in an excellent succinct review of this issue, *while creativity implies bringing something new into being, innovation implies bringing something new into use.*[17] In this section, we review the indicators of such a climate.

The findings of the studies on this topic are both mutually consistent and congruent with common sense expectations. Parmenter and Garber[18] asked a sample of U.S. scientists (whom their colleagues deemed creative) to rank order ten organizational factors which might encourage their creativity. Gerstenfeld[19] conducted a broadly similar study in which he asked a sample of 122 scientists to rank order goal priorities from a list of eight goals, while Kaplan[20] lists five important factors influencing organizational creativity. The seminal study of Burns and Stalker[21] provides an excellent background to the field. The overall conclusions of their studies were as follows:

1. Creative persons value autonomy and challenge. They desire freedom to choose challenging but realistic work assignments. However, provided they are involved in organizational research planning, they will seek to reconcile personal and organizational research goals. Some also welcome external *pressure to produce,* possibly because it helps stimulate the psychic energy required for creative effort. That is, to some, good R&D is like good journalism—*content under pressure.*

2. This means that the organization should be receptive and responsive to the individual's ideas, showing him appropriate recognition and appreciation. Despite a

person's commitment to internally set goals and indifference to formal performance appraisal, his sense of equity or natural justice dictates that the organization should implicitly sanction his own sense of worthiness through the renumeration and fringe benefits it accords him. Creative people will risk failure in order to explore novel ideas; therefore, the organization should not just reward error-free performance.

3. Puzzlement and wonderment, as well as errors, should also be nurtured in the continued search for invention. These are child-like traits which, as we have already seen, are typical of creative people. Furthermore, as we shall see in the next section, they are qualities which are encouraged in some creativity-enhancing techniques. Management convention tends to depreciate such traits as being symptomatic of indecisiveness, woolly-headedness, and immaturity—that is, devastating management sins! Such traits are very probably sinful for a man in a sensitive line-management position, but for the researcher, they may be virtues and should be accepted as part of the R&D climate. Indeed, tolerance of the nonconformist, provided he is productive in the long-term, should be axiomatic in a creative climate.

4. Together with tolerance of eccentricity should go tolerance of conflict. Creativity thrives on conflicts of ideas and opinion, and it need hardly be stated that scientific advance is often carried forward on the wings of controversy. Provided the conflicts are substantively about ideas that generate creative problem-solutions and are not centered on personal territorial disputes, they should be accepted as one of the colorful features of the creative climate.

5. We saw in the previous section that creative people are generally eager to discuss ideas with others. Indeed, the above conflict reflects on the Hegelian approach to intellectual discourse in the pursuit of *truth*. This consideration dictates that there should be minimal barriers to both intra- and interorganizational communication among individuals, even at some risk to commercial security. It is believed that, in general, creative organizations gain more than they lose by the ready willingness to exchange ideas and information. This has implications for the staffing of the R&D function, discussed in Chapter 7. These research findings provide useful management guidelines for creativity-squared, but we should conclude the section with two caveats.

First, although the psychological nature of the creative act may be invariant with respect to field, its contextual manifestation does vary throughout the innovation process. The creativity of a nondirected research group in a company following an offensive/defensive strategy should itself be *nondirected* since ideally its desired output is a continued but unspecified flow of novel inventive ideas. On the other hand, as Johne argues, once an invention reaches the prototype development phase of the innovation process, a more mechanistic directive structure is required.[22] Similarly, the success of a company following an applications-engineering strategy is dependent upon the continued ability of its development engineers to provide creative solutions to particular user problems *in a timely manner*. The first group can afford to employ a Kepler or a Newton who can deliberate indefinitely before publishing a specific novel insight, although they may continuously generate new insights throughout their creative careers. In contrast, the other two groups cannot afford this approach. Each must employ an Archimedes who can devise timely creative solutions to specific problems posed by an outside sponsor, especially when he knows that his personal survival may depend upon doing so! Although (as illustrated in the World War II careers of some eminent scientists) some creative individuals can be successful in either situations, it is important that the organization defines its manifest creative needs and seeks to recruit creative staff who can satisfy these needs. Too often, companies are unsuccessful because they recruit the wrong sort of creative people, as suggested by Ansoff and Stewart in their seminal article.[23] More recent research supports this observation. Ginn[24,25] has studied facilitators and barriers to successful innovation including an organizational culture factor, and suggests a *multidextrous* model for managing the creative aspects of the process. That is, innovation managers should choose between an organic versus mechanistic culture depending upon the needs of the specific situation. Management should also recognize the potentially critical role that creative thinking can play in all aspects of the innovation process. Quite often, the success of technological innovation is ensured through the timely creative solutions of "bugs" by technical and shop-floor staff. R&D laboratories usually establish a cadre of good technical and craft personnel, but an innovation may be delayed or aborted through the absence of a similar cadre at the manufacturing stage.

Second, although creative staff are vital, they are obviously not the sole ingredients required for innovation to occur. Indeed, they might be viewed as the yeast which ferments the "innovation brew" since successful innovation requires the cooperative participation of all corporate elements. This remark is valid even in the R&D phases of the innovation process because, as discussed in Chapter 7, the successful development of a project through to the pilot or prototype production stage requires a mixture of skills.

A7.6 CREATIVITY STIMULATING AND ENHANCING TECHNIQUES

Given that a company has hired creative R&D staff and provided a supportive organizational climate, the final issue which we consider in this Appendix is whether there are other means available for stimulating and enhancing individual and group creativity. Despite their commitment to internal standards, creative individuals are usually active members of the scientific communities in their fields. Their enthusiasms and curiosities make them ready to communicate ideas and information with colleagues, and they see this as a stimulus to their own as well as other people's thinking. This observation suggests that the creative process can be stimulated at the individual and group level. This is indeed so.

In Section A7.4 we distinguished between convergent and divergent thinking as being roughly equivalent to Green's systematic and intuitive thinking. We also quoted Thomas Kuhn's view that an individual's scientific creativity is based upon an "essential tension" between his convergent and divergent thinking. Thus, the creative process may be viewed as dynamic or psychic interplay between convergent and divergent modes of thinking, and there are specific analytical and psychological techniques in use which stimulate these modes of thinking at both the individual and group level. These may be conveniently classified as techniques for stimulating convergent and divergent thinking. Useful, quite detailed reviews of these techniques are given in Prince,[26] Rickards,[27] and Whitfield[28]; and we outline the more important ones in the remainder of this Appendix.

A7.7 TECHNIQUES FOR STIMULATING CONVERGENT THINKING

In Chapter 4, Appendix 4.1 (Section 4.1.11) we stated that TF is often conducted as a creative problem-solving process using creativity stimulating and enhancing techniques. In Sections 4.1.12–4.1.13 we discussed two such techniques—morphological analysis and relevance trees in the TF context. Both of these techniques are analytical and logical and stimulate the creation of problem solutions by rational thought processes or convergent thinking, and other simpler techniques are also used.

Two of the simplest analytical approaches, which owe their origins to method study, are attribute listing and value analysis. Attribute listing requires:

1. The description of each component.
2. The definition of the function of each component.
3. The evaluation of all possible ways in which each attribute of every component may be changed.

Attribute listing may be viewed as an incomplete form of morphological analysis, and also as the basis of another technique used in method study—value analysis. This latter technique compares the cost of each component with the value of the function it performs. Its application can often identify ways of improving the cost-effectiveness of a capability.

Earlier in this Appendix we stressed that creative skills are vital throughout the innovation process. Given the nature of this process, particularly in the development phases onwards, workers face a problem requiring a remedial as opposed to an inventive solution perhaps before a critical time deadline. For example, the performance of an innovation may fail to meet technical specifications at development, pilot/prototype, or full production stages. Kepner and Tregoe[29] are management consultants who pioneered an approach to analyzing management and organizational problems requiring the restoration of a performance attribute which has departed from some predetermined standard. The deviation is precisely defined, and alternative causes for this deviation are rigorously diagnosed or defined. Problem solving then focuses on remedying the most likely cause of the performance deviation.

A7.8   TECHNIQUES FOR STIMULATING DIVERGENT THINKING

Techniques for stimulating divergent thinking can be described as nonrational in that they essentially seek to simulate an individual's unconscious and preconscious nonrational cognitions. These techniques are most effective when performed in a small group setting (rather than by a lone individual) and, in light of Kuhn's stress on "essential tension," cyclically with the rational techniques we have already discussed. Because they seek to identify nonrational bisociations which normally occur at a prerational level of mentation, they may appear bizarre and frivolous to readers who have not met them before. Therefore, we stress that these techniques have been used effectively to stimulate creativity in many business organizations.

Brainstorming

Brainstorming is probably the technique most familiar to managers and the public at large. Indeed, it is possibly too well-known, since it is used loosely to describe any free-for-all group problem-solving or idea-generating session. The originator of the approach, Alex Osborne, points out that the verb to brainstorm means *to practice a conference technique by which a group attempts to find a solution for a specific problem by amassing all the ideas spontaneously contributed by its members.* Osborne, describes his brainstorming approach in *Applied Imagination,*[30] and the approach and applications are described in numerous publications. Our outline of the approach is mainly based upon Rickard's description.

The approach, usually conducted by one or two group leaders or organizers, is divided into a sequence of sessions. In the present context, the topic to be considered is likely to be an R&D problem to be solved. The brainstorming organizer and researchers involved can be expected first to define the problem. Once the problem is identified, the next task is to select the group of six to ten people who will attempt to solve it, and his criteria for group selection are:

1. *Prior Experience.* The group should include at least two people who have prior experience of brainstorming sessions and, preferably, at least two people without prior experience.
2. *Personalities.* Members should be restricted to people who participate constructively in committees and are sympathetic to the brainstorming approach.
3. *Disciplinary and Corporate Backgrounds.* Membership should embrace the disciplinary and functional backgrounds potentially relevant to the problem, subject to the overall size limitation. Seating arrangements should be informal—preferably a semicircle.

A sequence of several sessions is conducted, each with a specific purpose.

1. *Warm-up.* The organizer may encourage members to give some preliminary thought to the problem, but brainstorming will begin with a warm-up session to

encourage personal interaction and group awareness. The brainstorming approach will be described, and the session may end with a discussion and (possibly) a redefinition of the problem.

2. *Idea Generation*. This roughly corresponds to Koestler's cogitation process and seeks to encourage the *uncritical* generation of as many ideas as possible, while postponing critical evaluation and judgment. Each member is encouraged to suggest an idea which might be relevant, however "crazy," as it "comes into his/her head." The organizer lists ideas as they are generated, usually on a flip-chart at the center of the semicircle, so that they may be readily viewed by the whole group. Members are encouraged to improve or "piggy-back" on each others ideas.

Typically, at the beginning, ideas are generated fairly rapidly as members put forth their ideas. One session on increasing efficiency in an R&D contract research organization produced 55 ideas in ten minutes. There then follows a slump in ideas generation and possibly relatively long periods of silence. This period can be the most fruitful because it encourages members to cogitate on each other's ideas and generate useful combinations of piggy-backing. If the slump proves unproductive, continuing for too long so that members become bored, the organizer may stimulate the atmosphere by asking questions or introducing his or her own ideas from which to continue brainstorming. Richards quotes the example of a group session on improving the works canteen which appeared to have exhausted its capacity to generate ideas. Someone suggested the wild idea of topless waitresses, which led to the practical suggestion to install beverage vending machines. Specific techniques for generating ideas have also been used. "Trigger Sessions" is one approach in which each member works alone to produce an ideas list and then reads it out to the group as a stimulus. Another approach is "Recorded Round Robin," a variation on the party game of consequences. Each member lists an idea on a card and passes it on to another member, who attempts to piggy-back on it, then passes it on to the next member, and so on. If each of a six-member team generates three ideas on three separate cards, 18 ideas can quickly be explored.

3. *Idea Evaluation*. The idea generation session usually lasts up to two hours and produces a substantial list of ideas of varying usefulness. At least one member of the research team will have been included in the group. It is the researcher(s) who will be primarily concerned with the rational evaluation of the ideas generated. Further evaluation may be conducted with the participation of all group members or just the research team. This rational analysis will further explore the ideas which look promising, and may incorporate rational techniques such as morphological analysis. Other techniques for evaluating ideas are discussed under synectics below.

Synectics

The word Synectics, from the Greek, means the joining together of different and apparently irrelevant elements. With this sentence, William Gordon, its inventor, begins the introduction to his book describing the approach.[31] The words *synectics* and *bisociation* are virtually synonyms, and Gordon's approach presents a more explicit procedure than brainstorming for the group stimulation of nonrational cogitations. Indeed, it is fruitful to compare Koestler's[2] and Gordon's treatments of the creative process by reading both books together.

Synectics is quite similar to the brainstorming procedure in that an organizer or chairman conducts a small group (typically four to six) through a sequence of different sessions in an informal social setting. The criteria for group membership are also similar. The sequence of sessions is summarized as follows:

1. *Making the Strange Familiar.* Although we may suppose that most team members are already cognizant of the problem being considered, given the differences between individual perceptions, a group discussion is likely to uncover unexpected problem features and ramifications. This process of clarification or familiarization of the group with the "strange" problem yields the *problem as understood.* Group members may also suggest problem solutions at this stage which are evaluated or *purged,* and if one is acceptable, the procedure may be terminated. The term purging is used to encourage individuals to forget unacceptable solutions and not allow them to create psychological blockages to future constructive thought. Purged solutions are recorded, however, in case the group wishes to reconsider them later.

2. *Making the Familiar Strange.* Gordon contends that making the familiar strange or viewing the problem in an unfamiliar manner can generate creative solutions. He believes that such solutions are generated by metaphorical (as opposed to analytical) thinking and the development and exploration of analogies, and the synectics process is based upon a person's capacity to learn this metaphorical cast of mind. Gordon views biological analogies as being generally the most fruitful. He defines four types of analogies[31] for making the familiar strange, and we consider each in turn.

First, *personal analogy* involves the individual in an empathetic personal identification with the elements of the problem—that is, he thinks of himself as a dancing molecule, or in the case of Kekule when inventing the benzene ring, as a snake swallowing its own tail. Gordon also quotes the writings of Faraday and Einstein to show that they exploited personal analogies in their creative thinking. If the problem concerns the development of a tire material for use on vehicles in a desert terrain, a person may imagine himself as a tire traveling at 50 mph over hot sand. The exploration of personal analogies enables the individual to get inside the problem and provides a novel perspective from which a solution may be generated. Second, *direct analogy* involves the direct comparison of parallel observations or systems. We have already indicated Gordon's emphasis on biological analogies. He quotes several examples, including Kettering's addition of a red dye (tetraethyl lead) to kerosene to make it vaporize quickly, after observing that a wild flower (the trailing arbutus) blooms early in spring because its red-backed leaves more readily absorb heat. Twiss[32] quotes the use of a direct analogy with the homing pigeon to develop a method for a novice bricklayer to build a straight brick wall. Third, *symbolic analogy* is impersonal and objective and aethestically pleasing rather than technically accurate. Gordon says that James Clark Maxwell used it extensively, and quotes a synetics group using the symbolic analogy of the Indian rope trick to generate the solution to the problem of designing a jacking mechanism which fits into a four-inch cube, but would extend to three feet and support four tons. Fourth, *fantasy analogy*. Freud viewed creativity as a perpetuation of childhood fantasy, and Gordon views fantasy analogies as applying Freud's wish fulfillment theory of art to technical problem solving. It is similar to the wildest-idea approach in brainstorming, in which the group is invited to develop and explore any wild fantasy, which may incorporate the judicious suspension of the laws of nature and common sense to the problem. Gordon quotes an example applied to the invention of a vapor-proof enclosure for spacesuits. The group developed fantasy analogies of trained insects closing and opening the seal to military-style drill orders, and then a flea sewing a spider's thread in and out through closure holes! These fantasies led to a practical solution involving a wire pulling together overlapping steel spiral-springs embedded in the two rubber surfaces which had to be "stitched" together.

The organizer encourages the group to explore potentially fruitful analogies by diverting its thinking (or taking a vacation) from the problem to another (apparently) unrelated subject. This subject will be conceptually distant from the problem, but should, in the opinion of the organizer, offer scope for the development of fruitful analogies. Gordon recommends that a fantasy analogy is best tried first (presumably biologically based, given his preference for biological analogies), but the choice is probably best left to the discretion and judgment of the organizer. Further analogies may be introduced at appropriate stages in the discussion until an acceptable solution is generated. The whole session may typically take 60 to 90 minutes before a solution is suggested which offers practical promise. This solution (or solutions) may then be explored further, possibly using the other techniques we have already described.

Lateral Thinking

This approach has been developed by Edward de Bono and described by him in several books.[33,34] His concepts of *lateral* versus *vertical* thinking roughly correspond to divergent and convergent thinking introduced earlier. Vertical thinking is conventional logical thinking in which knowledge progresses in a sequence of analytical and logical steps. De Bono uses the analogy of water flowing down well-defined channels for this thinking, noting that the more water flows down these channels, the more it will continue to do so. In contrast, lateral thinking focuses on overall patterns and interrelationships between the elements of a problem and seeks to discover new patterns of association, using nonlogical approaches, which will lead to a problem solution. De Bono argues that deliberately departing from the conventional logical or vertical approach to the problem by the use of lateral thinking is equivalent to the damming-up of existing water flow channels and exploring the new flow patterns thereby generated to see if they offer a more fruitful irrigation catchment. As de Bono expresses it—in vertical thinking, logic is in control of the mind, whereas in lateral thinking, the mind is in control of logic. It is clear that the lateral thinking process is broadly similar to the synetics process and Gordon's metaphorical thinking in seeking out creative bisociations.

It follows that de Bono's approach to creativity stimulation is broadly similar to those brainstorming and synetics, and may be used in similar group settings. He stresses that first one should critically examine the ideas associated with the vertical thinking approach to the problem. One should identify:

1. *Dominant ideas and boundaries* which unduly constrain and influence solution approaches.
2. *Tethering factors and assumptions* which are implicitly included (without reevaluation) in every solution approach.
3. *Polarizing tendencies* which discourage the adoption of a position between two extremes.

This critical process can free the mind from the thought habits and preconceptions of previous approaches and set the stage for lateral thinking. As with synetics, specific techniques for stimulating lateral thinking are recommended:

1. *The Intermediate Impossible.* This technique seeks to break logical constraints and resembles the wildest idea and fantasy analogy in brainstorming and synetics. Although rationally untenable, the intermediate impossible idea provides a start for potentially fruitful explorations. Whitfield considers the problem of making road transport more efficient, and starts from the intermediate impossible idea of requiring road vehicles to be stationary.[35] The idea of stationary road vehicles leads logically to moving roadways onto which vehicles park—an idea which has yet to be implemented. However, it also leads to practical solutions which are in widespread use—motorail services (in which cars are carried on flat rail wagons behind passenger trains), peak-hour traffic restrictions (which constrain the directions of traffic flow by time of day), park-and-ride schemes (whereby commuters park their cars at suburban stations and complete their journeys by mass transit) or tele/video-conferencing techniques (which obviate the necessity to travel from A to B anyway).

2. *Random Juxtaposition.* This resembles the vacation approach in synetics, in which an apparently unrelated concept is introduced in the hope that it will trigger a sequence of associations which will lead to a problem solution. Whitfield considers the problem of providing a cheap domestic central-heating fuel.[35] He randomly selects the word "nonsense" which evokes the synonyms absurd, garbage, meaningless, trash, rubbish, waste. Garbage, trash, and rubbish? Why not develop a domestic furnace which will burn household garbage and trash? Waste? Why not control the decomposition of plant, animal, and human organic waste to convert it into fuel? Whitfield reports that someone is running a car on chicken manure and that sewer gas is being used for street lighting.

Clearly, the brainstorming, synetics, and lateral thinking approaches all stimulate successive periods of group convergent and divergent thinking, as illustrated in Figure A7.3. Cycles of convergent and divergent thinking may be repeated several times using these techniques (and perhaps others) until a solution is found.

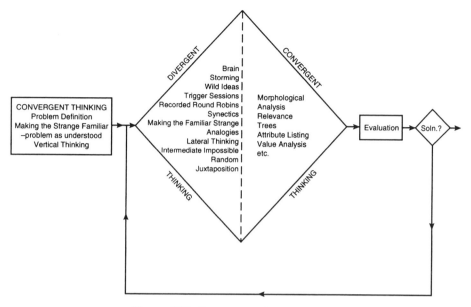

Figure A7.3. Convergent–divergent cycles.

Group creativity stimulating techniques, increasingly supported by PC software packages, are now used quite extensively in business and have been incorporated into overall systems for stimulating innovation and new product development in companies. One such system is described by Carson.[36]

FURTHER READING

1. Margaret Boden, *The Creative Mind.*   London: Sphere Books, 1992.
2. Arthur Koestler, *The Act of Creation.*   London: Picador, 1975.
3. Tudor Rickards, *Creativity at Work.*   Brookfield, VT: Gower Publishing Company.

REFERENCES

1. E. T. Smith, "Are You Creative?" *Business Week,* September 30 (1985).
2. A. Koestler, *The Sleepwalkers.*   Harmondsworth, Middlesex: Penguin Books, 1964.
3. A. Koestler, *The Act of Creation.*   London: Hutchinson, 1969.
4. W. Kohler, *The Mentality of Apes.*   London: Pelican Books, 1957, p. 35.
5. E. I. Green, "Creative Thinking in Scientific Work." In J. R. Bright (Ed.), *Research, Development and Technological Innovation: An Introduction.* Homewood, IL: Richard D. Irwin, 1964, pp. 118–128.
6. K. Lorenz, *Behind the Mirror.*   New York: Harcourt Brace Jovanovich, 1977.
7. M. Boden, *The Creative Mind.*   London: Sphere Books, 1992.

8. D. Hofstadter, *Nature (London)* **349,** (378) (1991).

9. L. Hudson, *Contrary Imaginations.* Harmondsworth Middlesex: Penguin Books, 1966.

10. T. S. Kuhn, "The Essential Tension: Tradition and Innovation in Scientific Research." In *Scientific Creativity: Its Recognition and Development.* New York: Wiley, 1963, Chapter 28.

11. J. W. Getzel and P. W. Jackson, *Creativity and Intelligence: Explorations with Gifted Students.* New York: Wiley, 1962.

12. E. P. Torrance and E. Hansen, "The Questing-Asking Behaviour of Highly Creative-Less Creative Basic Business Teachers." *Psychological Reports* **17,** 815–818 (1965).

13. D. MacKinnon, "Personality and the Realisation of Creative Potential." *American Psychologist* **20**(4), 273–281 (1965).

14. A. Roe, *The Making of a Scientist.* New York: Dodd, Mead, 1953.

15. J. A. Chambers, "Relating Personality and Biographical Factors to Scientific Creativity." *Psychological Monographs: General and Applied* **78**(7) (No. 584) (1964).

16. F. Barron, "The Disposition Towards Originality." *Journal of Abnormal Social Psychology* **51**(3), 478–485 (1955).

17. M. K. Badaway, "How to Prevent Creativity Mismanagement." *Research Management* **29**(4), 28–35 (1986).

18. S. M. Parmenter and J. D. Garber, "Creative Scientists Rate Creativity Factors." *Research Management* **14**(6), 478–485 (1955).

19. A. Gerstenfeld, *Effective Management of R&D.* Reading, MA: Addison-Wesley, 1970, p. 83 et seq.

20. N. Kaplan, "The Relation of Creativity to Sociological Variables in Research Organizations." In C. W. Taylor and F. Barron (Eds.), *Scientific Creativity: Its Recognition and Development.* New York: Wiley, 1953, pp. 44–52.

21. T. Burns and C. M. Stalker, *The Management of Innovation.* London: Tavistock Publications, 1961.

22. F. A. Johne, "How Experienced Product Innovators Organize." *Journal of Product Innovation* **1**(4), 210–223 (1984).

23. H. I. Ansoff and J. M. Stewart, "Strategies for Technology Based Business." *Harvard Business Review* **45**(6), 71–83 (1967).

24. M. E. Ginn, "Creativity Management: Systems and Contingencies from a Literature Review." *IEEE Transactions on Engineering Management* **33**(2), 96–101 (1986).

25. M. E. Ginn, "Creativity Management: Facilitators/Barriers in New Technology Development." Unpublished presentation. Manchester: University of Manchester, Manchester Business School Silver Anniversary Conference on R&D, Entrepreneurship and Innovation, July 1990.

26. G. M. Prince, *The Practice of Creativity.* New York: Harper & Row, 1970.

27. T. Rickards, *Problem Solving Through Creative Analysis.* Epping, Essex: Gower Press, 1974.

28. P. R. Whitfield, *Creativity in Industry.* Harmondsworth, Middlesex; Penguin Books, 1975.

29. C. H. Kepner and B. B. Tregoe, *The Rational Manager.* New York: McGraw-Hill, 1965.

30. A. Osborne, *Applied Imagination.*   New York; Scribner's, 1957.

31. W. J. J. Gordon, *Synectics: The Development of Creative Capacity.*   New York: Collier-Macmillan, 1961.

32. B. C. Twiss, *Managing Technological Innovation,* 2nd ed.   London: Longman Group 1980, pp. 86–87.

33. E. de Bono, *Lateral Thinking for Management.*   New York: McGraw-Hill, 1971.

34. E. de Bono, *The Use of Lateral Thinking.*   Harmondsworth, Middlesex; Penguin Books, 1971.

35. J. Carson and T. Rickards, *Industrial New Product Development.*   Epping, Essex: Gower Press, 1979.

36. J. W. Carson, *Innovation: A Battleplan for the 1990s.*   Farnborough, UK: Gower Press, 1990.

———8
PROJECT MANAGEMENT

The moral of the story is this: The inherent preferences of organizations are clarity, certainty, and perfection. The inherent nature of human relationships involves ambiguity, uncertainty, and imperfection. How one honors, balances, and integrates the needs of both is the real trick of management.

A. J. GAMBINO AND M. GARTENBERG,
INDUSTRIAL R&D MANAGEMENT

8.1   INTRODUCTION

In Chapter 6 and its Appendix we examined some approaches and techniques for evaluating and selecting R&D projects and here we turn our attention to the management of a project once it has been selected.

8.2   PROJECT PROGRESS AS AN UNCERTAINTY REDUCTION PROCESS

We may view the successive phases of the innovation process as a progressive reduction of technical, commercial, and marketing uncertainties, using a variation of the technology–market matrix suggested by Scott and Szany[1], shown in Figure 8.1. In the early phases of the innovation process, technical, commercial, and marketing uncertainties will be high, corresponding to a location in the upper right-hand cell of the matrix in Figure 8.1. As we move through the innovation evaluation described in Chapter 6, uncertainty is reduced, so we approach the bottom left-hand cell which corresponds to a commercially successful innovation when these uncertainties have been eliminated.

Thus, the early state of an R&D project evaluation corresponds to the upper-right cell in this matrix, and only fairly crude subjective estimates of projective outcomes can be made. We therefore initially examine the sources and magnitudes of estimation errors in R&D project evaluations.

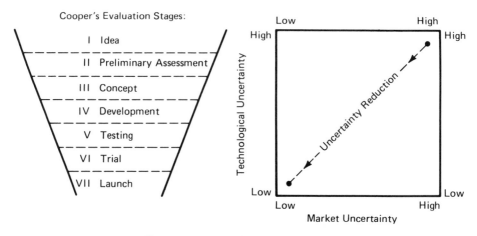

Figure 8.1. Uncertainty reduction process.

8.3   ESTIMATION PROCEDURES IN PROJECT EVALUATIONS

The innovation process consists of the progressive reduction of considerable uncertainty (particularly if the innovation proposes significant technological novelty), so that early estimates of outcomes must and should be subject to considerable errors. This is especially true in the early research (which really is characterized by ignorance rather than uncertainty) and primary development phases of the innovation process. Here, detailed estimations are of less consequence, if projects are subjected to peer group evaluations and selections and require a relatively small commitment of corporate resources. However, once a project moves into more advanced development, the increasing commitment of corporate resources coupled with the increasing competition between projects dictate that estimation procedure should facilitate realistic evaluations and comparisons between projects to be made. Often, estimation errors are large compared with the differences in yields between competing projects, so quite small differences in project estimates, which are well within the error ranges, could produce marked differences in project selections. For example, Meadows,[2] in one of the few reported studies of estimation errors, compared two projects and calculated that estimation errors of only 10% would have yielded a higher return on a project at present making a loss, as compared to one yielding a 230% ROI!

This being so, the emphasis in estimation procedures should be on the development of *consistent measures* of the most likely, pessimistic, and optimistic values of parameters between projects, rather than the pursuit of spurious accuracy. If consistent estimation methods are used for all projects, it should be possible to set up overall rankings with some confidence. The robustness of these rankings can then be subjected to some simple sensitivity analysis to see how they are maintained in the face of the variations of outcomes which may obtain in practice.

Identifying the Project Elements

Later in this chapter (Section 8.6) we shall cite network analytic techniques which are useful aids to project management. Both the application of these techniques and the requirements for consistency measures between projects dictated that a detailed identification of project elements is required. Obviously, also, this identification is a necessary requirement for the implementation of evaluation, assessment, and selection procedures discussed in Chapter 6 and Appendix. The following framework for identifying these elements is based upon Monteith[3]:

1. Define the project.
2. Divide the work into activities or tasks which may be independent or dependent upon the completion of predecessor tasks.
3. Decide upon all courses of action by which each task can be completed. This will identify the problems to be solved before the task can be completed.
4. Identify the alternative approaches which may be pursued, either in series or parallel, to solve these problems (note that the relevance tree techniques discussed in Chapter 4, Appendix 4.1, Section 4.1.14 may be of use here).
5. Select the most promising approaches for solving each of the problems, specify time and/or expenditure limits on these approaches, and the alternative approaches which will be used should the first ones selected be unsuccessful. That is, clarify the strategy for implementing alternative approaches.
6. Allocate priority to courses of action which are likely to yield the best technical and cost outcomes.
7. Delay the performance of any independent tasks which will prove to be worthless if another task or tasks proves to be infeasible.

Estimation Biases

The logical application of the above analysis should identify most-likely, optimistic, and pessimistic estimates of the cost time and resource (labor, equipment, etc.) requirements for a project, and should include appropriate overheads and contingency cost allowances. If applied consistently to all projects, it should enable realistic comparisons to be made. Although the framework is common-sensical, it would be utopian to expect unbiased estimates to be made in practice. There are two major sources for estimation errors: optimistic biases which individuals introduce into an evaluation to increase the chances of a project receiving continued support, and unconscious errors produced by individuals in their personal subjective judgments of outcomes.

Given that R&D projects are competing for funding, it is realistic to expect optimistic biasing of estimates by staff seeking support for "their" project. As Freeman expresses it, *"the social context of project 'estimation' is a process of political advocacy and clash of interest groups rather than a sober assessment of measurable probabilities."*[4] Realistically, we must expect such optimistic biasing to be present, and the few empirical studies in errors support this expectation. Thomas found this to be true in two scientific instruments firms he studied.[5] What tends to make matters worse is the way in which researchers compound cost errors. In a project which is commercially questionable but technically appealing, the estimator may reduce his cost estimations in two ways.[6] First, by removing elements from the project which are desirable rather than essential. Once the project has been funded, these elements may be reinstated and treated as cost overruns. Second, by reducing contingency allowances which, by definition, cannot be specified in detail. Because of the sunk costs incurred and the political and psychological trauma of terminating a project partially completed, the expenditure of these extra costs can be rationally justified at this later stage. Such actions are commonplace in some industries, and occasionally lead to disastrous cost overruns. It is reasonable to conjecture that the financial collapse of Rolls Royce in Britain in the early 1970s can be at least partially attributable to optimistic estimates of costs of developing the new technology in the RB211 engine.

Subjective Probability Estimates

Turning to unconscious errors, approaches for eliciting subjective probability estimates in management in general have been developed over the past three decades[7] and applied to R&D project evaluation.[8–10] Rubenstein and Schroder discuss some of the behavioral factors which may affect such estimates.[11] Within the R&D context, most studies focus on subjective estimates of the overall probability of technical success and have yet to be regularly applied in practice. Therefore, we concentrate our attention on empirical studies of estimation errors in R&D projects.

Despite the importance of the topic, there have been relatively few studies reported on the relationship between estimates and actual outcomes in projects. The most interesting one was performed by Norris.[12] Time overruns exceeded cost overruns in most of the projects he examined, from which he deduced one important conclusion. Time overruns often were due to delays in the starts of activities requiring laboratory-wide resources shared with other projects. Since an R&D function can be viewed as a job shop, the situation is analogous to job shop scheduling in manufacturing—to seek maximum machine and labor utilization more jobs are scheduled than the shop can handle, leading to work-in-progress (WIP) queues at successive stages in the manufacturing process and delayed completion dates for all job. Analogous projects-in-progress (PIP) queueing causes analogous delayed completion dates for R&D projects. As we shall see shortly, time delays in launching new products cause much greater reduction in their profitabilities than cost overruns, so firms are placing increasing emphasis on reducing new product development times. PIP queues must be reduced (or preferably eliminated) if these times are to be reduced.

Estimating Risk

Several project selection techniques described in the Chapter 6, Appendix incorporate risk terms which must be quantified as part of the overall estimation process.

Abetti and Stuart describe a useful approach to risk estimation.[13] They reason that there are three independent sources of risk (each measured on a 1–4 increasing risk scale) in a new product, plus its *innovation uniqueness* (measured on a 5–1 decreasing risk scale):

1. *Market risk* or the extent to which the customers and distributors of the product are new to the company.
2. *Functional risk* or the extent to which new functionality in the product: a) represents an incremental or major change, and b) is a new functionality in the market or a copy of a competitors product functions (that is, is a me-too product).
3. *Technology risk* or the extent to which it: a) incorporates an incremental engineering change or represents an application at the state of the art, and b) is new technology in the market or a copy of a competitor's technology.

Innovation uniqueness ranging from an incremental improvement to radical new products which render existing ones obsolete.

Risk should be estimated for each factor separately (remembering that me-too products usually fail, so can carry a high risk) and summed, with the uniqueness factor, to yield a score between 4 and 17. The lower the score, the lower the risk.

Three further points should be noted. First, the risks should be assessed after concept testing (Stage III in Figures 6.3 and 8.1) when significant technical, functional, and marketing features are known, but substantial resources have yet to be committed to the project. However, depending on the project, the assessment timings may vary with each, but they can be updated as the project progresses. Second, recalling the results of SAPPHO, NEWPROD, etc., reported in Chapter 6, innovation uniqueness should be perceived through the eyes of the user and the market. Third, although they are summed together to yield an overall score, again recalling the Chapter 6 results, the risk factors should be viewed as links in a chain, with its overall strength dependent upon the strength of the weakest link. Thus, a project that is very low risk on two factors should be rejected if it represents a high risk on the third factor.

8.4 PROJECT CONTROL TECHNIQUES

The Chapter 6 Appendix discusses the contribution of OR techniques to project selection and concludes that there are limited actual applications of sophisticated mathematical modeling to that decision problem. When we turn to project management and control, however, the situation is rather different. Network analysis techniques have been used quite extensively as decision aids to R&D project managers. Gantt or bar charts may be used, and project management is particularly associated with PERT/CPM, one of the most popular and effective of the battery of management aids which have been developed since World War II. PERT (or Program Evaluation and Review Technique) and CPM (Critical Path Method) use essentially simple network diagrams to represent complex projects and identify the timetable or schedule and critical-path which will minimize their completion time. CPM was developed by DuPont to schedule the maintenance of complex chemical plant, but PERT *was developed specifically to manage a complex R&D project.* It was originally developed to manage the Polaris missile development program, and it is claimed that it reduced the completion time for that project by two years. Since the original introduction of these techniques in the late 1950s, the scope of application to R&D project management has expanded quite considerably.

In its original and simplest form, PERT represents the logical interrelationships between the component activities or tasks in a project in a network, together with estimates of their "most-likely," "pessimistic," and "optimistic" durations. It then identifies the schedule which minimizes the expected completion time for the project. Such a network should be prepared within the framework suggested in Section 8.4 and become an integral part of the project management. In fact, since the 1960s, many U.S. government agencies (and some in other countries) have made the development of a PERT network mandatory for companies performing R&D contract work for them. Therefore, there is a considerable body of expertise on the use of PERT in R&D project management. In this simplest form, PERT may be used to monitor project progress to ensure that all tasks are being completed within schedule, or if the completions of tasks are delayed, the impact of these delays on the overall completion time of the project. As such, it is a valuable aid to project management, but it is possibly even more valuable in its extensions. In its earliest forms, it is most readily applicable to the tertiary design and development and prototype trial phases of the innovation process, but more recent extensions have extended its scope of application to earlier and later phases.[14,15]

The early phases are characterized by high levels of technical and commercial uncertainties and outcomes. The PERT network can handle uncertainties in activity duration times through the use of three time estimates, but cannot handle uncertainties in outcomes and the possibility that one or more of a number of alternative development paths may be followed. This is because it is a *deterministic* network in which no activity in the network can be started until all its predecessor activities have been completed and overall project completion is defined as the completion of the terminal activity or activities. To accommodate the possibility of alternative development paths, *probabilistic* or *stochastic* networks are required. These networks incorporate probabilistic nodes (or events) from which each of two of more mutually exclusive following branches (or activities) has a certain probability of being taken. The sum of these probabilities adds up to one, so one path is certain to be followed. The most widely used probabilistic network technique is GERT (Graphical Evaluation and Review Technique) which may incorporate both deterministic and probabilistic branching at nodes.[16] It can also incorporate recurring activities and cost information. More recently, another probabilistic network technique known as VERT (Venture Evaluation and Review Technique) has been developed which incorporates performance as well as time, cost, and probability measures.[17] It also incorporates the facility to calculate critical paths based upon time (as in PERT), cost, or performance considerations. GERT and VERT networks may be initially drawn up in an earlier development phase of the innovation process and be incorporated into repeated computer simulations to simulate a sample of alternative project outcomes. GERT simulations may be updated and used as the project progresses to simulate the costs of alternative manufacturing approaches. VERT simulations may be updated and used as it progresses to simulate both expected costs and financial returns from the project. Thus, a realistic and continually updated VERT network may be used as a tool in Cooper's ongoing evaluation process described in Chapter 6 (Section 6.5). Its approach is similar to the risk analysis models cited in the Chapter 6, Appendix.

Probabilistic network approaches offer considerable promise in project management, but, as yet, have not enjoyed the widespread application of deterministic approaches. We therefore turn to a discussion of the extensions of the deterministic approaches which have been applied quite widely over the past 10–20 years. The original approach was labeled PERT/TIME because only the duration times of activities were used, but subsequently, activity costs were incorporated into PERT/COST systems. Two pairs of duration times and costs may be estimated for some activities:

1. The NORMAL times and costs
2. The estimated duration times of some activities may be reduced by "crashing"—that is, by deploying more (or more expensive) resources on their completion. For example, workshop staff may be paid overtime rates to accelerate the assembly of a prototype item of equipment or faster and more expensive testing facilities may be used in an experiment. In such cases, CRASH times and costs are estimated which are less than the normal time and greater than the normal cost estimates, respectively.

This information may then be used to aid the control of the project. Project progress may be monitored in terms of "milestones." Successive milestones are typically separated by one to six month intervals. They are usually defined as groups of activities that should be completed by the given milestone date. Actual progress may then be compared with planned progress using the PERT/budgeting approaches described in standard texts, or a visually appealing progress chart approach suggested by Pearson.[18] As the project progresses, these approaches identify cost and/or time overruns. Time overruns may be eliminated or reduced by crashing selected uncompleted activities at an extra cost. By identifying the duration time–cost tradeoffs between activities, PERT/COST networks can be employed to identify alternative crash schedules by computer. The schedule chosen depends upon subjective judgments of the overall tradeoff of the disbenefits of time delays versus increased costs in particular situations. The overall outcome will be a time overrun or a cost overrun (or some combination of both), depending on the choice made. As we shall see in Section 8.6, increasing emphasis is being placed on reducing completion times, and delays can be very costly so, in general, cost overruns are preferable to time overruns.

In Section 8.4 we saw that time overruns are frequently caused by delays in gaining access to facilities shared with other users. Typically, certain laboratory resources (including labor and equipment) are shared by a number of projects, and such delays are to be expected if no attempt is made to schedule the overall assignment of these resources among projects. PERT/RAMPS (Resource and Multiple Project Scheduling) techniques have been developed to handle such situations. The individual project PERT networks may be aggregated into an overall multi-project PERT network, and alternative activity–resource schedules evaluated. As when considering activity crashing, even quite small networks can generate quite large numbers of alternative activity–resource loading and sequencing possibilities. Therefore, PERT/RAMPS software packages have been developed to identify and evaluate alternative activities–resources loading schedules by computer to determine the "best" overall allocation and scheduling of resources to projects.

Once a project progresses to the pilot production phase of the innovation process, there is often a requirement to manufacture, say, a few hundred units of a complex assembled product. That is, one is moving from scheduling and controlling an individual project to scheduling and controlling an ongoing assembly operation. PERT/LOB (Line of Balance) is particularly useful for scheduling such small-scale production. PERT/COST, PERT/RAMPS, and PERT/LOB are all described in texts on PERT methods.[19]

Provided they are conscientiously developed and updated at a realistic level of detail *by the project staff involved* in a participative management framework, networking techniques can provide a useful basis for planning, scheduling, and controlling activities in most phases of the innovation process. Note that they may readily incorporate the market planning aspects discussed in Chapter 6. Such networks can be prepared and implemented by the staff involved as an integral part of the financial planning and budgeting process discussed earlier in the chapter. PERT-based methods are conceptually straightforward and available in numerous inexpensive Project Management software packages, which may be used on desktop and notebook PCs.[20] GERT and VERT methods are more sophisticated mathematically, but most scientists and engineers should have little difficulty in understanding and applying them.

8.5 FAST INNOVATION: REDUCING NEW PRODUCT DEVELOPMENT TIMES

In Chapter 2 (Section 2.3) we indicated that there appear to be contrasting changes occurring in times taken to innovate or develop new products. More demanding regulative requirements (now affected by increasing "green" concerns, as discussed in Chapter 4, Appendix 4.2) appear to be increasing the time it takes to develop and launch new products, notably in the pharmaceutical, chemicals, and agricultural industries. In contrast, in the microelectronics and automobile industries, for instance, new product development times are shortening for several reasons. First, IC technology is still in the steep performance maximizing part of its overall S-curve so that the rapid introduction of successive generations of IC-based products continues. Second, these products themselves enhance the power of computer-aided design and manufacturing (CAD/CAM) which, in turn, enables the new product development process to be accelerated. For example, Hewlett-Packard has used an expert system to analyze a new product design and suggest changes to improve its manufacturability, which yielded 84% reductions in both failure rates and manufacturing times for 36 products.[21] Computer simulation techniques have been used quite extensively in the aerospace industry to accelerate new product development times. Third, Japanese companies have demonstrated that process may be accelerated by improved project management practices. For example, a Boston Consulting Group (BCG) study cites the example of a Western marine transmission manufacturer that took 36 to 48 months to develop a new product. In contrast, its Japanese affiliate could develop an equivalent product in 12 to 16 months at a 30% lower cost.[22,23] Clark and Fujimoto[24] also reported shorter new product development times in Japanese automobile companies, as compared with their U.S. and European counterparts. Moreover, as pointed out by Werner Niefer (Chairman of the Managing Board of Mercedes-Benz), these practices are also being adopted by companies in the newly industrialized countries (NICs) such as Hong Kong, Singapore, South Korea, and Taiwan as they seek to emulate the Japanese example.[25]

Although important, cost reduction may well not be the most important benefit of fast innovation. First-to-market often does enhance long-run market share and profit. Miller Lite was launched ahead of Schlitz Light and Anheuser-Busch Natural Light by one and two years, respectively—a fact that probably enabled it to dominate the light beer market.[26] A recent McKinsey & Co. study supports this observation. It showed that a product that stays within its budgeted costs but is six months late to market will forgo one-third of its potential profit over its lifetime. In contrast, one which incurs actual cost that are 50% higher than those budgeted but is launched on time, only loses 4% of lifetime profit.[27] This observation relates to the debate on offensive versus defensive strategies in Chapter 4 (Section 4.9), suggesting that, given that the company has access to the complementary assets needed, an offensive strategy is preferable. However, it also implies a need for a sustained fast innovation strategy, especially in industrial markets, where customers select a product based upon its performance, rather than its associated advertising and brand loyalty as in the beer and other consumer markets. If the offensive innovator wishes to sustain the advantage secured by being first-to-market, it should follow with a succession of incremental product improvements as quickly as possible to overwhelm any defensive innovator's fast-follower responses. This strategy of sustained offensive innovation has been followed by Japanese carmakers to take market share from their North American and European rivals and, within Japan and elsewhere, by Honda in securing leadership from Yamaha in the global motorcycle market.[22,23] Given the benefits of fast innovation, we now discuss how it is achieved.

Relay Races versus Rugby: Concurrent Engineering

In Chapter 2, for simplicity of exposition, the innovation process was primarily represented as a linear sequence of activities through research–development–production–marketing—a sequence which quite accurately reflected the practice of innovation in most, if not all, companies. That is, an innovation was first conceived as an invention in research and progressed as an innovation through successive technology transfers to development, production, and marketing functions. That is, at each stage the technology was transferred from one function to the next one, without any prior warning or discussion between the two functions. As it is sometimes expressed: *"When they had finished with it, Research tossed the technology over the wall to Development. When they had finished with it, Development tossed the technology over the wall to Production, and finally Production to Marketing."* Although this is an oversimplification of the truth (since, apart from anything else, other subfunctions such as design and pilot/prototype production are involved), some of us know from personal experience that it is by no means an inaccurate representation of the actuality that has applied in many firms until recently. In fact, it still applies in some firms today.

The basic principle of fast innovation management can be grasped by a simple analogy. The above innovation process resembles a team running a relay race. Each member runs their part of the race, then passes the baton onto their successor, with the last one completing the race. If we replace team member by function, baton by technology, and completion by market launch, it accurately represents the above *linear* innovation process. In contrast, fast innovation resembles rugby football or rugger. Readers familiar with rugger will know that forward passes are illegal. Therefore, the ball is best advanced down the field by its lateral interpassing between players as they run forward, often simultaneously interweaving together, evading, or brushing off the tackles of their opponents, until they reach the latter's goal line. This way, the team can score a touchdown without the ball being passed forward. If we replace player by function and ball by technology, it accurately represents the *fast* innovation process. That is, all the functions and subfunctions whose participation is required to complete the innovation process are involved in it as soon as possible, so they are involved *in parallel* or *concurrently,* rather than in sequence. It is sometimes described as concurrent, parallel, or simultaneous engineering because, as the new product is developed, individuals from all of the functions involved exchange ideas and suggestions, like rugger players interpassing the ball. The difference between sequential and fast innovation or concurrent engineering is shown in Figure 8.2.

In a rugger passing play as described above, not all players are initially involved. Typically, such a play is begun by the scrum-half who passes the ball to the fly-half before the three quarters and other players participate. Also typically, all the functions in Figure 8.2*b* may not be involved form the very beginning. A new product concept may emerge from initially informal discussions between R&D and marketing people. The key feature of the concurrent engineering approach is that all pertinent functions are incorporated into the New Product Development (NPD) Team *as soon as possible* in the innovation process, rather than as late as possible as in sequential innovation. This means that there is a large amount of overlapping participation among their members, which facilitates technology transfer. In addition, the timings of their involvements, as well as the titles and roles of the functions involved, will be technology-dependent. The titles in Figure 8.2*b* were used by

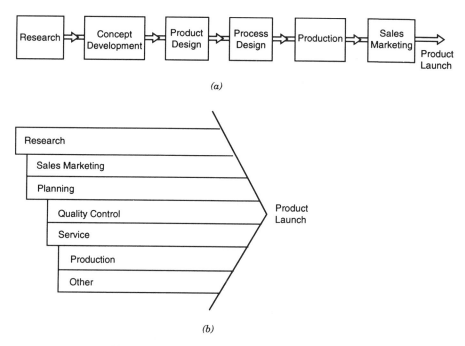

Figure 8.2. Sequential versus parallel innovation.

Nonaka in a study of fast innovation in seven Japanese companies (Fuji-Xerox, Honda, NEC, Epson, Canon, Mazda, and Matsushita).[29] The size of the NPD teams in the sample varied from 11 to 30 members. Although these examples are in Japan, which has espoused fast innovation, similar approaches are being successfully used in North America and Europe. For example, Ford used a concurrent engineering approach in developing its Taurus and Sable models.

The creation of a multifunctional NPD team is only the first step in a fast innovation approach, and several writers have discussed its other requirements. They can be summarized as follows:

1. *Top Management Support.* As in many other corporate endeavors, recognition by top management of the concurrent engineering approach is vital to ensure that the NPD team has timely access to needed resources. One survey in U.S. companies found that 42% of the respondents cited "lack of senior management support" and "lack of resources" as causes of NPD delays.[27]

2. *Collocation Plus Reliance on Outside Resources.* All members of the NPD team should be collocated, but also adjured to engage in networking with outside resource persons, particularly prospective customers, to clarify product definition as quickly as possible. The above-cited study found that *"poor definition of product"* with a 71% respondent citation was the leading cause of product delays. Hardly surprisingly, the second most cited cause was *"technological uncertainty,"* which

suggests that the NPD team should see its current and potential suppliers as technology, as well as component, sources. Both customers and suppliers should be viewed as associate members of the NPD team. If externally acquired technology looks as if it will do a quicker but equally effective job as its internally generated counterpart, then it should be used. The NIH syndrome should be a definite taboo in the fast innovation team! These considerations relate to the critical functions roles discussed in Chapter 7.

3. *Project Championship and Management.* A single NPD team will probably be one of several pursuing fast innovations and competing for corporate resources. The project champion must possess the diplomatic and political skills to secure resources, as and when needed, for the team. The project champion or manager (who may not be the same person, as noted in the previous chapter) should also have the participative style and the conciliatory skills needed to resolve the intrateam conflicts that often arise in multifunctional groups.

4. *Organizational Structure.* The relative merits of alternative structures (functional, matrix, and project) for implementing the innovation process were discussed in Chapter 7 and will be reconsidered in Chapter 9, so only brief comments are made here. Regardless of the structure used, each team may face competition for corporate resources from functional units as well as from its peers. This implies that functional units and NPD teams should report to a common higher-level manager, who will be required to reconcile competing demands within an overall priority ranking of projects (based upon methods discussed in the Chapter 6, Appendix) and functional needs. This role may be performed by a Chief Technology Officer, perhaps responsible for the IMF of the organization (Chapter 4, Section 4.2).[30]

5. *Fast and Flexible Innovations Strategy.* Honda's approach to capturing leadership of the motorcycle market from Yamaha illustrates the fast and flexible innovations strategy. That is, the company should rapidly follow one fast innovation with others, which provide a succession of incremental improvements as well as variations to satisfy a variety of market needs. Note that robust designs (as discussed in Chapter 6, Section 6.8) provide a sound base for this strategy. Stalk and Hout point out that it is analogous to fast and flexible manufacturing approaches. Frequent incremental product improvement corresponds to small manufacturing lot sizes, and tightly knit concurrent engineering NPD teams correspond to the tightly knit shop floor linkages needed to implement JIT. In fact, the evidence suggests that the factory of the future will combine flexible short-cycle time manufacturing with flexible short-cycle time innovation.

6. *Time Benchmarking.* Since time is the measure of all things and the calendar is king, time benchmarks are set for all activities, based upon those of competitors or technical feasibility. If a choice has to be made between taking longer or increasing costs, costs should be increased.

7. *Minimal Design Changes.* Once a product specification has been defined, it should, as far as possible, remain frozen. Only minimal changes should be made, and these should be fully documented and disseminated within the team and to other interested parties. The original product specification and design changes should

conform to the quality and environmental standards the company is maintaining and improvements it is seeking.

The rugby game analogy and the earliest possible involvement of relevant organizational functions in new projects, implicit in both some of the Chapter 6 materials as well as the above discussions, is now accepted by many firms. Nevertheless, if a project is to be both effectively and expeditiously transferred from an R&D to an operational setting, important technology transfer issues must be addressed. These are discussed in the next chapter.

FURTHER READING

1. P. G. Smith and D. G. Reinertsen, *Developing Products in Half the Time.*   New York: Van Nostrand-Reinhold, 1991.
2. G. Stalk, Jr. and T. M. Hout, *Competing Against Time.*   New York: Free Press, 1990.

REFERENCES

1. D. S. Scott and A. J. Szanyi, "Evaluating a New Venture." In D. S. Scott and R. M. Blair (Eds.), *The Technical Entrepreneur.*   Ontario: Press Porcepic, 1979, pp. 77–78.
2. D. L. Meadows, "Estimate, Accuracy and Project Selection Models in Industrial Research." *Industrial Management Review* **9**(3), 105–121 (1968).
3. G. S. Monteith, *R&D Administration.*   London: Iliffe, 1969, pp. 69–70.
4. C. Freeman, *The Economics of Innovation,* 2nd ed.   London: Francis Pinter, 1982, p. 151.
5. H. Thomas, "Econometric and Decision Analysis: Studies in R&D in the Electronics Industry," Unpublished Ph.D. thesis.   Edinburgh: University of Edinburgh, 1970.
6. B. C. Twiss, *Managing Technological Innovation,* 2nd ed.   London: Longman Group, 1980, p. 130.
7. D. Kahneman, P. Slovic, and A. Tversky, *"Judgement Under Uncertainty, Heuristics and Biases."* Cambridge, UK: Cambridge University Press, 1982.
8. R. J. Ebert, "Methodology for Improving Subjective R&D Estimates." *IEEE Transactions on Engineering Management* **EM-17**(3), 108–116 (1970).
9. W. E. Souder, "The Validity of Subjective Probability of Success Forecasts by R&D Project Managers." *IEEE Transactions on Engineering Management* **EM-16**(1), 35–49 (1969).
10. W. E. Souder, "The Quality of Subjective Probabilities of Technical Success in R&D." *R&D Management* **6**(1), 15–22 (1975).
11. A. Rubenstein and H. Schroder, "Managerial Difference in Assessing Probabilities of Technical Success for R&D Projects." *Management Science* **24**(2), 137–148 (1977).
12. K. P. Norris, "The Accuracy of Project Cost and Duration Estimates in Industrial R&D." *R&D Management* **2**(1), 25–36 (1971).

13. P. A. Abetti and R. A. Stuart, "Evaluating New Product Risk." *Research Technology Management* **31**(3), 40–43 (1988).

14. B. V. Dean and A. K. Chaudhurak, "Project Scheduling—A Critical Review." In B. V. Dean and J. L. Godhar (Eds.), *Management of Research and Innovation.* New York: North-Holland, 1980, pp. 215–233.

15. L. A. Digman and G. I. Dean, "A Framework for Evaluating Network Planning and Control Techniques." *Research Management* **24**(1), 10–17 (1981).

16. L. J. Moore and E. R. Clayton, *GERT Modeling and Simulation: Fundamentals and Applications.* New York: Petrocelli/Charter, 1976.

17. G. L. Moeller and L. A. Digman, "Operational Planning with VERT." *Operations Research* **29**(4), 676–697 (1981).

18. A. W. Pearson, "Planning and Control in Research and Development." *Omega* **18**(6), 573–581 (1990).

19. J. J. Moder and C. P. Phillips, *Project Management with CPM and PERT,* 2nd ed. New York: Van Nostrand-Reinhold, 1970.

20. K. O'Neal, "Project Management Computer Software Buyer's Guide." *Industrial Engineering* **19**(1) (1987).

21. Hewlett Packard Company, "The Promise and Measure of CAD." *I.Co Graphics Seminar,* Milan, Italy *1989.*

22. G. Stalk, Jr. and T. M. Hout, "Competing Against Time." *Research Technology Management* **33**(3), 19–24 (1990).

23. G. Stalk, Jr. and T. M. Hout, *Competing Against Time.* New York: Free Press, 1990.

24. K. B. Clark and T. Fujimoto, *Product Development Performance.* Boston: Harvard Business School Press, 1991.

25. W. Niefer, "Technological Expertise and International Competitiveness: Why New Strategies are Needed." *Siemens Review* 4, 4–9 (1990).

26. J. T. Vesey, "Speed to Market Distinguishes the New Competitors," *Research Technology Management* **34**(6), 33–38 (1991).

27. A. K. Gupta and D. L. Wilemon, "Accelerating the Development of Technology-Based New Products." *California Management Review* **32**(2), 24–44 (1990).

28. R. B. Kennard, "From Experience: Japanese Product Development Process." *Journal of Product Innovation Management* **8**, 184–188 (1991).

29. I. Nonaka, "Redundant, Overlapping Organization: A Japanese Approach to Managing the Innovation Process." *California Management Review* **31**(4), 27–38 (1990).

30. P. S. Adler and K. Ferdows, *California Management Review* **32**(3), 55–62 (1990).

PART IV
THE OPERATIONAL SETTING

In Chapter 9 we will discuss barriers to the transfer of an innovation from an R&D to a production facility, the organizational frameworks for facilitating such transfers, and the role of the "learning curve" in innovation.

____9
TRANSFERRING THE
PROJECT FROM R&D
TO OPERATIONS

*It must be considered that there is nothing more difficult to carry out,
nor more doubtful of success, nor dangerous to handle, than to
initiate a new order of things. For the reformer has enemies in all
those who profit by the old order, and only lukewarm defenders in all
those who would profit by the new order, this lukewarmness arising
partly from fear of their adversaries, who have the laws in their
favour, and partly from the incredulity of mankind, who do not truly
believe in anything new until they have had actual experience of it.*

MACHIAVELLI

In this chapter we will examine the issues and problems faced in the transition from
the R&D to the production phase in the innovation process. In a large offen-
sive/defensive innovative company, this typically involves the transfer of the inventive knowhow from a centralized R&D function to an operating division on production facility which is organizationally and (probably) geographically separate. The
problems endemic to this transfer process lead naturally to the consideration of
alternative organizational formats for the development of technological entrepreneurship and new ventures—an issue which will be discussed in the next three
chapters.

If a successful offensive/defensive innovator has established strong downstream
coupling (Chapter 4, Section 4.6) by using the project evaluation process described
in Chapter 6 (Section 6.5), together with the fast-innovation approach and a project
team staff based upon critical functions criteria (Chapter 7, Section 7.4), this transfer could be comparatively straightforward! In practice, for the reasons discussed in
this chapter, it rarely is. Rather than a one-step transfer from R&D to production
operations, it may also occur in two steps; the first from a central R&D laboratory to
a divisional engineering development facility and the second from that facility to
production. Whichever situation occurs, it is important to recognize that the differences in values and outlooks of the individuals involved can create barriers to the
transfer process, and we begin the chapter by discussing these.

9.1   SPECIFIC PROBLEMS OF TECHNOLOGY TRANSFER

Technology transfer has been researched and discussed extensively in the literature during the last 20 years.[1,2] Much of the research and discussion has focused upon international (particularly from developed to developing countries) and inter-organizational (particular from government and university R&D institutions to private industry) rather than intraorganizational transfers, and it is only the latter which are of specific interest here. Steele suggests the barriers to such transfers are as follows.[3]

Technical Barriers

The transfer of the technical knowhow of the project and the replication of performance and results in a scaled-up production rather than R&D environment is the essential purpose of the exercise, so we discuss technical barriers initially. First of all, what will already be obvious to most experienced scientists, engineers, and technology managers in industry is that technical knowhow cannot be transferred purely on paper. It is virtually impossible to completely document exhaustive detailed, unambiguous, and error-free specifications for a project. Much of the experience and insight built up by solving the problems and overcoming the "bugs" endemic to successful project progression can never be meaningfully documented on paper. That is, much of the knowledge is tacit rather than explicit, and duplication of learning will be required. This duplication will be minimized if the concurrent engineering approach is used. Although technology must be transferred by people, this does not mean that formal paper documentation can be ignored. The maintenance of an as complete as possible historical record of the genesis of the project, together with an up-to-date record of the product specifications, ensures that no explicit knowledge, which could facilitate the speedy solution of future expected problems, is lost.

Even with a multitechnical and multifunctional fast innovation project team, other technical barriers need to be overcome. The process of scaling up product output to the pilot/prototype level and beyond, within acceptable cost limits, almost inevitably involves technical design and performance changes and the modification of the original inventive concept. This process can frequently generate pride of ownership disputes between R&D staff and production engineers in operations. R&D staff inventors are sometimes reluctant to accept the technical compromises required for the large scale manufacture of their product at a price which the market will accept. As one technology manager expressed it to this author in the context of similar transfers: *"The scientist who developed the concept didn't give us much assistance. It was his baby. He didn't want any interference."*[4] A related technical barrier is the attitudinal difference between R&D staff and operations managers towards workability. To the former, an invention or concept works if it can be produced on a laboratory scale of manufacture, probably with a technician in virtually permanent attendance ready to repair or modify it. To the latter, an invention only works when it can be manufactured at the required full-scale output and cost level and is sufficiently robust to operate continuously in its perhaps rugged end-use market. Steele quotes a general manager's definition of working as something that works 24 hours a day, 7 days a week, 365 days a year! There is a considerable difference between these two concepts of workability.

First, the process of scaling up production will probably require considerations of mass production methods, tool-up cost (frequently a major cost element), learning effects and the lower standards of assembly operators performances and skills, quality assurance, the economics of shared components and modules with other related products, short- and long-term reliability, and after-sales service systems, all within acceptable cost constraints. Second, R&D staff sometimes lack insights into the end-use environment for a product. They may fail to recognize that the final user, who may be an unskilled worker, will not have the time, ability, or inclination to take much care in reading results from a piece of equipment. Furthermore, this environment may expose the equipment to considerable abuse. The final design of a product must accommodate these considerations. Such an accommodation must often degrade the technical performance of the invention, to the chagrin of its inventor. Some R&D personnel suffer from a professional hubris which equates scientific competence with manufacturing and commercial competence. The following comments made to the author by technology managers illustrate these points:

> Scientists don't have to worry about marketing a product. They often don't realize that the guy who will have to operate a gadget won't have a Ph.D. They don't know the manufacturing problems of trying to fit a million wires into a saleable package.

> The scientist had devised a perfect instrument which was difficult, if not impossible, for a layman to use. I took 20 minutes to take a reading, which is alright for a scientist, but not for a business that considers time a real cost. The lab model needs considerable re-designing for quicker readings and must be redesigned to manufacture at a lower price. One of the big problems is that the scientist has no market orientation. He is strictly tied up in his scientific toys.

> Also, when they (scientists) do become involved, they are reluctant to be a follower and accept the leadership and direction of the commercial entrepreneur. It seems that if they think themselves scientifically and technically competent, then they believe themselves to be commercially competent.

> I think the scientists and manufacturer should work together to build a commercial product that takes into consideration the scientific principle, design engineering, and quality that the customer wants and at an acceptable price.

The above discuss and quotations reflect differences in perceptions between people in R&D and operations functions, but it is also important to recognize that perceptions differ among different technical disciplines. The technology embodied in most (if not all) products derives from two or more engineering disciplines, so that its development and transfer involves intertechnical communications. That is, communications between, say, electronic and mechanical engineers or hardware and software engineers. Engineers (or scientists) with different professional backgrounds also have different explicit and tacit knowledge bases and mindsets, so they view a problem from different perspectives. If an electronic engineer is transferring information and knowhow to a mechanical engineer, some of it may be inadvertently lost through their differences in backgrounds. It is like seeking directions from someone in both English and the local language when lost in a foreign city. Although communication is apparently possible, because neither is really fluent in the other's language, some of the information exchange is lost. The communication is unknowingly and unwittingly distorted by both parties, so one ends up getting lost for a second time!

This means that, as well as the functional overlaps, technical overlaps may be required in the NPD team of Figure 9.1. That is, each relevant engineering discipline will need to be represented in each of the functional rows to ensure accurate and comprehensive technology transfers. Readers with electronic engineering backgrounds may view the representative of each discipline as a transceiver which transmits and receives signals at a frequency band unique to that discipline. Thus, separate transceivers are required for all the disciplines represented to ensure complete signal transfers. Alternatively, these readers may view arrangements for cross-functional technology transfer as analogous to impedance matching circuits in electronic systems. Although this requirement apparently creates duplications of effort or (in engineering design terms) redundancies, it does facilitate the fast innovation process. Nonaka reports that fast innovation or the *sashimi* approach used by Japanese companies favors such redundancies.[5] Although functions and technical disciplines are fully represented, labor is not rigorously divided between them. Rather, fast innovation involves "a shared division of labor" in which functional and technical staff share information and fluidly interact, as appropriate, throughout the successive phases of the innovation process.

Nontechnical Attitudinal Barriers

The above barriers are all to some degree technical impediments to technology transfer, but there are cultural and value differences between research-oriented and operations-oriented staff which impede transfer. First of all, there is a significant difference in attitudes towards time. As Steele points out, researchers view an event

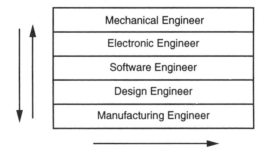

Figure 9.1. The NPD team.

or outcome as the independent variable and time as the dependent variable. That is, a researcher may be able to plan the study of a particular research problem, but he cannot guarantee that the expenditure of a given amount of time will generate the event or outcome of a successful problem solution, discovery, or invention. In contrast, the manager views time as the independent variable and an event or outcome as the dependent variable. He may plan and budget for a given interval of time and then identify and guarantee the sequence of events or outcomes which may be completed in that interval. Furthermore, because a manager typically has to produce results against unremitting time deadlines working from the annual budgeting and reporting cycle downwards, he is time-oriented and seeks to digitalize his time—that is, break it down into discrete units within which he completes specific preplanned tasks. On the other hand, a researcher produces results through discoveries or events which are unpredictable in time. Therefore, he is event-oriented and treats time as a continuum which cannot be divided between different tasks. Because of these orientation differences, researchers almost invariably appear to be poor time performers to managers.

The progression of a project from the R&D phases to the operations phases of the innovation process reflects a progressive reduction in technological uncertainty. Project management and control can thus be increasingly dictated by time-oriented rather than event-oriented considerations. This shift in emphasis is reinforced by the typical dominance of a time orientation in an operations facility. Failure to recognize the shift can erect a barrier to the transfer process. Ideally, projects should not be transferred to operations until technological uncertainties have been resolved to the extent that any further technological bugs can be solved in a time- rather than event-oriented management mode. Ideally, also, R&D staff remaining associated with the project should recognize this shift and change their own orientation accordingly.

Closely related to the event versus time orientation differences is the knowledge versus action orientations of researchers as opposed to managers. Research scientists place high emphasis on the importance of obtaining a theoretical understanding of the mechanism underlying the invention. They argue that the more comprehensive is this understanding, the less is the likelihood of technical bugs occurring later, and if they do occur, the more likely are they to be overcome. They are therefore prone to delaying the progress of the project while a satisfying theoretical explanation for unexpected results is derived, or what may be worse, become readily sidetracked into peripheral investigations which follow from these unexpected results. While this attitude may be desirable in a pure or undirected researcher, it has less appeal when a project has moved into the later phases of the innovation process. In contrast to the researcher, the manager is action-oriented. His job requires choosing between alternative courses of action, in a timely manner, with the best information which is available. If all the information pertinent to the decision is available, all well and good, but it not, he will choose anyway since he must take some action. Once a project has moved to an operations facility, it is likely and, on the whole legitimate, that a more action-oriented approach should be adopted. The technology of the project should be largely proven by then, and the increased costs and time delays associated with too exhaustive investigations will critically affect the profitability and the chances of commercial success of the innovation. However, it should be stressed, that each situation must be treated on its own merits. For example, in the aircraft and pharmaceutical industries, the risks to the consumer from the premature introduction of a product whose safety is subject to doubt may create situations where the knowledge orientation of the researcher takes a moral precedence over the action orientation of the manager.

Third, in private industry, innovations must ultimately be judged by commercial profit/ROI criteria. As a project moves into the operations phases of the innovation process, both decision makers and decisions are increasingly likely to be made against bottom-line and product launch considerations. To most operations managers, this criterion appears morally legitimate (and if does not, they would appear to be in the wrong jobs!). As we discussed in the previous chapter, some researchers view the profit motive with disinterest, arguing that their role as scientists is to generate knowledge for its own sake or for the good of society. These two viewpoints are not mutually antithetical since, following no less an authority than Peter Drucker, we may argue that today's profits are the costs which will finance tomorrow's innovations—an argument which is certainly congruent with our overall approach to corporate technological strategy formulation and implementation. However, some researchers may be either unaware of or disbelieve such arguments and, in any case, are unlikely to value profit as a criterion of success. Therefore, they are likely to measure internal rewards from their contributions to an innovation in terms of technical rather than commercial success.

All these differences in attitude and values may inhibit communication between the various professional groups in technology transfer. On the other hand, a recognition that such factors may inhibit transfers does enable management to design fast and effective transfer mechanisms in line with the earlier discussions.

9.2 MATRIX STRUCTURES

So far in this chapter we have focused on the barriers to transferring innovations from R&D to operations without regard to how the project team (however it is compromised) is incorporated into the operating management structure. We now consider this issue.

In Chapter 7 (Section 7.8) we argued that for offensive/defensive innovators, the research end of the R&D function, particularly nondirected basic research activities, is best performed in an organic organizational climate within a management structure based upon the disciplines appropriate to the technological base of the company. Once the invention moves into the development phases of the innovation process, we argued for the establishment of a team based upon the technological needs of the project, which must cut across functional disciplines in a matrix structure (recall Figure 7.5). The same considerations apply when the project is moved into the operational setting. The project team and its complement now cut across management functions and the requirements of ongoing product lines, and again, a matrix structure would appear to be desirable, as shown in Figure 9.2. As we stated

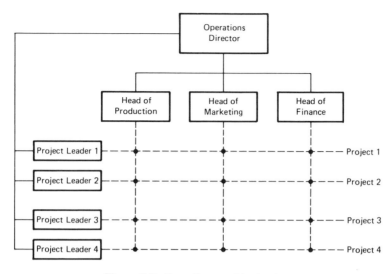

Figure 9.2. Operations matrix structure.

in Chapter 7, the efficiency of the matrix structure is disputed. So as well as discussing matrix structures in operations, here we will continue the discussion begun in Section 7.5 and consider the benefits of matrix structures in general.

The benefits of a matrix structure are perhaps best judged in the context of a compromise between two alternative extremes. One possibility is the pure project organization, where operations are organized as an aggregate of largely autonomous projects, typified by the construction industry. At least in theory, it might be possible to manage the operations of any high-technology company as an aggregate of projects in various stages of the innovation cycle. In this framework, the composition of a project team would change with the varying critical function requirements of the successive phases of the innovation process. This framework is most nearly approached, in industries involved in the assembly of large, complex high-technology products such as air transports and ships.

Most innovations are, however, incremental, and are more typically introduced into production operations in mature industries which have reached the cost-minimizing stage of their industry life cycles. Part of this cost-minimizing rationalization process involves the establishment of a functional management structure in which potential economies of scale are exploited by manufacturing or processing a limited range of closely related products in mature, established divisions. Even in the aircraft and shipbuilding industries, a project management structure is only partially realized since much of the subassembly work may be performed in functionally managed organizations. Thus, realistically, it can be expected that most incrementally innovative projects will be transferred from R&D to an operations facility which is mainly functionally managed. This disadvantage of the latter structure, in the context of innovation management, is that it creates an environment which is probably resistant to technological change. Operations staff will identify their loyalties, career aspirations, and goals with the dominant operations functions and respective departmental heads. For an innovative project to thrive in the operations facility, it will require able team members from that facility. If the management, leadership, and reward structures are dominated by a functional orientation, such staff will not wish to join an innovative project team since they will perceive it as penalizing their career prospects. If they do accept such an assignment, they may do so reluctantly and on a part-time basis, sharing their time between the project and an ongoing functional responsibility. In a functionally dominant organizational climate, they will probably assign higher priorities to their functional responsibilities and a relatively low commitment to the project. The project team will thereby fail to secure its legitimate share of organizational resources and suffer a decline in morale, cohesiveness, and entrepreneurial drive. Therefore, there are sound arguments for introducing a new product into operations through a matrix structure, with project team members reporting to both project and functional managers. Once the new product has achieved market acceptance, its production may be transferred to a conventional functional operations management structure.

The benefits of a matrix structure can be summarized quoting the words of Kolodny, who has made a specialist study of the subject:[6,7]

> Matrix organization is designed to have the best of both project and functional worlds. It overcomes the weakness of poor task responsiveness on the part of the functional organization by channelling the knowledge of the specialists through project and program teams. It overcomes the weakness of limited long term specialist competence development on the part of the project organization by keeping specialists closely connected to functional homes.

Matrix management structure in operations is similar to that in R&D (compare Figures 7.5 and 9.1), except that the vertical columns of the matrix are now management functions rather than subject disciplines. The main disadvantage is that the individual team member is a two-boss person since he has reporting responsibilities to both the project manager and his functional manager. He may therefore suffer from the role uncertainty and conflict of loyalty that this creates. Equally, the project and functional manages may experience discomforts with the ambiguities created for them, particularly if they are competing for an individual team member's time, as well as other organizational resources. Such potential conflicts can be avoided or resolved, provided that top management understands its leadership role in a matrix organization.

Lawrence, Kolodny, and Davis point out that a matrix structure is really a diamond structure (Figure 9.3) with both project and functional managers reporting to a common superior.[8] Successful matrix leadership behavior requires this person to initiate dual control, evaluation, and reward systems which balance project and functional needs and create a matrix culture which fosters open discussion and conflict resolution. These views are echoed by Klimstra and Potts, who emphasize a commitment to the norm of "no surprises" and the "proper management" of conflict.[9] They emphasize the importance of informal discussions of problems within

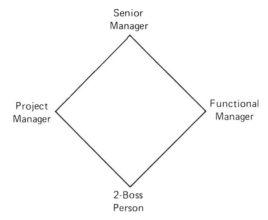

Figure 9.3. Matrix diamond.

project teams, between project teams, and between project teams and line functions. Only when such problems cannot be resolved informally should they be taken up through the formal management hierarchy to the senior manager in the diamond structure. Even when conflict does arise, it is important that it is about issues (rather than personalities) and that both line and project management play constructive roles in seeking its productive resolution.

Given the undoubted difficulties of implementing an effective matrix organization, it is pertinent to comment briefly on the research which has evaluated the approach. Kolodny provides most valuable reviews in the references cited earlier, and two investigations are specifically revealing. Marquis studies 37 projects in the aerospace industry in which he identified four organizational formats.[10] They are: (1) functional, (2) matrix with large functional areas and small project teams, (3) matrix with small functional areas and large project teams, and (4) project. He found that matrix structure (2) was the most effective format for achieving technical success. Jermakowicz studied 100 projects spread across seven organizational formats with functional and project formats at each extreme and the matrix format in the middle.[11] He used two measures of effectiveness for evaluating these formats— number of new products introduced in the organizational system and originality of new products. He found that while the highest level of originality occurred in a project format, the highest implementation rate was achieved in a matrix structure. Given that the implementation of a profitable innovation is the ultimate measure of success, it could be argued that the two studies yield broadly identical conclusions.

The overall evidence suggests that often the best framework for transferring projects from R&D to operations is via a mixed transfer team, possibly under the sponsorship of a senior coordinator, from a R&D management matrix structure to an operations management matrix structure. When the project becomes a full-fledged new product, it may be transferred to a functionally structured production facility. Given that the original invention began in a discipline structured research facility, the overall sequence of organizational structures in the innovation process is

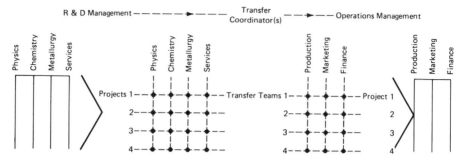

Figure 9.4. Overall transfer process.

functional–matrix–functional, as illustrated in Figure 9.4. This sequence has been achieved through intuitive trial and error organizational learning by many offensive/defensive innovators.

9.3 LEARNING CURVES AND THE TECHNOLOGICAL PROGRESS FUNCTION

Once a full-fledged new product is transferred to functionally managed mainstream operations, its production and marketing may be integrated into operations. Once it becomes an operations (as opposed to innovation) management responsibility, it could be said to be no longer the concern of this text. Nevertheless, there is one important property of technological innovations in the full-scale production setting which remains to be considered. This is the impact of learning effects on unit production costs and pricing of an innovative product. In Chapter 2 we viewed technology as progressing though the cumulative trial and error elimination impacts of successive innovations. It is important to recognize that the trial and error elimination learning occurs not only in the inventive phases, but throughout the innovation process, including the production and diffusion phases when an innovative new product has been adopted by mainstream operations. In production, this property is reflected in the learning curve, which was first observed in the aircraft industry and subsequently in numerous other manufacturing and process operations.

Such learning patterns are documented by Hirschmann.[12] All readers will know from personal experience that the time required to perform a repetitive task decreases with repetition. That is, it can be performed more quickly the second time than the first, the third than the second, and so on. What is less well-known is that the rate of improvement is consistent and appears to continue indefinitely. For example, if the improvement rate is 80%, the man-hours required to produce one unit will be reduced by 20% each time output is doubled. That is, if it takes 100 man-hours to produce the first unit, it will take $100 \times 0.80 = 80$ hours to produce the second, $80 \times 0.80 = 64$ hours to produce the fourth, as shown in Table 9.1. It is expressed mathematically in the learning function shown in Table 9.2, where the parameter b is the improvement rate ($b = 0.80$ above). It follows from the form of

TABLE 9.1 Learning Effect

Total number of units produced	1	2	4	8	16	32
Manhours required for last unit	100	80	64	51.2	40.96	32.77

this function that when we plot manhours per unit against cumulative output, we obtain the curve shown in Figure 9.5a, and when we plot the logarithms of these measures, we obtain the straight line shown in Figure 9.5b. The slope of this line measures the improvement rate b. The ubiquity of this production learning effect has been amply demonstrated in numerous industries, including those cited by Hirschmann. The value of the improvement rate b varies between different production technologies, and is particularly dependent upon the relative proportions of manual assembly versus machine-paced work requirement. It typically lies between 70% and 90%. However, what is more important to recognize is that such learning effects are not confined to unit production, but also influence a broad range of other costs. The term experience curve is sometimes used to describe these broader learning effects which may apply to the total costs of successive versions of a product which incorporate a succession of incremental innovations. For example, Abernathy and Wayne show that the price of the dominant design, the Model T Ford, followed a 85% slope experience curve over the years 1909–1923.[13]

The consistency of experiental learning effects may be exploited by companies in their pricing strategies when launching innovations. Companies in, for example, the aircraft and electronics industries (such as Boeing and Texas Instruments) may launch an innovative new product at a price below current production costs, anticipating that the levels of future sales will ensure that costs will fall below price from learning effects as output increases. This strategy of pricing ahead down the learning curve enables them to reap the benefits of large market share as well as learning effects. We must also stress that such exploitation must be based upon prudent management judgment. First, if continued learning improvements are to be maintained, the organization must continually seek ways of improving production methods and products. That is, it must maintain a continued positive climate for improvement from all its members. Second, it is obvious that pricing down the learning curve can be financially disastrous if anticipated sales and therefore outputs are not realized. Third, the continued exploitation of the learning effects from a successful product range should not be allowed to constrain technological flex-

TABLE 9.2 Learning/Improvement Function

$$y = ax^{-b}$$

where y is man-hours required to produce x-units
    x is the number of units produced
    a is man-hours required to produce the first unit
    b is the learning or improvement parameter

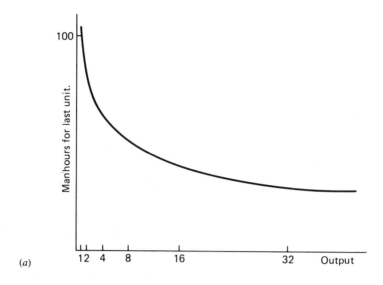

(a)

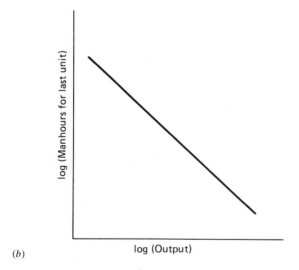

(b)

Figure 9.5. Learning curve.

ibility. An excessive dependence on its benefits may lead to undue rigidity and the inability to respond to the future opportunities and threats of technological change. The limitations of the learning curve are discussed in the article by Abernathy and Wayne cited above.

9.4 NEW VENTURE OPERATIONS

So far in this chapter we have concentrated upon the transfer of technology from R&D to the mainstream operations in a company. Since revolutionary innovations are based upon new technological paradigms, it is often technologically infeasible to transfer them to existing operations as they require the establishment of new businesses and new industries to exploit them commercially. Moreover, as suggested in Chapter 6 (Section 6.3), some micro-radical innovations, while falling short of inducing a Kuhnian paradigm-shift, are sufficiently novel to require the introduction of radical new production technology. Also, even if an innovation possesses more modest technological novelty, it may have to struggle for continued survival in face of the opposition from operations personnel who believe (rationally or otherwise) that they have a vested interest in maintaining the technological *status quo,* particularly if the company suffers from the technological rigidity indicated above.

The causes of technological conservatism among operations personnel are discussed in Chapter 11 later, but if an otherwise promising innovation remains unexploited by mainstream operations, the project team will be frustrated and may set up their own new company to do so (see next chapter). If this happens, the innovation will be lost to the present company. To avoid this outcome, an alternative approach to transferring technology to mainstream operations is to set up a separate business venture to manufacture and market it. Since entrepreneurship is a vital element in the innovation chain equation, it is hardly surprising that some companies have adopted a separate venture approach to exploiting innovations commercially. This separate ventures approach will be discussed in some detail in Chapter 11, after we have examined another aspect of technological entrepreneurship—the spin-off phenomenon and the process for establishing new technology businesses, in Chapters 9 and 10.

FURTHER READING

S. R. Rosenthal, *Effective Product Design and Development.* Homewood, IL: Richard D. Irwin, 1992.

REFERENCES

1. W. H. Gruber and D. G. Marquis (Eds.), *Factors in the Transfer of Technology.* Cambridge, MA: MIT Press, 1969.

2. H. F. Davidson, M. J. Cetron, and J. D. Goldhar, (Eds.), *Technology Transfer.* Leiden, Netherlands: Noordhof International Publishing, 1974.

3. L. W. Steele, *Innovation in Big Business.* New York: American Elsevier, 1975.

4. M. J. C. Martin, J. H. Scheibelhut, and R. Clements, *Transfer of Technology from Government Laboratories in Industry,* Report No. 53, Technological Innovation Studies Program. Ottawa, Ont: Department of Industry Trade and Commerce, November 1978.

5. I. Nonaka, "Redundant, Overlapping Organization: A Japanese Approach to Managing the Innovation Process." *California Management Review* **31**(3), 27–28 (1990).

6. H. F. Kolodny, "Matrix Organisation Designs and New Product Success." *Research Management* **23**(5), 29–33 (1980).

7. H. F. Kolodny, "The Evolution to a Matrix Organization." *Academy of Management Review* **4**(4), 543–553 (1974).

8. P. R. Lawrence, H. F. Kolodny, and S. M. Davis, "The Human Side of the Matrix." In M. L. Tushman and W. L. Moore (Eds.), *Readings in the Management of Innovation.* Marshfield, MA: Pitman, 1982, pp. 504–519.

9. P. D. Klimstra and J. Potts, "Managing R&D Projects." *Research Technology Management* **31**(3), 23–39 (1988).

10. D. C. Marquis, "A Project Team + PERT = Success, Or Does It?" *Innovation* May, No. 7 (1969).

11. W. Jermakowicz, "Organizational Structures in the R&D Sphere." *R&D Management* **8**(Spec. Issue), 107–113 (1978).

12. W. B. Hirschmann, "Profit From the Learning Curve." *Harvard Business Review* **42**(1), 125–139 (1964).

13. W. J. Abernathy and K. Wayne, "Limits of the Learning Curve." *Harvard Business Review* **52**(4), 109–119 (1974).

PART V
THE ENTREPRENEURIAL SETTING

Part V considers the entrepreneurial aspects of the technological innovation–entrepreneurship process. It begins by considering the key role played by small spin-off companies in pioneering innovation and the personal characteristics of technological entrepreneurs. Some researchers may be contemplating launching their own technological ventures, so Chapter 11 extensively discusses such start-up operations. In contrast, Chapter 12 discusses frameworks for stimulating intra-preneurship, that is, technological entrepreneurship, in larger companies.

___10
SMALL IS BEAUTIFUL: SPIN-OFFS, INNOVATION, AND ENTREPRENEURSHIP IN SMALL BUSINESSES

> *The successful entrepreneur is driven not so much by greed as by the desire to create. He has more in common with the artist than with the business manager.*
>
> PAUL JOHNSON, "GEORGE GILDER PRAISES CAPITALISM"
> WALL STREET JOURNAL

10.1 INTRODUCTION

The establishment and growth of new high-technology industries in the nineteenth and early twentieth century was particularly associated with the pioneering efforts of inventor–entrepreneurs. These were individuals who were both creative scientists and/or engineers *and creative businessmen.* Their technological and commercial vision, drive, and judgment created what sometimes ultimately became major high-technology corporations, often built around revolutionary innovations. Examples of such innovation giants can be found in most Western countries. They include Nobel (Swedish and dynamite), Linde (Germany and liquid air distillation), Brins (France and industrial oxygen), Solvay (Belgium and ammonia–soda), Perkin (England and anilene dyes), Bell (Scotland–North America and the telephone), Edison (United States and electric light), and Marconi (Italy–United Kingdom and radio), to name but a few. In Chapter 1 we surveyed the experiences of the last of them. As the twentieth century progressed, however, with the increasing institutionalization and professionalism of R&D in university, government, and industrial laboratories (often founded by such individuals), the role of these individual inventor–entrepreneurs appears to have declined, although Land (polaroid camera) and Carlson (xerography) are recent examples of the breed.

In Chapter 2 (Section 2.7) we hypothesized that creative young technological entrepreneurs pledge adherence to revolutionary innovations just as creative young research scientists pledge adherence to new paradigms following Kuhnian scientific revolutions. During the Abernathy–Utterback fluid stage following a revolutionary innovation, new companies are often founded by such individuals to seek to exploit the new technology. Such new companies are usually founded or spun off by individuals from larger companies, government laboratories, and universities. This spin-off phenomenon is notably identified with specific industries (solid-state electronics and computers) and localities (Silicon Valley in California and Route 128 in Massachusetts). It might be conjectured that this phenomenon characterizes the establishment and growth of new high-technology industries at the present time as did the individual inventor–entrepreneur in the past. The success of a new high-technology company is partially dependent upon the entrepreneurial aptitudes of its founders, and entrepreneurship may be viewed as the commercial expression of the creative process described, in the R&D context, in Chapter 7, Appendix. Moreover, such successful small companies usually generate economic growth and employment opportunities in a region. Therefore, in this chapter, we briefly comment on the role of small business technological entrepreneurship in economic growth, then examine the spin-off phenomenon–the personal characteristics of technological entrepreneurs and new venture initiation.

10.2   THE ECONOMIC ROLE OF SMALL HIGH-TECHNOLOGY BUSINESS

As Gee and Tyler indicate, innovative companies have much better records for generating economic growth and employment than others, and an industrialized nation's economic ability to grow and maintain international competitiveness is largely determined by technological innovation.[1] This observation, and the contribution that smaller companies can make to innovation, is further supported when seen in combination with the results of other studies. Mansfield has found that the social return (that is, the benefits to society) of an innovation is typically double the private return (that is, the ROI to the company which introduced the innovation).[2] This finding is reflected in EMI's experience with the CAT scanner described in Chapter 1.

Furthermore, Kamien and Schwartz[3] point out that smaller companies play a vital role in the process of technical change, while Shapero[4] points out that the 1000 largest U.S. firms are outperformed by smaller firms. These contentions are supported by the results from a comparative study of large mature, large innovative, and young high-technology companies made between 1969 and 1974.[5] It showed that the young companies have far more impressive sales, employment, and tax compounded growth rates than large innovative and mature companies. Perhaps more impressively, between 1969 and 1974, they created more absolute employment opportunities (34,369 versus 25,558) than their mature counterparts from an employment base which was less than 1% of the latter's (7579 as opposed to 786,793 employees in 1969)! New company formations also increased considerably over that time. Birch reports that annual new company formations in the United States increased from 90,000 in 1950 to 700,000 in 1985.[6] The majority of these new companies will not, of course, have been New Technology-Based Firms (NTBFs). Even in Silicon Valley and Route 128, high-technology firms of any size account for less than 20% of local jobs.[7] What is more important is that a core of burgeoning high-technology firms stimulates the regional economy and culture and many more jobs through multiplier effects.

The superior growth records of these small companies suggest that they may demonstrate superior innovative records as supported by U.K. studies. Gibbons and Watkins, discussing the competitive ability of small firms, argue that large firms find it more difficult to react to technological change, so that small firms benefit from their faster reaction time.[8] Freeman found that small firms made significant innovative contributions in many industries.[9] The Bolton Report also concluded that small firms make an important contribution to innovation in the United Kingdom.[9]

At first sight, it might be argued that the unemployment created by frequent small business failures must be set against the growth records of successful companies. This argument is of only limited validity, however, for two reasons. First, the mortality rate of emergent high-technology business (20–30%) is much lower than all small business (over 80%). Second, just as Mansfield demonstrates that the social benefits of innovation outweigh their private benefits, Shapero argues that a business failure generates social as well as individual learning. He also cites Henry Ford as an example of entrepreneurs who experienced several failures (in Ford's case, two) before achieving business success. Barron also gives several examples of Silicon Valley technological entrepreneurs who have achieved success after initial failure.[10] Thus, technological entrepreneurship, like technological evolution, is a trial-and-error, learning-by-doing process. This technological and entrepreneurial learning process is reflected in the spin-off phenomenon, which we now discuss.

10.3 THE SPIN-OFF PHENOMENON

Earlier, we have referred to *incubator organizations,*–that is, institutions which appear particularly prone to employing individuals who later establish (that is, spin off) successful new companies. Government R&D facilities, universities, as well as private companies have been credited with the propensity to spin off such companies. Stanford University can claim to have "hatched" some very successful post World War II high-technology companies. William Hewlett, David Packard, and Russel Varian graduated from Stanford, and later established Hewlett-Packard and Varian Associates, while in 1972, the Stanford Industrial Park contained 65 companies employing 16,000 people. The evidence suggests, however, that private companies are more effective incubators than the other institutions, and we focus on these. There are two factors, one transient and the other permanent, which have stimulated the spin-off phenomenon. First, the Apollo space program launched by President Kennedy in 1960, and second, the personal frustrations experienced by individuals in the larger companies, both coupled with the business opportunities created by microelectronics technology. We will examine each of these factors, in turn, before studying the motivations of individuals who have spun off.

The influence of the Apollo Program on the creation of new companies is described by Osborne.[11] This program generated the need for numerous industry-based, limited-period R&D contracts funded by the U.S. government. Government policy allowed companies fulfilling such contracts to charge profits based upon a fixed percentage of total salaries paid. This policy encouraged some companies to hire more professional staff and pay them higher salaries than the R&D needs and labor market rates warranted—that is, to pad their payrolls to increase their profits. While such contracts lasted, the arrangement was fine for R&D staff. They were well-paid and underemployed, so they could learn new subjects and develop new ideas on the government's time. But they also knew that they enjoyed no job security. Once a government R&D contract ended, they could expect to be laid off. Therefore, they changed jobs frequently, hopping from one R&D contract to the next one in another company, becoming professional gypsies. If such footloose individuals develop an idea or invention which looks commercially promising, it is less than likely that they will seek to exploit it with their current employers. They are more likely to set up a company to exploit it themselves. The burgeoning expansion of the solid-state electronics and computer industries in the 1960s and 1970s created many business opportunities of this kind, which Osborne compares with the California gold rush, particularly in Silicon Valley. Since the 1970s, this ten-mile stretch of real estate between the Santa Cruz mountains and San Francisco Bay has experienced substational fluctuations in its fortunes, associated with global changes in the semiconductor and computer industries, but we review the spin-off phenomenon in Northern California and elsewhere, as it occurred in the 1960s and 1970s.

We have already seen that the comparative fecundity of scientists and engineers to generate potentially profitable inventions, coupled with the rising costs of a project as successive phases of innovation are completed, usually dictate that all projects cannot be supported to completion. Furthermore, projects sometimes fail to survive once they have been transferred to the manufacturing setting through lack of enthusiastic support from operations and/or general management. Whenever a project is aborted (or subjected to malign neglect) for whatever reason, it is hardly surprising that members of the project team often experience frustration and dissatisfaction with the treatment of "their project." One reaction to this frustration and dissatisfaction is for such individuals to sever their association with their employers and set up a new company to manufacture and market the product they are developing. This is the second factor which has led to the spin-off phenomenon and the establishment of networks of small high-technology companies in specific localities. The Santa Clara Valley in California and the Greater Boston area in Massachusetts are the most notable of these, but others exist in the United States and elsewhere—for example, Silicon Glen in Central Scotland and Silicon Valley North in Ottawa, Canada. Danilov, in describing the spin-off phenomenon, lists the following examples of spun-off companies: Control Data Corporation, Digital Equipment Corporation, Teledyne Inc., Aerojet General Corporation, High Voltage Engineering, and Itek Corporation.[12] Such companies have subsequently spun off second- and third-generation offspring, including Spectra Physics from Varian Associates and Coherent Radiation Laboratories from Spectra Physics. More notably, perhaps, we have the RAND Corporation, which spun off from Douglas Aircraft and was the parent of System Development Corporation and the Hudson Institute.

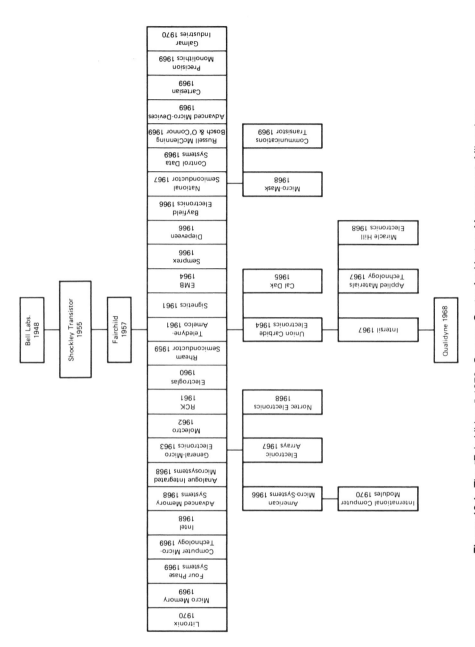

Figure 10.1. The "Fairchildren," 1970. *Source:* Center for Venture Management, Milwaukee, WI 53204.

Probably the most famous of these incubator organizations, however, is Fairchild Semiconductors, and the cascade of its spin-offs or progeny is shown in the "Begat Tree," Figure 10.1, based on Draheim.[13] This figure shows that between 1957 and 1970, Fairchild begat numerous "fairchildren, grandchildren," etc. Although Fairchild's fecundity is exceptional, being related to the rapid growth opportunities manifest in the "California gold rush" climate of the solid-state electronics industry and the "go-go" venture investment climate of the period (partially financed from the high profits from the Apollo program cited earlier), it is not unique. Miller and Cote display a comparable "Begat Tree" derived from Engineering Research Associates (ERA) in Minneapolis–St. Paul.[14] Formed in 1946, ERA had begotten 43 companies by 1983. These included Remington Rand, Sperry, Univac, Control Data Corporation, and Cray.

Some companies clearly act as better incubators than others. Cooper[15] performed a study to identify factors which influence a company's propensity to spin off NTBFs. He studied about 250 NTBFs which were founded in the San Francisco area in the period from 1960 to 1969 from incubator parents whose total employment exceeded 77,000. The results of his study, interspersed with some from other related studies, can be summarized as follows.

Founder's Expertise

The majority of the new firms seek to exploit their founder's expertise in the technology–market matrix of their parents. Of Cooper's new firms, 85.5% sought to exploit the same technologies and/or markets as their parents. Obviously, this makes prudent commercial sense since fledgling venture founders are likely to be more successful in technologies and markets which they already know well. Astute participants in a high-technology industry in the fluid growth stage of its life cycle will be able to identify profitable market niches or, as Cooper puts it, *pockets of opportunity*. Roberts has studied over 500 spin-offs from MIT, government laboratories, and private companies in the greater Boston area over almost 30 years.[16] Perhaps surprisingly, since MIT is the largest single incubator in his sample, founders more frequently have a background in development rather than research. He concludes that the key initial technologies of spin-off firms were transferred primarily from development projects carried out by their founders for their previous employers, where they excelled as high performers.

Location Choice

For several reasons, NTBFs tend to locate close to their parents. First, the founders may wish to exploit an intimate network of fairly local professional and social contracts, both to finance and develop their product(s) and to exploit the target market. Second, there may be a period of part-time activity or even moonlighting prior to the founder's final break with their parents (see Chapter 11, Section 11.2) during which they develop their venture plan and possibly secure financial support. Alternatively, they may build up local reputations as consultants, which they may wish to exploit by performing consulting work during the early phases of NTBF development to generate supportive income. Third, apart from the fact that the founders may have to take a second mortgage on the family home as collateral for start-up capital, they may be unwilling to disrupt the lives and education of their family. This observation supports the evidence which is cited later in Section 10.5, namely, that such venture founders have an average age of 33 years when founding their first NTBF and usually have supportive stable marriages. Roberts' work does not explicitly consider location choice since it focuses exclusively on spin-offs in the greater Boston area. However, it is reasonable to conjecture that most (if not the whole) of his sample chose to remain in that area.

In two U.K. studies, Dickinson and Watkins[17] and Watkins[18] reported broadly similar conclusions. In the former study, only two of a sample of 21 spin-off companies reported that the new entity was founded outside the founding proprietors' residential areas.

Maturity of Industry

Others of Cooper's observations suggest that industries in the fluid rather than mature stage of their technology life cycle are more likely to provide spin-off opportunities. At the mature stage, the surviving companies are typically pursuing cost-reducing process innovations with technologies offering fewer spin-off opportunities. However, it should be noted that contemporary developments and applications of CIM technologies in the automobile industry do offer spin-off opportunities.

Size of Incubator Firm

At first sight, it might be expected that larger firms should show greater spin-off propensities than smaller ones. Larger organizations are more likely to suffer from organizational senescence and a bureaucratic climate which frustrates younger staff. In fact, Cooper found that the spin-off rate, expressed as the inverse of the number of employees per spin-off, was ten times greater for companies employing less than 500, as compared with those employing more than 500. Note that the firms were classified as under 500 employees, over 500 employees, and subsidiaries under 500 employees, and that the spin-off rate of the latter was also eight times that of the larger firms. This result suggests that size rather than ownership affects spin-off propensity. The explanation for the higher spin-off rates in the smaller firms possibly lies in the positive attributes of their organizational climates as much as the negative attributes of those of the larger ones.

First, individuals with the longer-term aspiration of setting up their own small businesses are more likely to be attracted to a position in a smaller firm. There, they will probably be given broader responsibilities, in both range and substance, at an earlier age than they would in a larger company. Second, they will be gaining direct experience in the problems of operating and managing small businesses, having closer contacts with different functional managers, as well as familiarity with the markets in which they operate. Third, they will directly observe a small businessman in action—namely, their own employer, and this, together with the examples of colleagues who have spun off a business, may provide a positive "demonstration effect."

Cooper suggests other mechanisms which may increase the propensities of small business environments to spin off new ventures. He suggests that some of the small business subsidiaries studied may have been taken over quite recently, so that individuals spin off because they are unwilling to adjust to the loss of management independence associated with the take over. Furthermore, some of these may have had substantial stakes in the equity of the taken over company, and have achieved generous capital gains from the take over deal. They are then ready and able to finance an independent new venture. One behavioral trait of some technological entrepreneurs, which supports this contention, is to establish successfully and then sell at profit a succession of new ventures. Cooper also found that in the larger firms studied, the spin-off rate appeared greater from departments which constituted the "small business" of the firm. Apart from the reasons cited above, he suggests that such departments may be less well managed than the larger divisions of these firms, and have relatively weak bargaining power in securing corporate resources for new product development. This conjecture is congruent with the reported reasons individuals offer for "going it alone" (see below). In contrast, Watkins did not find a small business bias in spin-off propensities in Britain.[19] Such a difference may perhaps be attributed to the absence of regional developments comparable in their economic impacts to the Silicon Valley/Route 128 phenomenon and the lower number (in relative and absolute terms) of small high-technology businesses in the Britain.

Motivation Factors

The personal frustration with his employers obviously may be a significant factor in inducing an individual to found an NTBF, and this observation is borne out by Cooper's results. He subjected 30 individuals to in-depth interviews to elicit their reasons for leaving their incubator-organizations. Seventy percent reported themselves to be highly frustrated in their previous jobs, 30% said they quit without specific alternative employment, and 40% said they would have left anyway. Cooper reports that the major areas of concern among these people are poor selection and development of managers and poor investment in products and technologies. Other writers suggest the importance of precipitating events or major displacements, such as bankruptcies, dismissals, or extensive lay-offs and major company policy changes, in stimulating spin-offs. Watkins quotes the example of a policy change in Plessey leading to 12 spin-offs in Britain.[20] Such a displacement, together with a sympathetic economic climate, may well trigger an individual with entrepreneurial leanings to found his NTBF, so it is pertinent to our discussion to examine the personal characteristics of the technological entrepreneur.

10.4 CHARACTERISTICS OF TECHNOLOGICAL ENTREPRENEURS

Before discussing entrepreneurial characteristics, it is useful to establish a working definition of the noun *entrepreneur* since it is a word which has considerably broadened its meaning over the last 100 years. The 1897 edition of the *Oxford*

English Dictionary defined an entrepreneur as: *the manager or director of a public musical institution: one who "set up" entertainments, specially musical performances.* But the 1975 edition of *Webster* defines one as someone who undertakes to: *organize, manage, and assume the risk of business.* This latter definition is a realistic semantic approximation of its present general usage. Presumably, its etymology is the French verb *entreprendre* (literally, between taker), meaning to undertake or assume a responsibility or task. Clearly, an entrepreneur undertakes or assumes business responsibilities and tasks, and in France, it is often used to describe a small businessman or trader, particularly a building contractor. Its German equivalent, *unternehmer,* does literally mean undertaker, and it is perhaps unfortunate that this word has acquired a funereal colloquial usage in English. The entrepreneur typically assumes responsibility for *creating* and developing new living businesses rather than burying old dead ones, and it is this quality of creativeness which is important in technological entrepreneurship! In the context of technological evolution, this quality converts inventions or "technological mutations" into innovations which occupy useful social as well as technological niches.

As well as experiencing changing meanings, the word has acquired an unfortunate perjorative association since some equate the entrepreneur with a "fast buck artist" or "huckster." For this reason, some companies prefer to use the terms project or product champions to describe the role of the individual whose drive and tenacity sustains an innovation to fruition. This misconception is both unfortunate and unfair as well since the evidence suggests that, if anything, technical entrepreneurs have somewhat higher ethical standards than the average person.

As with creativity, different behavioral schools of thought have offered different explanations for an entrepreneur's behavior, and a very useful review of them is provided by Sexton and Bowman-Upton.[21] However, it appears that, at present anyway, there is no agreement from the results of the various studies on the psychological profiles of entrepreneurs, although the differences may in part be attributable to differing biases in the samples studied. Some have studied technological entrepreneurs (that is, people who have set up a high-technology business) as opposed to all entrepreneurs (that is, people who have set up any kind of business, including small traders, Mom and Pop storekeepers, etc.), while others have studied business students or have included managers in their samples. Consequently, it is difficult to draw firm overall conclusions from these numerous studies. A useful recent review of such work by Brockhouse and Gasse[22] is provided, and some overall generalizations can be made.

McCelland, in *The Achieving Society,*[23] has been an active contributor to the field and he summarizes his ideas in a *Harvard Business Review* article.[24] He argues that a society's economic development is stimulated and maintained by its entrepreneurs who are motivated by a high need for achievement (n-Ach). He describes a high need for achievement as:

> the desire to do something better, faster, more efficiently, with less effort. It is not a generalized desire to succeed, nor is it related to doing well at all sorts of enterprise . . . Rather it is peculiarly associated with moderate risk-taking because any task

which allows one to choose the level of difficulty at which he works also permits him to figure out how to be more efficient at it, how to get the benefit (utility) for the least cost. And business, cross-culturally, is the specific activity which most encourages or demands using the calculus of cost-benefit.

He characterizes an individual with a high n-Ach as someone:

1. Who likes situations in which he takes personal responsibility for finding solutions to problems.
2. Who has a tendency to set himself challenging but realistic achievement goals and to take calculated risks.
3. Who also wants concrete feedback as to how well he is doing.

He observes that a society which offers equal opportunities to people regardless of origin is likely to show a high n-Ach. Hagan also argues that such development occurs when creative problem solving ability is directed towards industrial innovation.[25] In fact, both argue that economic development is stimulated by individuals with personal attributes and goals (whether they be labeled high in n-Ach or creative problem solvers) which set them ahead of the prevailing norms. Such individuals may be somewhat alienated from the prevailing social values, but they have curious enquiring natures, rich stores of ideas, and confidence in their own evaluation of problem-situations. They provide much of the driving force of economic development, and may be archetypes for the many inventor–entrepreneurs, some of whom we listed at the beginning of this chapter. Thus, they may also represent the archetype for the technological entrepreneur we are discussing here. Both the scientist and the technological entrepreneur have a strong need to reach high self-set goals. Some individuals may set themselves goals which embrace R&D inventiveness and entrepreneurial aspirations, and may thus become inventor–entrepreneurs. Associated with a high n-Ach is a desire for feedback on how well they are doing, and an anxiety that such feedback will reveal failure to achieve the desired goals. This need for feedback is also consistent with the entrepreneur's attitude towards risk. To the lay person, an entrepreneur may be viewed as a gambler or irrational risk taker. However, as Roberts and Wainer indicate, the technical entrepreneur sees himself as a moderate risk taker who would not take the gamble of founding a new business venture until he had estimated that it had a reasonable choice of success.[26]

An alternative and possibly complementary explanation of entrepreneurial behavior is offered by Rotter's locus-of-control theory.[27] According to this theory, a person perceives the outcome of events to be controlled by chance, powerful other persons (or factors), or himself. One who believes that he exercises personal control over outcomes is described as believing in an internal locus-of-control. Clearly, a person is unlikely to set up in business unless he believes in an internal locus-of-control. Such "internal" individuals are more striving and competent than their "external" counterparts (that is, those who believe that chance or powerful others control outcomes) and extract more information from ambiguous situations or environments. Rotter conjectures that a high n-Ach should be related to a belief in an internal locus-of-control, and this conjecture is supported by other studies.[28–30] In another study, Borland asked over 300 business students to indicate their future expectancy of starting a business, as well as determining their n-Ach scores and locus-of-control beliefs.[31] She found that students with high n-Ach scores and internal loci-of-control expressed high expectancies, and for students with low n-Ach scores, an increasing internal locus-of-control correlated with an increased expectancy of starting a business. Her results suggest that an internal locus-of-control is better than a high n-Ach in measuring entrepreneurial aspirations. Brock-house[32] also found that the owners of a business which had survived for three years held stronger internal locus-of-control beliefs than those businesses which had failed in the same period. In contrast, Hall, Basley, and Udell[33] failed to find a correlation between locus-of-control scores and entrepreneurial activity among business school alumni.

Both high n-Ach and internal locus-of-control scores appear intuitively to be reasonable measures of entrepreneurial aptitudes and, apart from the differing sample frames chosen, conflicting results may be attributable to inadequate psychometric techniques rather than invalid concepts. Sexton and Bowman[34] suggest that entrepreneurs "possess more intense levels of growth-oriented traits than do managers in general." This growth orientation distinguishes the entrepreneur from the "Mom and Pop" small-business owners or franchisees. All three measures can be perceived in the personality of David Thomas, the founder of Wendy's International. He began his entrepreneurial career as a "company scrounge" in the U.S. Army. All readers with military experience (whether in Uncle Sam's or any other army) will recognize "company scrounge" as an archetypal figure who clearly possesses an internal locus-of-control in the face of chance and powerful other persons of the military hierarchy! After leaving the Army, David Thomas worked for Colonel Sanders and Kentucky Fried Chicken before deciding to launch his Cadillac Hamburger business, which has grown into a major international fast-food chain. Presumably, he possesses an intense level of growth-oriented personality traits as well as an internal locus-of-control; otherwise, he might have been satisfied with a Kentucky Fried Chicken or McDonald's franchise.

Turning from psychological attributes, it is useful to look for more pragmatic measures of entrepreneurial aptitudes. Probably the best summary of these aptitudes was proposed by Williamson.[35] He does not suggest that *all* successful entrepreneurs will possess *all* these personality characteristics, but rather that

> the probabilities of entrepreneurial success may be expected to be proportional to the degree to which the aspiring entrepreneur possesses those characteristics which appear common to individuals who have started and successfully operated a new business.

He suggests ten characteristics:

1. *Good physical health* since establishing a new business requires the physical stamina to sustain long hours of hard work.
2. *Superior conceptual and problem solving abilities* since the entrepreneur

must learn from his mistakes and resolve complex technical and commercial problem-situations quickly.

3. *Broad generalist thinking* since the above problem solving skills must be based upon the ability to maintain continued overviews of situations and to relate and integrate diverse technical and commercial factors into the overall thrust of the venture.

4. *High self-confidence and tolerance of anxiety* so that the inevitable set-backs and adversities can be overcome and "defeats" turned into "victories."

5. *Strong drive* since persistence and tenacity plus a strong sense of urgency will ensure that "things get done" and that the venture maintains its momentum.

6. *A basic need to control and direct.* The successful entrepreneur wants to maintain overall control of the situation and rejects higher authority and externally imposed bureaucratic structures. He wishes to establish and maintain his own control mechanisms and fully accept responsibility and accountability for his decisions.

7. *Willingness to take moderate risks* based upon a rational analysis of alternative actions the calculation of consequent risks before decision making.

8. *Very realistic* since he perceives and accepts situations as they are, seeks to monitor them continuously, and solves problems pragmatically. Although both cautious and suspicious at times, he is honest and dependable and expects these virtues in others.

9. *Moderate interpersonal skills.* The technical entrepreneur wants to "run his own show," so is uninterested in delegation and a participative approach to decision making. While the NTBF remains small, he does not need the interpersonal skills of the professional manager, although if he does possess them, they will be personal assets as the firm grows larger.

10. *Sufficient emotional stability.* The successful establishment of an NTBF imposes physical, emotional, and time pressures which in turn impose a mandatory rugged ability to "keep one's cool" in crisis situations.

When we turn to the social and domestic (that is, what might be called biographical as opposed to psychological backgrounds of technological entrepreneurs), we find that studies yield remarkably consistent results. First, McClelland and others have amply demonstrated that entrepreneurs frequently belong to ethnic or sectarian refugee minorities in a society of which the Jews (in Western Societies) and Chinese (in both S.E. Asian and Western societies) are obvious examples. The experience of minority status may predispose people towards entrepreneurial initiatives regardless of their ethnic or sectarian backgrounds. Litvak and Maule, in studies of technological entrepreneurs in predominantly Catholic Quebec, found a lower proportion of Catholics and a higher proportion of first generation Canadian immigrants in their sample than chance would indicate. In contrast, Catholic entrepreneurial minorities are to be found in the United States. Second, as Shapero and Sokol indicate, numerous studies in different countries show that a self-employed parent strongly influences an individual to set up in business.[36] In all of the studies these authors cited, 50–89% of the entrepreneurs involved had at least one self-employed parent. This percentage is much higher than the corresponding population-wide values (12% in the United States). Such children initiate business ventures regardless of their parent's success or failure in self-employment.

Having reviewed some of the personal characteristics of entrepreneurs, we turn to some specific examples.

10.5 ILLUSTRATIONS OF TECHNOLOGICAL ENTREPRENEURS

We have already cited Cooper's, Roberts', and Watkin's detailed studies of NTBFs in the United States and United Kingdom, respectively. Others were performed by Shapero in the United States, France, Italy, India, and South Africa[37] and by Litvak and Maule in Canada. Shapero's work provides a partial basis for the new venture initiation framework we discuss in the next section. Probably the best summary of such studies is provided by Litvak and Maule within the context of their study of a sample of entrepreneurs who launched NTBFs in Canada.[38,39]

They found that 65% of their respondents were Canadian born and 35% non-Canadian born. In 1971, the population-wide proportion of heads of household who were non-Canadian born was 23%, a much lower percentage. The relatively high proportion of nonnative respondents tends to support the "stranger hypothesis"— that is, that immigrants in a society may be more predisposed towards entrepreneurial initiatives because of their feelings of being "outsiders" in their present society and of the need to justify their traumatic act of emigration from their native heath. This hypothesis is supported, at least impressionistically, by the number of successful entrepreneurs who have been immigrants from other countries in the United States and United Kingdom as well as Canada. However, it should also be pointed out that the proportion of immigrants among scientists and engineers in Canada is high due to the sustained growth of the Canadian economy (until recently) which has created a continued demand for technologists in excess of the native supply available, and the continued "brain-drain" of Canadian scientists and engineers to the United States. It seems likely that all the above factors were contributing to produce a relatively large proportion of nonnative respondents.

Although the mean age of the respondents was 47.7 years at the time of their response, their mean age at the time that they incorporated their first firm was 33 years. This agrees with the U.S. results of Roberts and Wainer[26], and Cooper[40] who states:

The firm is started by two founders, both of whom are in the middle thirties. One usually can be described as the driving force. He conceives the idea and enlists the other founder. They come from the same established organization, which is where they got to know each other. Either both are in engineering development, or one is in engineering and the other is a product manager, or in marketing. Often they have achieved significant prior success, with titles such as *Section Head,* or *Director of Engineering,* being common.

The U.S., British, and Canadian studies all found that technological entrepreneurs were much more likely to have, or have had, a self-employed parent, reflecting a "parental" as opposed to a "collegial" demonstration effect. The Canadian researchers also found that a relatively larger proportion of the population with Jewish religious backgrounds are likely to become technological entrepreneurs, as opposed to those with Protestant and Catholic backgrounds. Roberts obtained almost identical results from his sample of technological entrepreneurs *whose parents were not self-employed*. The proportions with Jewish backgrounds, together with the population-wide proportions in brackets, were almost identical in the Canadian and Roberts' "not self-employed parent" subsamples, being 10% (1.3%) and 10% (2.0%), respectively. In contrast, he found that for the proportions with Jewish, Protestant, and Catholic backgrounds, in "self-employed parent," subsamples were almost identical with the population-wide proportions.

Most Canadian technological entrepreneurs held two (although a few up to ten) jobs before incorporating their first firm, as compared with an equimodal distribution of one, two, and three jobs in the British study.[41] One interesting observation was that the proportions of Canadian respondents holding managerial positions increased from 4.6% in their first job to 38.7% in their last job before incorporation. Hardly surprisingly, 54% held university degrees (mainly in science or engineering) or technical diplomas, which is probably at least ten times the population-wide statistic for the corresponding age group.

Their responses to questions asking their reasons for starting a business supported the conclusions cited earlier in this chapter. Seventy-four percent cited "challenge" and 50% cited "being one's own boss" and "freedom to explore new ideas" as prime reasons for starting up on their own. These results were broadly consistent with the British study which listed "desire for independence" and "desire for increased job satisfaction" as the most frequent reasons cited, followed by "a release of creative urges" and "financial motivation" significantly less frequently. When asked for the prime reason for incorporating a new firm, 51.8% cited the development of an existing product for a new market or new product for an existing or new market, 20.5% the acquisition of partners, and 19.7% the acquisition of financial support. Of the respondents who cited new markets or product as their reason, 83.7% believed that their previous employers would have refused them permission to develop their ideas.

Fifty-six percent of the respondents founded their first company with part-timers—a figure which corresponds closely with U.S. and British findings. Most first companies were financed through personal savings, bank loans, and borrowings from friends and relations (in that order), with venture capitalists providing little funding. Fifty-two percent of the respondents received federal or provincial government grants, and 38% of respondents (as compared to 33% in Britain and only 6–15% in the United States) cited financing as a major problem. The next most frequently cited problems were selling and managing personnel. Seventy-eight percent of the respondents formed more than one company. The mean number of firms formed per respondent was 3.25, with 2.87 still in operation at the time of the study. These results are also consistent with U.S. findings, and suggest that such individuals learn to become better technological entrepreneurs through trial and error. About one third of the sample expanded by mini-conglomerate and a second third by horizontal integration.

10.6 ENTREPRENEURIAL VENTURE INITIATION

Liles[42] and Shapero[43] provide frameworks for viewing the entrepreneurial venture initiation process based upon their own extensive studies of entrepreneurship. The work we have cited in the previous two sections indicates that certain individuals have a predisposition or readiness to initiate an NTBF. However, this *readiness to act* of prospective technological entrepreneurs is initially restrained by their need to acquire professional education and business experience and then by increasing family responsibilities. Such individuals typically enter the work force in their early- or middle-twenties, and gain progressive experience and responsibility, as shown in Figure 10.2, leveling off as they approach their fortieth birthday. During this time, they also typically accept the increasing financial commitments (including heavily mortgaged houses) of matrimony and family life with a reorientation of goals from career to familial and other concerns (again shown in Figure 10.2). Therefore, during their thirties, they experience a *free choice period* when they still retain a readiness to act and are, as yet, unrestrained by noncareer concerns.

Some people never experience this free choice period because the other commitments become too large before they have the experience and confidence to set up a company. Many others pass through such a period but do not set up a company

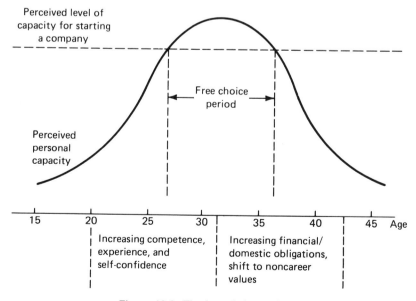

Figure 10.2. The free choice period.

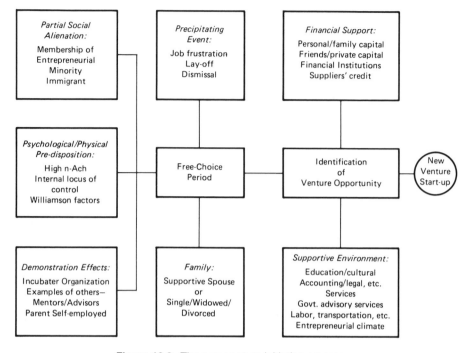

Figure 10.3. The new venture initiation process.

because they fail to experience a *precipitating event* which induces them to do so. It is a minority, who experience precipitating events during their free choice periods, who become technological entrepreneurs. We therefore now focus on the characteristics of such precipitating events.

The interactions of a readiness to act, precipitating events, etc., is illustrated in Figure 10.3. If a significant proportion of these factors is present in an individual's personal situation, it is reasonable to conjecture that there is a substantial chance that they will launch a new venture. Individual readinesses to act is determined by psychological make up, social background, and the experience of a demonstrative models in a parent and or a colleague/mentor in an incubator organization. If individuals experience precipitating events (such as job dissatisfaction, dismissal, or lay off), they will be predisposed to launching their own companies. This predisposition will be reinforced if their spouses are supportive (a crucial factor), see next chapter, (Section 11.3) or conversely, if they are single, divorced, or widowed without significant family commitments. Clearly, they should only launch a new company if they perceive a suitable opportunity which they are confident of exploiting (probably with one or more partners) and they can assemble the financial and other required resources from a supportive environment. The launch and establishment of such a new venture is a challenging process which we discuss in detail in the next chapter.

FURTHER READING

R.-E. Miller and M. Cote, *Growing the Next Silicon Valley: A Guide for Successful Regional Planning.* Lexington, MA: Lexington Books, 1987.

D. L. Sexton and N. B. Bowman-Upton, *Entrepreneurship Creativity and Growth.* New York: Macmillan, 1991.

REFERENCES

1. E. A. Gee and C. Tyler, *Managing Innovation.* New York: Wiley, 1976.
2. E. Mansfild, "Returns from Industrial Innovation." In A. Gerstenfeld (Ed.), *Technological Innovation: Government/Industry Co-operation.* New York: Wiley, 1979, pp. 18–19.
3. M. I. Kamien and N. L. Schwartz, "Market Structure and Innovation." In H. R. Clauser (Ed.), *Progress in Assessing Technological Innovation—1974.* Westport, CT: Technomic Publishing Co., 1975, pp. 32–35.
4. A. Shapero, "Entrepreneurship and Economic Development." In W. Naumes (Ed.), *The Entrepreneurial Manager in The Small Business.* Reading, MA: Addison-Wesley, 1978, pp. 183–202.
5. *Emerging Innovative Companies: An Endangered Species.* Chicago: National Venture Capital Association, 1976.
6. D. L. Birch, *INC.* **21** (March) (1987).
7. R.-E. Miller and M. Cote, *Growing the Next Silicon Valley: A Guide for Successful Regional Planning.* Lexington, MA: Lexington Books, 1987.
8. M. Gibbons and D. S. Watkins, "Innovation and the Small Firm." *R&D Management* **1**(1), 10–13 (1970).
9. C. Freeman, "The Role of Small Firms in Innovation in the UK since 1945." *Bolton Committee of Enquiry on Small Firms,* Research Report No. 6. London: H.M. Stationery Office Cmnd 4811, 1971.
10. C. A. Barron, "Silicon Valley Phoenixes." *Fortune,* November 22 (1987).
11. A. Osborne, *Running Wild: The Next Industrial Revolution.* Berkeley, CA: Osborne/McGraw-Hill, 1979.
12. V. J. Danilov, "The Spin-Off Phenomenon." *Industrial Research* **11**(5), 54 (1969).
13. K. P. Draheim, "Factors Influencing the Formation of Technical Companies." In A. C. Cooper and J. L. Komives (Eds.), *Technical Entrepreneurship: A Symposium.* Milwaukee, WI: Center for Venture Management, 1972, K. P. Draheim and R. P. Howell, "Comparative Profile—Entrepreneurs versus the Hired Executive: San Francisco Peninsula Semiconductor Industry." *Ibid.*
14. R.-E. Miller and M. Cote, *Growing the Next Silicon Valley: A Guide for Successful Regional Planning.* Lexington, MA: Lexington Books, 1987, p. 52.
15. A. C. Cooper, "Incubator Organizations and Technical Entrepreneurship." In A. C. Cooper and J. L. Komives (Eds.), *Technical Entrepreneurship: A Symposium.* Milwaukee, WI: Center for Venture Management, 1972.

16. E. B. Roberts, *Entrepreneurs in High Technology.* New York: Oxford University Press, 1991.

17. P. J. Dickinson and D. S. Watkins, *Initial Report on Some Financing Characteristics of Small Technology-Based Companies and their Relation to Location,* R&D Research Unit Internal Report. Manchester: Manchester Business School, 1971.

18. D. S. Watkins (Ed.), *Founding Your Own Business,* Conference Proceedings, R&D Research Unit. Manchester: Manchester Business School, 1972.

19. D. S. Watkins, *The Role of the Small Firm in Innovation Since 1945: Extension and Critique,* R&D Research Unit Internal Report. Manchester: Manchester Business School, 1972.

20. D. S. Watkins, "Technical Entrepreneurship: A cis-Atlantic View." *R&D Management* **3**(2), 65–70 (1973).

21. D. L. Sexton and N. B. Bowman-Upton, *Entrepreneurship Creativity and Growth.* New York: Macmillan, 1991, Chapter 1.

22. R. H. Brockhouse, Sr., "The Psychology of the Entrepreneur." In C. A. Kent, D. L. Sexton, and K. H. Vesper (Eds.), *Encyclopedia of Entrepreneurship.* Englewood Cliffs, NJ: Prentice-Hall, 1982, pp. 39–71; Sr. H. Brockhouse, R. and Y. Gasse, "Elaborations on the Psychology of the Entrepreneur." *Ibid.*

23. D. C. McCelland, *The Achieving Society.* New York: Irvington Publishers/Wiley (Halstead Press), 1976.

24. D. C. McCelland, "Business Drive and National Achievements." *Harvard Business Review* **40**(4), 99–112 (1962).

25. E. E. Hagan, *On the Theory of Social Change: How Economic Growth Begins.* Homewood, IL: Dorsey Press, 1962.

26. E. B. Roberts and H. A. Wainer, "Some Characteristics of Technical Entrepreneurs." *IEEE Transactions on Engineering Management* **EM-18**(3), 100–109 (1971).

27. J. B. Rotter, "Generalized Expectancies for Internal Versus External Control of Reinforcement." *Psychological Monographs: General and Applied* **80**(1), 1–28 (1966).

28. P. E. McGhee and V. C. Crandall, "Beliefs in Internal-External Control Reinforcement and Academic Performance." *Child Development* **39**(1), 91–102 (1968).

29. P. Gurin *et al.,* "Internal-External Control in the Motivational Dynamics of Negro Youth." *Journal of Social Issues* **25**(3), 29–53 (1969).

30. R. C. Loo, "Internal-External Control and Competent and Innovative Behaviour Among Negro College Students." *Journal of Personality and Social Psychology* **14**(3), 263–270 (1970).

31. C. Borland, "Locus of Control, Need for Achievement and Entrepreneurship," Unpublished doctoral dissertation. Austin: University of Texas, 1974.

32. R. H. Brockhouse, "Psychological and Environmental Factors Which Distinguish the Successful from the Unsuccessful Entrepreneur: A Longitudinal Study," Unpublished proceedings. Academy of Management Meeting, August 1980.

33. D. L. Hall, J. J. Basley, and G. G. Udell, "Renewing the Hunt for the Heffalump: Identifying Potential Entrepreneurs by Personality Characteristics." *Journal of Small Business* **18**(1), 11–18 (1980).

34. D. L. Sexton and N. B. Bowman, "Validation of a Personality Index: Comparative Psychological Characteristics Analysis of Female Entrepreneurs, Executives, Entrepre-

neurship Students, and Business Students." In J. Hornaday *et al.* (Eds.), *Frontiers of Entrepreneurship.* Wellesley, MA: Babson College, 1984, pp. 40–51.

35. B. Williamson, Address to Seminar, Life Planning Center, Vail, CO: SMU School of Business, March 1974.

36. A. Shapero and L. Sokol, "The Social Dimensions of Entrepreneurship." In C. A. Kent, D. L. Sexton, and K. H. Vesper, (Eds.), *Encyclopedia of Entrepreneurship.* Englewood Cliffs, NJ: Prentice-Hall, 1982, pp. 72–90.

37. A. Shapero, *An Action Program for Entrepreneurship,* Austin, TX: Multi-Disciplinary Research Press, 1971.

38. I. A. Litvak and C. J. Maule, "Some Characteristics of Successful Technical Entrepreneurs in Canada." *IEEE Transactions on Engineering Management* **EM-20**(3), 62–68 (1973).

39. I. A. Litvak and C. J. Maule, *Policies and Programmes for the Promotion of Technological Entrepreneurship in the US and UK: Perspectives for Canada,* Report No. 27, Technological Innovation Studies Program. Ottawa, Ont.: Department of Industry, Trade and Commerce, May 1975.

40. A. C. Cooper, "The Palo Alto Experience." *Industrial Research* **12**(5), 58 (1970).

41. D. S. Watkins, *Founding Your Own Business,* R&D Research Unit Internal Report. Manchester: Manchester Business School, 1972.

42. P. R. Liles, "Who are the Entrepreneurs?" In W. Naumes (Ed.), *The Entrepreneurial Manager in the Small Business.* Reading, MA: Addison-Wesley, 1978, pp. 10–21.

43. A. Shapero, "Engineering and Economic Development." In W. Naumes (Ed.), *The Entrepreneurial Manager in the Small Business.* Reading, MA: Addison-Wesley, 1978, p. 31.

11
CREATING THE NEW
TECHNOLOGICAL VENTURE

The day will come, as it must to 99% of all technical professionals,
when the thought of developing your own ideas for your benefit
crosses your mind. If you let this seed lodge and grow, then you begin
to realize that the resulting fruit might require the establishment of
your own business.

D. S. SCOTT, "SO YOU WANT TO RUN YOUR
OWN BUSINESS" IN *THE TECHNICAL ENTREPRENEUR:*
INVENTIONS, INNOVATIONS AND BUSINESS

11.1 INTRODUCTION

The qualities of individual creativity and entrepreneurship have much in common.
In fact, the inventor–entrepreneurs of the nineteenth century could equally well be
called *technologically creative entrepreneurs*–that is, individuals who were twice-
blessed by Providence with the flair and drive to generate both new technological
ideas and the new companies and industries to exploit them. In Chapter 10, we
reviewed the relatively recent resurgence of creative entrepreneurship through the
spin-offs of new technological ventures from parent incubator organizations—a
resurgence most visible in the microelectronics and computer industries, notably
around Route 128 and Silicon Valley. Although the enthusiasm for new venture
creation has waxed and waned through successive cycles, individuals are still
launching new autonomous ventures. In this chapter, therefore, we discuss the
problems and pitfalls of new venture creation. The issues of sponsored ventureship
in larger organizations are discussed in Chapter 12.

Many R&D professionals and technology managers employed in larger organizations have an unfulfilled dream to launch their own NTBFs. Whether this dream is
turned into a reality is dependent upon a combination of circumstances illustrated in
Figure 9.3 of the previous chapter. This chapter provides such individuals with an
overview of the factors which they should consider before undertaking the undoubt-
ably hazardous step of launching their own businesses. There are numerous books
on approaches to setting up and managing a small business, and we shall not attempt
to review such a vast literature here. One of the best of these is by Timmons,
Smollen, and Dingee, and is strongly recommended reading for all aspiring new
venture creators.[1] It is based upon new business creation programs implemented by

the authors and their colleagues in the United States, Canada, Sweden, and Britain which, by 1979, had generated ventures valued at $70–80 million. Therefore, we shall draw upon their experiences as the major individual source material for this chapter. Keirulf[2] have also provides useful brief reviews of the new venture creation literature which may also be of interest to readers.

11.2 STAGES IN THE NEW VENTURE CREATION AND GROWTH PROCESS

We will discuss the process, assuming that the venture is founded by the growth-oriented entrepreneurs of the previous chapter. There are a number of multistage models of the process extant in the literature and, for the purposes of exposition, we define five stages in the new venture creation process. Each stage is defined now to give the reader an overview of the process, and each will be examined in some detail in subsequent sections of this chapter.

Preliminary Moonlighting

Autonomous ventures are spun off by individuals from incubator organizations which may be private firms, government, or industry agencies (such as the British research associations), or academic institutions. Unless a person is laid off or fired by his employer, he should not decide to launch an autonomous venture (even if he can get the money to do so!) without careful thought and planning, typically with one or more partners from the same organization. This gestation process will occur while they are currently employed (hence the term *incubator organization*) when they are, in effect, moonlighting. Timmons *et al.* suggest that 200–300 hours of moonlighting work is needed to produce a viable new venture concept, but this figure could be too conservative. Roberts reports an average moonlighting period of 30 months in a sample of biomedical start-up firms he studied.[3] During this time, the partners formulate the product concept, typically engage in prototype development and testing, and plan how the product should be made and sold. They will thereby identify the initial capital needed to launch the new venture. They may also find the further partners they need to create a viable technical and entrepreneurial team, as discussed shortly.

Launch and Test Operations: First Round Financing

A successful output from the moonlighting stage could be a venture concept which deserves financial support. Professional financial bodies rarely fund proposals at this stage. Capital requirements for the initial launch often amount to tens or low-hundreds of thousands rather than millions of dollars. The founders often find that they can raise such a sum from their own personal savings plus support from families, friends, and possibly a "business angel."[4] The latter is an affluent individual (perhaps a successful technological entrepreneur) who enjoys investing in promising new technology ventures and can afford to take the risks involved. As one successful technological entrepreneur seeking promising start-ups put it to this author: "He was looking for 'frogs' whom he could kiss and turn into 'princes'!"

The development of this proposal and plan will identify many of the problems and pitfalls to be expected in initial operations. However, problems of plant construction and initial manufacturing operations, which may be unpredictable and uncertain beforehand, will have to be overcome. The product may initially be sold to a few, what might be called beta site customers, who showed an interest in the product during the moonlighting stage. This can be seen as test marketing, but obviously the ultimate test of the marketability of the new product must await the full market launch. If these customers give the product favorable report cards, the scene is now set for a full market launch. This launch is likely to require a more substantial multi-million dollar capital investment for further production, as well as sales and distribution resources. The founders will look to the professional financial community for this support.

Initial Growth: Second Round Financing

The acquisition of second round funding from professional investors may well prove to be the most time-consuming and frustrating stage of the overall process. Prospective investors such as banks and venture capitalists are unlikely to be enthusiastically supportive at this stage. Many bank managers are reluctant to invest in new technology start-ups because of the intangible nature and risk of new technology—at least in the bank manager's eyes! It is much safer to support a proven concept, such as a new franchisee of an established fast-food chain. Venture capitalists prefer to support third stage financing, when the venture has a proven track record, for the reasons outlined in Section 11.4.

These observations suggest that many new venture founders will have to expend significant efforts in first identifying potential investors and then securing their support. Professional investors will assess the attractiveness of a proposed new venture as an investment opportunity based upon the strengths and personalities of the individuals founding the proposed venture, the business plan submitted to them, and the commercial and technical merits of the new product, *probably in that order.* They are knowledgeable, experienced, and astute evaluators of new venture proposals. In contrast, scientists and engineers seeking to exploit their own inventions may be commercially and managerially naive. Such individuals can find that the development of a venture proposal and business plan which will survive the critical scrutiny of investment professionals to be a time-consuming, traumatic, but *ultimately rewarding learning experience.* Such a critical scrutiny may well identify errors in thinking which might well have proved costly, or even disastrous, if they had been carried over into the fully-developed venture.

Once second round financing has been secured, output can be scaled up and the product concept fully market tested. It leads to the fourth stage.

Consolidation and Growth: Third Round and Subsequent Financing

Once the venture has established itself as, at least, an initially viable commercial entity, its founders will look to consolidation and probably expansion of both product lines and markets. This expansion will almost certainly require further financing, but given the venture's success to date, this should be forthcoming. Venture capitalists will, in particular, view it favorably now because it possesses the risk–return characteristics that they are seeking. Indeed, the founders' major concerns in third round financing may be retention of control. First, a second round financing may well have required some external equity and management participation in the venture. Parties investing orders of millions of dollars will require substantial equity and probably management participation in the venture. Furthermore, given that a steady growth is sustained, the company may need repeated cash infusions. Drucker suggests that growing small high-technology companies require financial restructuring every three years.[5] The founders may therefore limit their growth plans and consequent financing requirements to ensure that they can retain overall control of "their" company.

This stage presents other opportunities and threats. If the venture is generating a solid and growing sales revenue, procedures for manufacturing and marketing may have to be standardized and the original venture team expanded to include functional management and specialist personnel, without creating undue red tape or lowering organizational morale. It will require the institution of one or more levels of middle management and some delegation of decision-making. One of the dangers faced by venture teams at this stage is the risk of *following success with failure*. It goes without saying that the venture teams we are discussing here will be both technologically and entrepreneurially oriented. If they enjoy commercial success with their first innovative project, they may become overconfident, and rather than consolidating their position, pursue more ambitious product ideas which are beyond their capabilities or market needs. These issues will be discussed in Section 11.7, and we now come to the final stage.

Maturity: "Go Public" or Sell

By this stage, the venture should be enjoying revenues of tens of millions of dollars and faces two growth alternatives. First, the venture can "go public," thereby securing extra cash for further expansion. In return for surrendering further equity, the founders often enjoy substantial capital gains from its sale. There are definite disadvantages to becoming a publicly-traded company, however, which will also be discussed in Section 11.7. The second alternative is for the founders to sell the company, making larger capital gains, and maybe starting further ventures or becoming business angels. This author knows of one successful technological entrepreneur who sold his company to a large multinational corporation when he had taken its annual sales to about $50 million. Now, still in his mid-forties, he acts as a business angel and advisor to start-ups, but spends much of his time backpacking and mountaining.

11.3   THE FORMATION OF THE VENTURE TEAM

As we stated above, much of the planning required for the launching of a successful new technological venture is typically conducted in the moonlighting stage before its founders sever their associations with their incubator organizations. Timmons *et al.* contend that success depends above all else upon three essentials. First, a capable lead entrepreneur supported by a cohesive entrepreneurial team; second, a feasible business idea; and third, appropriate financing, in that order.[6] This ranking is supported by Knight's survey of venture capitalists, cited in the previous section, as well as other writers concerned specifically with NTBFs. As C. Gordon Bell, a very experienced technology manager and entrepreneur, puts it: "the three most important factors in the formation of start-up companies are 'people, people, and people'."[7] General George Doriot, who is regarded as the father of venture capital institutions (and who helped launch many high-technology companies, including Digital Equipment Corporation), is reputed to have preferred a Grade A team with a Grade B idea to a Grade B team with a Grade A idea.[8] Therefore, although the decision to try to launch an NTBF may be triggered by the desire to exploit a specific invention, a detailed self-analysis of the strengths and weaknesses of the proposed venture team should be a first consideration.

A new venture should coalesce around the technological entrepreneurial aspirations of one or a very small group of individuals, whom we may call the lead entrepreneur(s). The core of the venture must be this team's entrepreneurial and technological inventive capabilities and vision. If the business idea is to be launched as a new venture, it must, however, incorporate minimum functional management capabilities to support or, as shown in Figure 11.1, surround this core. One of the first tasks of this group, therefore, should be to complete a self-inventory of their collective management expertise to identify any skills which may be lacking. This stock-taking should, of course, be conducted in the context of developing the venture idea and business plan, as discussed in the next section. Different industries and products have different specialist functional skill requirements, particularly in production and marketing. If the venture is to succeed, the group should be knowledgeable of the industry and possess the insight to identify the key determinants of business success within it. If it lacks this knowledge and insight, it probably lacks the maturity and experience to launch the venture. The stock-taking should identify what, if any, further management skills are required. If further skills are required, they may be acquired either by coopting additional members with these skills into the founding team or hiring such individuals on a full- or part-time basis once the venture is launched. Bell recommends that if none of the original team is suited to be CEO, if possible, a suitable person should be coopted into the team to act as its leader.[9] Outside financial support may also be contingent upon the acquisition of such expertise. The outer layer of Figure 11.1 represents the specialist expertise which, although required, may be employed on a part-time consulting basis during the early life of the venture. For example, in one study of small high-technology businesses, this author found that most did not hire a full-time accountant until several years after they had been launched.[10]

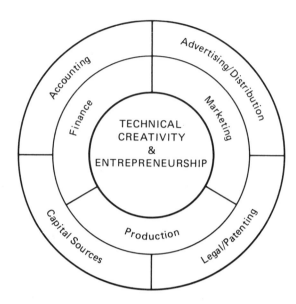

Figure 11.1. The venture base.

The group's self-assessment process must not be confined to only technological, entrepreneurial, and managerial considerations, but *most importantly,* must include behavioral aspects. The venture team should be cohesive, and once the venture is launched, practice a kind of "group marriage." The moonlighting period offers an opportunity for "courtship" during which the personal aspirations of individuals should be identified to determine their interpersonal compatibilities. They should be totally committed to the success of the venture, which can take priority over all other commitments, including (in a sense) their responsibilities to their families. At first sight, this requirement appears morally outrageous and certainly requires clarification. The launching of a new venture invariably requires the team to work long hours, probably on a seven-day week basis, which leaves members little time to exercise their domestic responsibilities. Furthermore, once the second stage of the process is entered, it will require the investment of personal capital and the foregoing of a regular monthly income as a salaried employee. Typically, team members can expect to pay themselves low or no salaries during the early days of the venture, and their families may have to live at least partially on accrued savings or the incomes of working spouses. Their families may experience a short-term drop in their material standards of living. A commitment of this magnitude should only be made with the agreement and support of the team's spouses and families, *on both moral and psychological grounds.* Apart from the fact that both husband and wife must make such a commitment jointly if they have a viable marriage, it is unlikely that an individual would be able to withstand the psychic strain of new venture creation without familial support. This issue is discussed by Jolson and Gannon[11] and was confirmed by the author's study cited above. the NTBF founders interviewed in that study confirmed that, for the married, the successful development of the venture required the indulgent support of their families. Moreover, many of the men interviewed said that their wives worked during the initial phase of the venture development to provide the family income. Typical comments were as follows:

> Had not the wives of my partner and myself been prepared to work in order to support our homes we would never have been able to get this company off the ground. We made no profit, not even enough to pay ourselves salaries during the first two years of our operation.

> Our wives worked. There is no way we would have done without them.

Roberts makes similar observations, based upon his studies of technological entrepreneurs over many years.[12]

The founders also need the support of others in their reference group (former work colleagues, other friends, and the community in general) if they are to succeed. We pointed out in Chapter 10 that both paternal and incubator organizational demonstration effects influence an individual's decision to become a technological entrepreneur, and community attitudes towards entrepreneurship are also strongly influential.

Both Bell and Timmonds *et al.* stress the importance of team building and corporate culture. "The two key aspects of that must be defined at the outset are how the company will treat its employees and how it will manage its cash" is how Bell puts it.[13] The latter authors suggest that a cohesive venture team should be built around the concepts of organizational climate and interpersonal and helping skills.[14] The climate should emphasize and reward responsibility and commitment to high-performance standards through teamwork. It should therefore reflect some of the features of a creative climate, discussed in the Appendix to Chapter 7, but be focused upon the goals of the venture and teamwork rather than on individual R&D creativity. The establishment and maintenance of a purposeful organizational climate is dependent upon the interpersonal and helping skills of team members. They should readily articulate problems, consciously develop their listening and participation skills, and be willing to help and accept help from others. This willingness to give and receive help is particularly important in a new venture situation since individual team members will typically lack the range of expertise and experience which the venture demands, and they will have to do a lot of learning on the job. Leadership is also likely to be flexible since different situations may require the expertise of different team members. It is almost certain that the team will have to pay deliberate attention to these behavioral factors, especially as they are in conflict with the individualistic personality attributes of creative R&D personnel and entrepreneurs identified in the Appendix to Chapter 7 and in Chapter 10. This potential conflict must be faced and accommodated if the team is to be cohesive.

This process of what could be called group introspection should enable the team members to clarify their personal goals in joining the venture. It is most desirable that individuals should identify and make explicit their personal goals to ensure that they are mutually congruent given the requirement for managerial expertise in the venture team. The research evidence suggests that inventors and entrepreneurs are achievement-oriented, in contrast to managers who are more power- and status-oriented. The former individuals seek self-actualization through R&D or business creativity. In contrast, the latter derive satisfaction from outward manifestations of managerial success. These latter manifestations are reflected in status symbols such as titles, office sizes and furnishing, company cars, expense accounts, etc. An NTBF offers little scope for "empire building" of this nature, and an individual who seeks to accrue such external status symbols will squander the scarce financial resources of the venture and disrupt team cohesiveness. Therefore, it is vital that all team members should recognize that their external status rewards can be expected to be modest during the first few years of the life of the venture.

This consideration brings us to the important issue of the reward structure for team members, and indeed all employees. The design of the business plan, which is to be discussed in the next section, should identify specific business goals (expressed in terms of sales growth, return on capital employed, etc.). Individuals will join the team, almost certainly making immediate material sacrifices, in the expectation of future psychological and material rewards. They will be coopted into participation because they appear to be personally acceptable and they offer the required expertise, commitment, and financial investment. Obviously, the split of the initial equity and other shareholdings must be agreed upon between the members, together with some rules for future divisions and buy-back agreements. A member may later wish to leave the venture, become sick, or die. Clearly, stock buy-back rules should be decided at the outset which, as far as is possible, protect the interests of the team and the departing individual (or his estate, if he dies). These rules should also allow for the contingency of inviting new members to join the team in the future, should the need arise. Some shares might be held as a pool available for purchase by new team members who may, of course, be employees. The team will also need to determine each of their individual salaries, based upon some group assessment of each individual's performance and contribution to the goals of the venture.

As the venture grows, it can be expected to take on a growing number of paid employees at all levels. Therefore, the team must plan a future reward structure which not only maintains its own cohesion, but also promotes a productive organizational climate. A pioneering spirit of a new venture can create a participative and performance-oriented organization climate for all personnel. The employee reward structure should therefore sustain its development and, as we suggested in the previous paragraph, it could allow for the longer-term contingency of offering the opportunity for share purchase to key and loyal employees.

11.4 SOURCES OF FINANCIAL SUPPORT

Interlocking with the development of the venture team will be the search for financial support, often the most frustrating aspect of new venture creation. It is therefore useful to comment on sources of funding for the launching of new technological ventures, which can be roughly classified as follows:

1. The venture team, plus their families and friends.
2. Independent private investors or business angels.
3. Banks.
4. Government business support programs, such as the SBIC in the United States.
5. Venture capitalists.

We discuss each of these in turn.

The Founders plus Families and Friends

The nucleus of financial support will virtually certainly have to be provided by the venture team, supplemented by whatever else is available from families and friends. Clearly, if the team members are unwilling to put their own money into the venture, no one else will. In a few instances, these immediate sources are sufficient to launch the venture, but these instances will be very much the exception rather than the rule. The research on the profiles of technological entrepreneurs reported in Chapter 10 suggested that they were typically in their thirties. Given their relatively short working lives to that age, the costs of education, and their domestic obligations, it is likely that such individuals will have limited savings to invest in the new venture. This seed capital may be supplemented by unsecured borrowings from families and friends (probably at low or zero interest rates), but since these individuals may also have modest savings, the venture team may need to seek support elsewhere.

Private Backers or Business Angels

The next step should be to look for other private backers or business angels. Between them, the team members may interact in quite diffuse professional and social networks, which will include individuals with relatively substantial private wealth that they have inherited or earned through their own professional or entrepreneurial activities. The team should engage injudicious and tactful word-of-mouth advertising of their need for financial support, in the hope that an, as yet unknown, backer might be forthcoming. Networking skills are important in all areas, including finance, and are of critical importance to technological entrepreneurs. These skills are described by Peterson and Ronstadt who stress that *knowwho* is as important as *knowhow*.[15]

Banks

Although alternatives 1 and 2 may provide adequate start-up support, they are unlikely to be sufficient for second round financing. The team will therefore need to approach professional financing agencies which are most visibly represented by the clearing banks. The attitude of the clearing banks towards NTBFs varies between countries, but some overall comments can be made. All clearing banks are in the business of lending money to attractive ventures, but many have a prudent and conservative concept of attractiveness. In Britain and Canada, the clearing bank business is largely performed by an oligopoly of nationally organized corporations who (rightly or wrongly) are viewed as highly conservative by technological entrepreneurs. In the United States, the local banking community is more sympathetic because clearing banks are not national institutions, but federally regulated local business, usually in intense competition with each other to lend money. However, in any country, a loan request to support an NTBF by a commercially inexperienced entrepreneurial team may well appear unattractive to the average bank manager. Unproven technology coupled with unproven entrepreneurs looks to be an unacceptably high risk situation. In practice, particularly if government financial support is also forthcoming (see immediately below), clearing banks are likely to provide medium-term loans or lines of credit at market interest rates to the extent of the realizable assets which can be offered by the venture, but the capital securable on these assets will probably be insufficient for its needs.

Governments

Governments of most developed nations recognize the important role played by the smaller high-technology businesses in stimulating growth and employment, so they offer a range of support programs to help technological entrepreneurs. Simultaneously with exploring alternatives 2 and 3, the team should identify and, if possible, exploit the government support programs for which they qualify. In the 1970s, the U.S. National Science Foundation and Small Business Administration introduced the innovation center concept, and similar centers have been set up in other countries. Such centers offer financial support and a range of management and technical advisory services to prospective or existing NTBFs. These are often associated with universities, recognizing that they have an obligation to assist in the deployment as well as the generation of new technology.[16] Some governments are also prepared to act as guarantors of bank loans to NTBFs provided certain conditions are fulfilled. Individuals who do seek financial support from nontechnical government institutions may experience exasperation. This exasperation arises from the differences in personalities and attitudes between entrepreneurially-oriented individuals and civil servants. Civil servants are likely to have managerial rather than entrepreneurial attitudes since they have chosen to work in large bureaucratic organizations. Furthermore, like bank managers, they must exercise responsible stewardship in managing taxpayers' money. Through temperament and a legitimate sense of fiscal stewardship, like bank managers, they may be reluctant to invest in new technology that they do not understand.

A complementary avenue for seeking government grant money is the pursuit of R&D or manufacturing contracts from government or private institutions. Some NTBFs have begun as contract R&D service companies *before* moving into manufacturing. This approach normally requires considerably less initial capital, but it is a less attractive growth strategy since it delays the development of production and marketing capabilities. R&D contracting is used quite extensively, *in combination with* other strategies which enhance the other capabilities, by new biotechnology firms.

Venture Capitalists

The term venture capitalist or venture capital institution can, of course, be literally applied to any individual or institution providing financial capital to any new or existing business venture. However, it is more often used to describe financial institutions which aim to invest money in a particular type of venture. These institutions try to invest their money in ventures, including those which embrace novel business ideas, which may offer a *moderate* (but not high) commercial risk and a *high growth potential.* They want to invest equity or a mixture of equity and debt capital in a venture which, if it fulfills its promise, will enable them to sell off their investment after about five years for a capital gain of five times or more. By their nature, such agencies want to invest in promising technological innovations in industries, such as microelectronics and computers, in the fluid-growth stages of their life cycles. Venture capitalist institutions have proliferated in the United States to help fund the proliferation of NTBFs in such industries.

Technological entrepreneurs can again expect to experience frustrations in their dealings with venture capitalists for several reasons. First, the availability of venture capital, like other institutional capital, fluctuates with economic and stock-market conditions. Thus, the chances of funding are markedly affected by the economic conditions at the time it is being requested. Second, even with venture capital fairly accessible, technological entrepreneurs often find it difficult to attract it for first and second round funding. Hardly surprisingly, several studies show that the failure rates of NTBFs rapidly fall as they grow. Sadly surprisingly are the results from a comparative study of funding criteria of venture capitalists in the United States, Canada, Pacific Asia, and Europe.[17] Out of 24 criteria used by venture capitalists to evaluate a venture, Knight found that "may be described as high tech" came 18th in the rank order of importance in the United States and 23rd elsewhere! Several European venture capitalists even suggested that a new technology venture was viewed negatively there.

These results possibly follow from the objectives of venture capitalists outlined above, namely, that they are looking for opportunities to inject capital into a company and then withdraw it (hopefully with a substantial capital gain) *after about five years.* If an NTBF receiving first or second funding from a venture capitalist in exchange for equity participation fails, then the venture capitalist loses most or all of his investment. If, as he hopes, the venture flourishes and his investment appreciates in the manner required, he may still find it difficult to realize the expected return on this investment within a few years. The logical purchasers of his equity are the venture's founders or others who have joined it. However, it is unlikely that these people will have the financial resources to do so. Their own assets must be tied up in the venture (or they would not have sought equity financing in the first place), and, since they have been initially successful, they will be seeking further capital to finance expansion. Thus, his best hope for realizing his desired capital gain is to sell his equity to a third party or to persuade the firm to allow a public stock issue. But since original investment provided first or second financing, the venture may not yet be mature enough to exercise this option. These considerations, plus the fact that they, like everyone else, prefer to invest money in a company with a visible track record, mean that venture capitalists invest a relatively small portion of their portfolios in launch or start-up situations. They prefer to invest in an existing firm which has a few years of successful operations or track record completed and needs third or further round financing to expand and consolidate its success.

At the beginning of this section, we stated that prospective technological entrepreneurs may find the acquisition of start-up capital to be the most frustrating aspect of launching an NTBF. By now, some readers may be wondering how start-up capital for independent NTBF ever gets raised at all! The truth is that if the proposed venture looks to be technically and commercially viable, and the venture team has the entrepreneurial or project championing determination to launch and prove their innovation, they will raise the required money *some how*. Raising capital, like inventive genius, also requires the infinite capacity for taking pains. A technically creative entrepreneurial venture team which has the infinitely painstaking determination to generate an invention and innovate it into an NTBF will have the determination to find some way of raising the money which it requires. Somehow, it will put together a package of equity and debt capital from some combination of the sources listed above and manage to launch its venture. However, to achieve this goal, it will also have to have prepared a credible business plan.

11.5 THE BUSINESS PLAN

The previous two sections discussed the behavioral aspects of new venture formation and potential sources of financial support. The processes of team selection and drawing up of both written legal and abstract psychological contracts must clearly interlock with the process of formulating a business plan. Moreover, it is foolish to seek support from professional investors without a well thought-out and well-presented business plan. The formulation of a detailed and rigorous business plan, as the situation allows, is also desirable for two psychological reasons.

First, a person often desires to launch an NTBF out of frustration with working as a paid employee and a healthy desire to become his own boss. If this need for personal autonomy and independence is strong—and it must be if a person is willing to hazard the economic security of his family in order to launch his own business—there may well be an element of unconscious self-deception in evaluation the venture idea. The individual *wants* to become his own boss and he *wants* the business idea to look commercially viable, so that its pursuit can be viewed as an excellent opportunity rather than as an unacceptable risk to both his family and career. The wish is indeed often father to the thought, so that he runs the risk of unconsciously repressing critical aspects of the business idea in order to justify to himself the decision to attempt the venture.

Second, in this context, we are discussing the launching of an NTBF to make and sell a new product, based upon an invention or discovery which was probably made by one or more members of the venture team. The psychology of the act of creation often appears to dictate that the inventor has a strong "parental" attachment to his invention or "baby"—a parental instinct which can easily favorably bias his assessment of its intrinsic economic and social worth and make him resist the modification and/or degraduation of the inventive idea in the interests of commercial expediency. The untoward assertion of this instinct risks entrapping the new venture into a technology-push mode of development leading to commercial fiascoes of the kind quoted in Chapter 6. This danger is reinforced by the prevalence of "mousetrap myopia." Emerson's claim that the world will beat a path to your door if you invent a better mousetrap is, quite simply, untrue. The world will only buy it after it has been promoted and marketed—so the world knows of its availability, and if it proves to be superior to others in the market. Because of his emotional attachment to his invention and probable commercial inexperience, the inventor is particularly prone to mousetrap myopia. Although this emotional attachment is understandable, it is by no means forgivable, and the venture team should ensure that their assessment of their product concept is subject to critical commercial self-scrutiny.

Fortunately, even though the team may fail to recognize its vulnerability to the two forms of self-deception listed above, it will almost certainly be forced to develop a critically formulated business plan for the third imperative reason. Most NTBFs will require first or second round financial investment over and above that which can be provided by venture team members, their families, and friends, and thus must seek other sources of financing. Such financial support, even from a business angel, will only be provided after a critical evaluation of at least an informal business plan formulated by the venture team. Thus, in most circumstances, the venture team will be forced to formulate an informal business plan which will survive critical independent scrutiny if the proposed venture is to progress from the moonlighting to the launch stage.

The formulation a detailed business plan will certainly be required once the venture team seeks second round financing from professional investors. Its format could correspond roughly with the procedure for project evaluation outlined in Chapter 6, but should incorporate more detail since it constitutes a plan for developing a business operation in its entirety. Whatever the degree of novelty in the proposed innovation, it will virtually constitute a revolutionary innovation to the venture team so they are likely to find White's conceptual framework useful when developing the plan. Prescriptions for composing such plans are provided in numerous texts. Both Bell (see **Further Readings**) and Timmons *et al.* describe how to prepare one in some detail and provide useful illustrative examples. Funding agencies also sometimes provide guidelines for the preparation of new business proposals. Readers who are considering attempting to launch a new venture are recommended to obtain a copy of one of the above texts or another of their own choosing. Therefore, we will only outline the main features and requirements of a well-prepared plan.

When preparing such a plan, two overriding points should be borne in mind. First, most funding agencies prefer to make their initial contact with proposers *on paper;* that is, through a copy of the business plan. They can then make a swift preliminary evaluation of the appeal of the proposal before discussing it further with the venture team. Over 50% of the proposals submitted to agencies fail to overcome this initial hurdle, so it is often cost-effective to complete this preliminary screening on paper rather than waste up to a whole day of everyone's time (plus traveling expenses) on face-to-face interviews. This means that the proposal should be well presented and be lucidly and succinctly written, omitting any irrelevant detail, so that the reader can readily identify its intrinsic merit. The proposers will be asking for some tens of thousands or hundreds of thousands of dollars, and no responsible agency is going to part with this amount of money without a thorough and critical evaluation of the merit of the venture. Nevertheless, it should not exceed ten pages in length (excluding appendices) and all material should be pertinent to the evaluation process.

As stated at the beginning of Section 11.3, whatever the intrinsic merit of the proposed innovation and business concept, the appeal of the venture will be highly dependent upon the credibility of the venture team. They must not only demonstrate that they possess the experience and expertise to launch and manage the venture, but also give no reason for the reviewers to question their integrity. This means that any expected financial performance for the venture or technical performance claims for products should be as specific and accurate as realism allows, but err on the side of conservatism. Only quantitative estimates or data and opinions which can be cited or defended should be quoted. If the funding agency is considering funding the proposal, the venture team will be subjected to searching interviews before financial support is forthcoming. These interviews can be expected to expose any claims and opinions which cannot be defended and justified. Failure to justify statements made in the written proposal will undermine credibility in the integrity and competence of the venture team. In fact, writing a business plan may be compared with a research student writing a master's or doctoral dissertation. Ultimately, his dissertation will be passed based upon its intrinsic quality and his ability to defend it before an examining committee. A wise research student never makes any claim or expresses any opinion in a dissertation which he is not prepared to defend in an oral examination. To do otherwise would be to risk losing his credibility in the examiner's eyes and having his dissertation rejected. A business plan should be seen in exactly the same light. If the venture team includes someone who has successfully completed a dissertation for a higher degree, the team should perhaps listen carefully to his opinions when drawing up their business plan!

As we stated above, we will not write a detailed treatise on the design of a business plan here. Timmons *et al.* suggest the following framework:

Company Name and Address

A company, like a baby, needs a name so that it can be discussed by all as a separate entity. Remember that, like a baby, providing that it survives, the entity will grow into maturity with an identity of its own. Therefore, give it a name and a "home" (that is, an address) as soon as possible. This suggests to others that the team has confidence in its offspring!

Executive Summary

This should be prepared last, but appear immediately after the company name, etc., summarizing the following: It should identify the business idea and the industry in which it will operate, indicate any inventive novelty which is being developed and the market opportunity it will exploit, as well as the competence of the founders to exploit this opportunity. Financial performance projections for the first few years of the venture should be summarized, together with the financing required and the equity and security being offered.

Technology–Market Niche

This should describe the industry (including its current economic climate) in which the company will compete and what the founders identify as key determinants of success in it. It should describe the products, their proprietary positions, and their competitive advantages.

Marketing

The market research and analysis should be as detailed and comprehensive as the current state of knowledge allows. The intended marketing strategy should be described, including distribution and selling methods, servicing and warranty procedures, promotion plans, advertising, and pricing. The target market share and sales should be estimated over the first two or three years of operation. The procedures for ongoing market monitoring should be described.

Technical Development and Manufacturing

The evolution and current state of development of the innovative idea should be first described. The program and timetable of future technical development should be specified, identifying the anticipated problems and how they may be overcome. A timetable of development tasks, in the form of a bar chart or PERT network, should be included, which preferably illustrates the tasks which have been completed to date to convey a sense of accomplishment to the reader. The manufacturing strategy should be described, including the production, location, facilities and layout, manufacturing methods, labor requirements, sources for raw materials and brought-in components, and the impact any sales seasonality will have on production planning, scheduling, and inventory policies.

Human Resources

This section of the plan should describe the backgrounds and past achievements of the venture team, the organizational structure, and role each will play in the venture. It should also specify what other key management and professional personnel are required, and how these personnel will be recruited. It will specify salaries to be paid to founders and staff. The venture team should plan to pay themselves relatively modest salaries during the start-up phase of the venture as a demonstration of their commitment. Requirements for part-time professional services should be specified, together with the likely sources of these services. The use of reputable professional advisors strengthens the appeal of the proposal.

Financial Plans

Financial forecasts for the first few years of intended operations should be prepared, including profit and loss statements, cash flows, balance sheets, and a break-even chart. The cost and credit control system should be specified.

Contingency Planning

The anticipated risks and threats to the proposed venture should be identified and contingencies for dealing with them described. The ability to anticipate potential problems and set up contingency plans for mitigating their impact demonstrates management foresight. Some writers argue that *three* business plans should be prepared, based upon optimistic, most likely, and pessimistic scenarios and contingencies.

Overall Timetable

The foregoing analysis should be incorporated into an overall timetable or schedule, probably expressed in the form of bar charts or a PERT network. As was stated when discussing development and manufacturing, this schedule of tasks can reflect the work that has already been completed to demonstrate a record of proven accomplishment.

Funding Requirements

The proposal can end (apart from any appendices) with an analysis of the funding that is being requested, broken down into individual years. If the proposal is being submitted to a venture capital institution, it will also specify the support which has been obtained from other sources—the venture team, families and friends, bank loans, etc. It will state the securities being offered and the proportion of equity the venture team is prepared to surrender in return for funding. The description of the organization will have specified the membership of the Board of Directors, including a nominee from the funding agency. The capital structure of the proposed venture should be specified. Last, but by no means least, the proposal should clearly specify why the money is needed and how it would be spent!

All of the above sections should be succinct. Possibly, more detailed information can be offered in supplementary appendices, which potential financial supporters have the option of examining if they are interested in a further exploration of the proposal.

Assuming that this goal has been achieved, we now discuss some of the more common problems and mistakes made by NTBFs during their first year or two of operation.

11.6   COMMON PITFALLS OF INITIAL OPERATIONS

Until now, we have discussed the requirements for launching and securing funding for a new technology venture. We now examine some of the problems and pitfalls of its early operations. Technological and commercial uncertainties will have to be successfully resolved and sound business practices established if the NTBF is to be consolidated on a sound basis. The venture team members should be able to resolve technical uncertainties, but they may well lack business management skills. Clarke gives a useful succinct overview of the managerial requirements of an NTBF.[18] As he points out, the majority of business failures can be attributed to poor management. The problems of the venture team's management inexperience are compounded by the need for each to accept a wider span of management responsibilities than would be necessary in a larger, established firm. It is therefore imperative that the team ensures that, between them, they accept responsibility for all aspects of the management of the firm, and that each member recognizes and accepts his individual managerial roles and responsibilities. Some of the problems NTBFs may experience through poor management are now discussed.

Technological Obsessions—Invention versus Innovation

Given that they have strong technical backgrounds, novice technological entrepreneurs may place too great an emphasis on vocational as opposed to management tasks, pursuing *invention* rather than *innovation*. The author is familiar with one potentially very successful new venture which failed because of its owners' technology obsessions and preferences for invention over innovation. The company founders had developed a good product concept which led to several years of profitable operations and funding, including support from government and venture capital. It had also won a valuable manufacturing contract in collaboration with a major multinational corporation, presenting it with a profitable learning opportunity. Its CEO, a very creative engineer, had numerous new product ideas, some of which appeared promising to potential customers. Unfortunately, neither he, nor anyone else in the company, appeared interested in making and selling its current products, let alone converting new invention ideas into saleable products. Needless to say, revenues fell and the company ceased trading after a few years of operations.

Ill-Considered Diversification

Earlier in the book we pointed out that technological innovations have frequently proved successful in product–market niches different from those originally intended. NTBFs have had such experiences. Roberts found that only 40% of his large sample of NTBFs started off with a specific product–market focus, and most of those without a specific market assessment.[19] One company known to the author planned to develop fiberglass-reinforced plastic materials for use in the fisheries industries, but found they were unsuited to this intended purpose. Rather, they proved to be excellent source materials for piping, etc., used in chemical plants, and the company pioneered its application in this market. If the venture team discovers, after a time, that its technological competence may be more profitably employed in a different product–market niche than originally intended, it should make the change. However, such a radical shift in product–market strategy should only be made after conscious deliberation. Too often, in their understandable eagerness to generate income, new companies seek to divert and diversify their efforts into different market niches in a willy-nilly, ill-considered manner, thereby manufacturing a poorly coordinated product ranged based upon a nonexistent venture strategy. The point will be considered further in the next section.

Poor Pricing

The price should be set to ensure an adequate cash-flow into the company and a satisfactory return on the capital employed after operating costs have been covered, plus a surplus that may be invested in future innovative growth. Bell recommends that price should be set at four times cost[20] and underpricing is foolish for two other reasons. First, often, during the first year or two of operations, venture teams fail to measure and control costs adequately, so they are underestimated. Therefore, although the team thinks it can sell a product at a lower price than the competition and still generate a sufficient cash flow, it may simply have underestimated costs. It may only discover this error after a year or more of operations when the firm is virtually bankrupt. Second, people who are naive about market behavior believe that pricing a product lower than the competition is bound to generate high sales. In fact, the converse may well occur. Some purchasers view price as a measure of quality, and an NTBF with no track record and an as yet unproven product will be treated cautiously in the marketplace. Prospective purchasers may view a lower price as indicative of an inferior product, and so pay a higher price for a competitive product with which they are already familiar.

Sales Equated with Orders Predicted

Customers, when discussing their future anticipated (as opposed to current actual) product orders with a company's sales representative, sometimes overestimate. The customer wants to guarantee adequate supply in the long-term. Customers' long-term predictions of product requirements versus current purchase orders should be treated as optimistic and discounted accordingly (possibly by 50% or more). Failure to discount sales estimates obtained in this manner may lead to an overproduction that can have dire consequences for a financially vulnerable new venture.

Overreliance on Part-Time Staff

In Section 11.3 it was pointed out that a new venture will probably have to employ some part-time people in its early years to ensure that it can call upon the entire range of expertise required for effective management. Because of both time pressures and a realistic recognition that they lack expertise in all aspects of running a business, the venture team risks delegating too much responsibility to part-timers who, by definition, cannot be expected to give total commitment to the new venture. Although part-time managerial and professional help can be immensely valuable, the venture team must ensure that it is retaining ultimate managerial accountability for running its own business and must not become overdependent on outside help. A venture capitalist or other investor may provide some expertise if only to protect his investment, but the team should always be learning new skills to ensure the long-term management autonomy of the company. This does not preclude the permanent retention of some part-time support and/or professional services or management consultants. In particular, a judiciously selected nonexecutive part-time director retained on a modest honorarium can often enact a valuable coaching or alter-ego role.

Financial and Cost Measurement and Control

In Section 11.3 we quoted Bell on the prime importance of cash management in the corporate culture, and nonexistent or inadequately designed monetary measurement and control systems are a common source of problems for a new venture.

During its first few years of operations, an NTBF may not require, or be able to afford, the full-time services of a professional accountant. Obviously, accounting skills will be a mandatory requirement in preparing the business plan, and once the venture is launched, in fulfilling auditing requirements. This, however, will not be adequate for measurement and control purposes. Moreover, auditors and other part-time accounting professionals are unlikely to be able to provide the services required to develop an adequate measurement and control system. The definition and measurement of the costs of developing, manufacturing, and marketing new high-technology products require significant sustained attention, and part-time professionals cannot be expected to undertake this task. One study of technological entrepreneurs found that many had developed their own financial measurement and control systems more or less from first principles (possibly after advice and/or reading an appropriate text).[10] The scientific and engineering training of these individuals made them particularly adept at designing such systems as an exercise in empirical observation, measurement, analysis, and design. The performance of the exercise also gave each of them a useful insight into the pattern of monetary flows in his company.

Amazingly, some companies sustain several years of operations without making serious attempts at cost measurement and control. If they are manufacturing several products or product-lines, they cannot be in a position to know which are their most profitable products and which (if any) are being manufactured and marketed at a loss. This means they cannot truly identify the business they should be in. Sloppy cost measurement and control can also go hand-in-hand with other management "sins of omission." Cost analysis should inevitably involve value analysis—the identification of the cost and value added for each bought or manufactured component which is incorporated in the final assembled product. It can also highlight faults in (or the absence of) quality control procedures. Finally, it can draw attention to excessive inventory levels and accounts receivable, both of which can tie up significant amounts of the scarce working capital required by the company. Failure to pay due regard to all these cost measuring and control aspects can make the difference between bankruptcy and survival for a small new company.

11.7 CONSOLIDATION AND GROWTH

Some NTBFs remain, through choice or necessity, small and beautiful. Others grow to become mature firms. Growth requires the continued judicious exploitation of technological entrepreneurial opportunities. It also requires an increasing strategic management orientation on the part of the entrepreneurial team, especially its CEO, if the venture is to minimize the risk of following success by failure. Hambrick and Crozier[21] suggest the following guidelines for managing the growing firm:

1. The CEO should establish a vision of the firm as a larger entity, without abandoning its original vision.
2. Staff to support future growth should be hired and trained. If not done so already (see Section 11.3), employees should be given financial stakes in the firm.
3. Further information systems to control growth should be introduced as needed, without creating a bureaucratic hierarchical structure.

There appears to be no consensus on strategies to promote growth which may vary between technologies and industries, but some brief observations can be made. One strategy is to enter into one or more strategic alliances with larger organizations. This approach has been used successfully in the computer industry, notably by MICROSOFT in its alliance with IBM. Moreover, because of the nature of recombinant DNA and genetic engineering, discussed in the next paragraph, it is probably an essential strategy for the over 500 new biotechnology firms (NBFs) in the United States. For example, Forrest and Martin[22] surveyed such alliances in a sample of 42 NBFs, mainly in the United States. They found that the firms had entered into a total of over 1100 alliances between them, for an average of more than 26 per firm. Because the promise of new biotechnology is still largely unfulfilled, it is probably premature to speculate on the effectiveness of such alliances in fueling the growths of NBFs, but it is not entirely surprising that they are so extensively used, and they are considered in Chapter 13.

In Chapter 2 we drew attention to key differences between two post World War II technological revolutions, the discoveries of the double-helix and transistor action. Each has led to new Kuhnian technological paradigms in the new biotechnology and semiconductor industries, respectively, but there are key differences between the two. Although the new biotechnology industry is based on both new scientific and technology paradigms, design and manufacturing in the new biotechnology need to be *substantially integrated into* the old technology paradigm to be effective. This has meant that most (if not all) NBFs have sought growth through strategic alliances with other organizations, including larger companies that provide access to the old technology. In contrast, although the mechanism of transistor action is not based upon a new scientific paradigm,* the semiconductor design and manufacturing technology paradigm *is very different* from that of the vacuum tubes it replaced, so that the semiconductor industry has evolved largely independently of its vacuum tube predecessor. Strategic alliances played a much more limited role in the growths of small firms during the formative years of the semiconductor industry because of incompatibility of semiconductor and vacuum tube technologies.

If an NTBF, either through choice or necessity, seeks growth without entering into significant alliances with larger partners, how should it do so? Again, Roberts makes some useful suggestions.[23] He has found that the more successful NTBFs follow what he calls a *focused strategy* from a focused competence in a key core technology and products initially targeted at a focused set of customer needs, sold to gradually broadening groups of end-users. Recall the technology–market matrix of Chapter 4, Figure 4.3. He delineates new technology rows into four levels:

1. *Minor improvements* in existing product technology, which can be developed quickly in three months to a year or so and can be frequently introduced as new products.

2. *Major enhancements* to existing product technology, which may incorporate new base technologies and components to achieve leverage from the existing key technology. These take longer to develop, and are only introduced as new products at about three-year intervals.

3. *New, related technology* or a new core technology related to the existing technology base.

4. *New, unrelated technology* or a new core technology unrelated to the existing technology base.

He delineates the new market columns three ways: by (1) new product functionality, (2) new end-user groups, and (3) new distribution channels.

Roberts defines the focused strategy in terms of the above technology–market delineations. A company following a focused strategy evolves and grows through periodic major enhancements of its current technology competency that are leveraged into products to provide new levels of functionality to its users. As well as

---

*Shockley's explanation of transistor action in his *Electrons, Holes and Semiconductors* was based upon the extant paradigm of the band theory of solids, whereas Crick and Watson invoked a new paradigm with their double-helix.

offering new levels of functionality to their customers, more successful companies also seek to add new customer groups and distribution channels to expand their markets. In fact, as it grows and evolves, the venture places increasing emphasis on strategic and marketing concerns to adopt a market-driven as opposed to technology-driven strategic vision of its future. This does not mean that an original pioneering spirit of technological innovation is abandoned. Rather, through its improved understanding of the needs of its growing customer base, it is better able to conceive and develop new products that satisfy these needs.

Roberts' focused strategy would appear to be attractive to the company that began operations by following a technology nichemanship strategy of Chapter 4 (Section 4.8) and seeks growth. Nautical Electronics of Maine and Nova Scotia cited there appears to have successfully followed this strategy. In contrast, MITEL Corporation, the Canadian telecommunications company, followed success with failure through developing new, related technology for new markets. MITEL began operations in the early 1970s, and enjoyed a phenomenal compounded growth of more than 9% *per month* over a decade through designing, manufacturing, and marketing analog Private Automatic Branch Exchanges (PABXs) for small- and medium-sized organizations. In 1980, it decided to pursue growth by developing a digital PABX for medium and large organizations. The development of this product took longer and cost much more than anticipated, and coinciding with the recession of the early 1980s, forced the company to post losses for the first time. Marketing the new product line to new customers also proved to be much more difficult than anticipated, particularly in the United States, so that it has never achieved its promised success.[24]

Linked to the issue of growth and the transition from a technology- to a market-driven strategic vision of its future is "founders disease" or the inability of the founding CEO to "grow" with his venture, leading to his replacement (voluntarily or otherwise). The most well-known example is the replacement of Steve Jobs by John Scully at Apple.[25] Many, but by no means all, new technology ventures are forced into this often traumatic action to protect future growth opportunities. Paul Cook, as founder and CEO, grew Raychem into a billion dollar corporation, and Ken Olsen steered Digital Equipment Corporation from a $70,000 spin-off of MIT's Lincoln Laboratories to a Fortune 100 company. Both are archetypal technological entrepreneurs, and Bell rates Ken Olsen as the best CEO he has ever met.[26] Some cannot quit, even when they want to. The founder–CEO (now in his sixties) of one publicly traded company with a turnover of about $200 million has twice stepped aside in favor of successors he has hired. He has had to twice return as CEO when his successors proved unsuited to the role and, at the time of this writing, was seeking his third successor!

11.8 MATURITY: "GO PUBLIC" OR SELL

Assuming that, by now, the venture team has acquired the commercial maturity and judgment to implement suitable growth opportunities and to avoid willy-nilly diversification, it could be described as having reached the age of majority and the threshold of mature growth. Again, the acquisition of further capital support for this mature growth often presents real problems. Venture capitalists are perhaps now most interested in realizing their past investments with substantial capital gains rather than making further more substantial investments in the company. They recognize the wisdom of the old show business adage—always quit when the audience is asking for more!

If the founders wish for their company to continue its growth, sooner or later, they face the issue of going public. Providing the desired growth can be maintained, it is probably best to delay this decision as long as possible since more mature companies are viewed more favorably, and hence valued more highly, on the open market. Going public is a time-consuming and hazardous process and, as Howard points out, a significant proportion of companies that do so regret the decision afterwards.[23] Some founders may also be experiencing "hi-tech burnout" from some years of the sustained family and personal stress that has been the real cost of this growth. That is why some founders, when faced with the decision, prefer to sell the company if a purchaser can be found. Sexton and Bowman-Upton suggest that this option is chosen when the individual entrepreneur has changed his priorities from growth as an end to growth as a means—a change which may also be dependent upon the individual's biological clock. All members of the venture team may not have similar biological clocks, so disagreements may arise. Once that disagreement has been resolved, it is reasonable to expect that those who wish to exit the venture will sell at least part of the holdings to the remaining partners, on the stock market, or to the purchaser. Having achieved financial independence, if not considerable riches, they can each then decide whether to try to launch another venture or whatever else strikes their fancies.

Whether under the ownership of the remainder of its original founders or outside purchaser, once the venture is set on a course for mature sustained growth, it must then find ways of institutionalizing innovation and entrepreneurship if it is to avoid stagnation. Some of these methods are discussed in the next chapter.

11.9 FINAL COMMENTS

At the beginning of this chapter, it was suggested that many readers had unfulfilled dreams to set up their own high-technology companies. Reading this chapter from beginning to end may have shattered some of these dreams. We hope not. Launching and nurturing a high-technology venture is undoubtedly a challenging, protracted, and hazardous enterprise, but history demonstrates *it can be done*. Many have demonstrated this fact. Most readers will hold (or be obtaining) academic or other professional qualifications. Most readers will agree that the acquisition of such qualifications is also a challenging, protracted, and (at times!) hazardous enterprise. If individuals have the ability and tenacity of purpose to obtain good qualifications from reputable academic or professional bodies, they have the inherent personal resources to succeed in life. If these resources are matched with personal aptitudes for technological entrepreneurship, they should be capable of participating in successful ventures.

FURTHER READING

C. G. Bell with J. E. McNamara, *High-Tech Ventures: The Guide For Entrepreneurial Success.* Reading, MA: Addison-Wesley, 1991.

E. B. Roberts, *Entrepreneurs in High Technology Lessons from MIT and Beyond.* New York: Oxford University Press, 1991.

REFERENCES

1. J. A. Timmons, L. E. Smollen, and A. L. M. Dingree, *New Venture Creation: A Guide to Small Business Development.* Homewood, IL: Richard D. Irwin, 1977.

2. H. Keirulf, *"Additional Thoughts on Modelling New Venture Creation." In C. A. Kent, D. L. Sexton, and K. H. Vesper, (Eds.), Encyclopedia of Entrepreneurship.* Englewood Cliffs, NJ: Prentice-Hall, 1982, pp. 126–39.

3. E. B. Roberts, *Entrepreneurs in High Technology.* New York: Oxford University Press, 1991, pp. 110–111.

4. E. B. Roberts, *Entrepreneurs in High Technology.* New York: Oxford University Press, 1991, Chapter 5.

5. P. F. Drucker, *Innovation & Entrepreneurship.* New York: Harper & Row, 1985.

6. J. A. Timmons, L. E. Smollen, and A. L. M. Dingee, *New Venture Creation: A Guide to Small Business Development.* Homewood, IL: Richard D. Irwin, 1977, p. 12.

7. C. G. Bell, *High-Tech Ventures: The Guide for Entrepreneurial Success.* Reading, MA: Addison-Wesley, 1991, p. 10.

8. G. Bylinski, *The Innovation Millionaires: How They Succeed.* New York: Scribner's, 1976, Chapter 1.

9. C. G. Bell, *High-Tech Ventures: The Guide for Entrepreneurial Success.* Reading, MA: Addison-Wesley, 1991, Chapter 2.

10. M. J. C. Martin, J. H. Scheibelhut, and R. C. Clements, *Technology Transfer from Government Laboratories to Industry,* Report No. 53, Technological Innovation Studies Program. Ottawa, Ont.: Department of Industry, Trade and Commerce, November 1978.

11. M. A. Jolson and M. J. Gannon, "Wives—A Critical Element in Career Decisions." *Business Horizons* **15**(1), 83–88 (1972).

12. E. B. Roberts, *Entrepreneurs in High Technology.* New York: Oxford University Press, 1991, pp. 184–185.

13. C. G. Bell, *High Tech. Ventures: The Guide for Entrepreneurial Success.* Reading, MA: Addison-Wesley, 1991, p. 26.

14. J. A. Timmons, L. E. Smollen, and A. L. M. Dingree, *New Venture Creation: A Guide to Small Business Development.* Homewood, IL: Richard D. Irwin, 1977, p. 368.

15. R. Peterson and R. Ronstadt, "Developing Your Entrepreneurial Knowwho." In Y. Gasse (Ed.), *Entrepreneurship: Into the 90s.* London, Ont.: National Centre for Management Research and Development, University of Western Ontario, pp. 12–16.

16. M. J. C. Martin and D. A. Othen, "Developing University Technology: Some Observations and Comments." *Management of Technology. III. The Key To Global Competitiveness,* Proceedings of the Third International Conference on Management of Technology. Norcross, GA: Industrial Engineering and Management Press, 1992, pp. 48–56.

17. R. M. Knight, "Criteria Used by Venture Capitalists: A Cross Cultural Analysis." *Management of Technology. III. The Key To Global Competitiveness,* Proceedings of the Third International Conference on Management Technology. Norcross, GA: Industrial Engineering and Management Press, 1992, pp. 574–583.

18. T. E. Clarke, "Managing Your Management Time." In D. S. Scott and R. M. Blair (Eds.), *The Technical Entrepreneur.* Ontario: Press Porcepic, 1979.

19. E. B. Roberts, *Entrepreneurs in High Technology.* New York: Oxford University Press, 1991, Chapter 6.

20. C. G. Bell, *High-Tech Ventures: The Guide for Entrepreneurial Success.* Reading, MA: Addison-Wesley, 1991, p. 36.

21. D. C. Hambrick and L. M. Crozier, "Stumblers and Stars in the Management of Rapid Growth." *Journal of Business Venturing* **1**(1), 31–45 (1985).

22. J. E. Forrest and M. J. C. Martin, "Strategic Alliances: Lessons from the Biotechnology Industry." *Engineering Management Journal* **2**(1), 13–21 (1990).

23. E. B. Roberts, *Entrepreneurs in High Technology.* New York: Oxford University Press, 1991. Chapter 10.

24. M. J. C. Martin and P. J. Rosson, "MITEL and the SX-2000." In *Further Cases in Managing Technological Innovation and Entrepreneurship.* Ottawa, Ont.: DRIE, 1986.

25. "Playboy Interview: John Sculley." *Playboy Magazine* pp. 51–66 (1987).

26. C. G. Bell, *High-Tech Ventures: The Guide for Entrepreneurial Success.* Reading, MA: Addison-Wesley, 1991, p. 14.

____12

SMALL IN LARGE IS ALSO BEAUTIFUL: STIMULATING INTRAPRENEURSHIP

To encourage innovation, industry must now re-discover the fact that any innovation starts with a dream, and only people have dreams. It must accommodate to the individual in its organizational planning, and allow some of those dreams to materialize.

JOSEPH W. SELDEN, "ORGANIZING FOR INNOVATION"
IN *INNOVATION AND U.S. RESEARCH:*
PROBLEMS AND RECOMMENDATIONS

In reading what has been said about major corporations and their innovative capabilities, one gets the impression that there is almost a love-hate relationship . . . few corporate leaders would admit to being anti-innovation. Yet many will wonder whether a relatively "free-wheeling" culture can coexist under the same corporate skin with traditional proceduralized business.

ROBERT A. BURGELMAN AND LEONARD R. SAYLES,
INSIDE CORPORATE INNOVATION

12.1   INTRODUCTION

In Chapter 10 we discussed the spin-off phenomenon and the personality characteristics of technical entrepreneurs. We saw that, in many instances, a key factor stimulating independent spin-offs from larger companies was a person's frustration with general management and the environment for entrepreneurship in the incubator organization. In the previous chapter, we examined the problems of launching and operating an independent spin-off or NTBF, concluding by suggesting that when it grows to maturity, it runs the risk of losing its innovative entrepreneurial drive. Since entrepreneurship is a vital constituent in the innovation chain-reaction, well-established high-technology firms which fail to stimulate and nurture entrepreneurship also run the risk of becoming technologically moribund, losing their capacity for innovation. At the end of Chapter 9, we briefly referred to the need to set up independent or quasi-independent ventures to exploit innovations and forge the entrepreneurial link in the innovation equation. At that point, we stated that we would return to this important topic in Chapter 12, having reviewed the literature on spin-offs and NTBF entrepreneurship.

12.2 DYNAMIC CONSERVATISM VERSUS ENTREPRENEURSHIP IN LARGE ORGANIZATIONS

Donald Schon is a leading thinker and writer who has explored many facets of technological and social change. He argues that social systems, including high-technology corporations, have a built-in tendency to seek to maintain a stable state and resist change and this tendency is much stronger than an inertial resistance to change:

> The system as a whole has the property of resistance to change. I would not call this property "inertia" a metaphor drawn from physics—the tendency of objects to move steadily along their present courses unless a contrary force is exerted on them. The resistance to change exhibited by social systems is much more nearly a form of 'dynamic conservatism'—that is to say, a tendency to fight to remain the same.[1]

We have already commented on some of the causes of dynamic conservatism in previous chapters, but it is useful to review them here. Collier[2] provides the following summary:

1. Financial resources are usually allocated by DCF criteria which favor short-term incremental investments in the company's present business. Other innovations require a longer development lead time (recall Chapter 2), before generating net positive cash flows and a significant ROI, which must be discounted over a longer time period. This observation is borne out by Biggadike in an analysis of new ventures undertaken by large corporations.[3] He found that, on average, they took eight years to become profitable and 10 to 12 years before their ROI's equaled those of the corporations' mature businesses.

2. Product engineering is required to produce designs which work reliably and safely in the user environment (recall Chapter 9). Product engineers are therefore typically conservative since they prefer to support new product (or process) designs which incorporate incremental changes in proven designs, rather than risk their professional reputations in unproven, significantly new product or process design concepts.

3. Manufacturing achieves highest efficiency in routine operations on an assembly-line basis. Quoting Collier: "*This the exact antithesis of what happens when a new product is introduced. To manufacturing, the new product is a monkey wrench tossed into the gears of a highly efficient operation.*"

4. Trade unions resist any process improvements which reduce labor requirements, and when new products are introduced, seek to loosen work content standards.

5. Salespeople, who are usually paid partially by commissions on sales revenue generated, know that time is money. New products require them to invest extra time in educating themselves and their customers, so they thereby receive less commission per hour of sales effort invested.

6. The reward system for general managers is typically based upon annual profits or ROI of corporate resources managed. They are therefore rewarded for achieving short- rather than long-term profit. Moreover, apart from the greater inherent risks involved, the rewards associated with the profits from any longer-term, more radical innovations are unlikely to accrue to the manager making the original investment since he is likely to have moved on to other responsibilities before they are achieved.

These factors largely explain the dynamic conservatism of many high-technology organizations which has to be overcome if they are to be technologically innovative. As we discussed in Chapter 4, such organizations must stimulate innovation as well as operations management if they are to remain commercially healthy. Fortunately (from the viewpoint of large corporations), not all entrepreneurially-motivated individuals wish to set up their own independent ventures. Some are attracted by the relatively high community status enjoyed by individuals holding middle or senior management positions in larger recognized corporations. Moreover, such individuals' ambitions may motivate them to seek advancement in larger organizations which offer rewards in terms of increased formal power and internal status. Thus, larger high-technology organizations can retain a reservoir of entrepreneurial talent, and we now consider alternative organizational mechanisms for stimulating intra-corporate entrepreneurship. Pinchot coined the apposite and more succinct word *intrapreneurship* to describe this activity, and suggested that the above barriers constitute a *corporate immune system* which must be held in check if intrapreneurial endeavors are to succeed.[4] Some companies have succeeded in doing this, while others have failed. Despite its initial failure to recognize the importance of the personal computer, IBM redeemed itself through an imaginative intrapreneurial project.[5] In contrast, Xerox's immune system apparently caused it to ignore the concept of the product in the first place.[6] The personal computer was obviously a revolutionary innovation, but some companies are better than others at managing a succession of less radical efforts. Later in the chapter, we shall review 3M's intrapreneurial achievements, which again contrasts with Exxon's less successful efforts.[7]

12.3 INTRAPRENEURIAL ORGANIZATIONAL STRUCTURES

Formal approaches to the stimulation of intrapreneurship can be traced back to at least the 1960s, and a number of different organizational structure approaches have been adopted. As Burgelman suggests, the required approach is dependent upon the proximity of the intrapreneurial new product concept to the corporation core business (especially its core technologies) and its strategic importance.[8] We shall see in the next section, despite the fact that both Texas Instruments (TI) and the Minnesota Mining & Manufacturing Company (3M) have fairly concentrated core technologies in semiconductor electronics and surface coatings, respectively, they have quite different approaches. This difference is at least partially due to differences in products and new product introduction strategies. Both are offensive innovators; but whereas TI concentrates on a limited range of high-volume products, and is reported to have withdrawn from the electronic watch market because it could not achieve the market share that it desired, 3M seeks to proliferate new products. For example, in 1990, it was reportedly making 60,000 different products.[9] Approaches vary from those of TI, where individuals are expected to enact both operational and intrapreneurial roles through ventures largely severed from their parent corporations. We discuss such alternative approaches, with illustrative examples, in the next section.

12.4 INTRAPRENEURIAL VIGNETTES

We begin with a series of vignettes which examine some of the alternative approaches to the stimulation of intrapreneurship adopted by major corporations in quite different industries. First, the TI approach which encourages managers and technologists to enact both innovation and operations roles, in quite considerable detail. Second, 3M, where the story of ArtFry and the *Post-It* note has become a legend. Third, Eastman-Kodak, another corporation that has a long history of such endeavors. We then make a less detailed overview of the approaches adopted by other corporations, and conclude the chapter by reviewing the issues and problems inherent in intrapreneurship.

Texas Instruments—Objectives, Strategy, and Tactics (OST)

Despite more recent downs-and-ups following the retirement and then sad, sudden death of Patrick Haggerty, it almost goes without saying that Texas Instruments (TI) has been one of the most successfully innovative high-technology corporations since World War II.[10] Founded in 1930 as a geophysical services company, it established laboratory and manufacturing facilities in 1946, and entered the then infant semiconductor industry in 1952. Since then, it appears to have followed a consistent offensive strategy, pioneering the development of silicon semiconductor technology, first with discrete semiconductor devices, then integrated circuits, and finally a range of products embracing this technology. Innovation is described as a way of life at TI, and this quotation of the words of one of the company's senior managers illustrates this philosophy:

> We are convinced that useful products and services as well as long-term profitability are the result of innovation. Further, we feel that profitability above the bare compensation for use of assets can come only from a superior rate of innovation and can no longer exist when innovation is routine. This is why our long-range planning system is fundamentally a system for managing innovation.[11]

TI's long-range planning system designed to institutionalize innovation is called Objectives, Strategies, and Tactics, or OST. As Jelinek states, it is difficult to capture on paper a true feeling for the organizational climate created by OST.[12] Indeed, OST appears to be not so much a system, but more a way of life at TI. Reading his own account of its conception and inception, it appears that the system owes much to the impressive personality and managerial insight and judgment of Patrick Haggerty, the sometime president and chairman of TI.[13] OST was conceived and instituted in TI in the 1960s, during what might be loosely labeled its adolescent, formative years, and now provides a framework for managing the corporation at virtually all levels. It is relevant to note that OST was instituted at a formative stage of organization development, and the installation of a similar system in a mature, larger company might be very difficult. Despite this reservation, a brief examination of its features is in order here since OST offers salutary ideas for managers in other high-technology firms.

When TI entered the semiconductor industry in 1952, virtually as a small novice company, it achieved considerable technological and marketing success through developing silicon transistor technology. By 1959–1960, top management recognized that as the organization grew out of its success, it risked losing its innovative drive and edge through the onset of what might be called *bureausis,* or the inhibiting effect of bureaucratic hierarchies and procedures that can degrade the performance of an organization, like arteriosclerosis degrades the performance of the cardiovascular system of the human body. Top management thinking echoed the view of Gardner, who argues that societies and institutions tend to lose their vitality as they mature, and that to avoid this decay, they must build in a capacity for self-renewal, based upon individual self-renewal.[14] TI's success was based upon successive innovations realized within well-conceived strategies and well-executed tactics in support of these strategies. This approach was therefore formalized in the company.

As Haggerty states: *"The objectives, strategies, and tactics system at Texas Instruments is just that—an attempt to create for one organization a system or framework within which continuous innovation, renewal, and rebirth can occur."*[15] Historically, TI's operations have been managed through a group divisional structure and Product Customer Centers (PCCs). In 1978 there were four groups, 32 divisions, and more than 80 PCCs with annual sales between $10 million and $100 million. The management of each center holds responsibility for marketing, manufacturing, and new product development and is judged annually on its profit and return on assets performance. Although the PCC concept should ensure that products are developed which satisfy market needs, it is vulnerable to stagnation and bureausis for the reasons we discussed earlier. If a manager is judged annually by his operating performance in terms of profits and returns on assets, he is less than likely to wish to promote innovations which require lead times longer than one year, incur significant development costs, and (if they are really innovative) carry a definite risk of commercial failure anyway. Furthermore, the administration of the innovation process becomes more cumbersome as an organization grows, and the industry maturing process may make profitable innovation more difficult. Therefore, a greater premium gets placed on administrative rather than innovative skills, so that managers seek to demonstrate the former rather than the latter skills in order to obtain promotion. These views were expressed by one of TI's senior managers as follows:

> OST constitutes a formalization and institutionalization of the informal approach to long range planning used by senior corporate management in the 1950s, but extended throughout the organization. Implemented in conjunction with PPCs it creates a form of matrix organization in which all staff can wear two hats as both innovation and operations managers.[11]

This matrix structure is illustrated in Figure 12.1. Individual PPCs are aggregated into divisions, and then groups, based upon generic technologies or markets, which report to corporate management. It is the OST structure, which constitutes a hierarchy extending downwards from corporate objectives to tactics, that requires some explanation.

The objectives are broadly stated goals at the corporate level and each business area. The strategies define long-term general courses of action in pursuit of these goals. The tactics represent the relatively short-term projects in support of the strategies, to which are allocated individuals with specific responsibilities and resources. It is the hierarchy of individuals who are responsible for the objectives, strategies, and tactics which collectively constitute the innovation management function. The purpose of this hierarchy is to create a climate which stimulates the

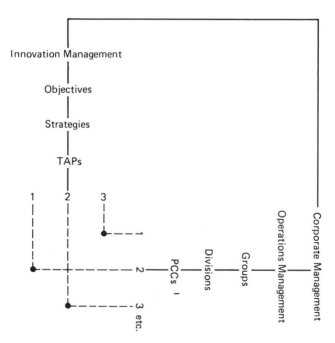

Figure 12.1. The Texas Instruments matrix.

generation of innovative ideas or proposals throughout the organization, and to evaluate these proposals in terms of corporate and group objectives and strategies.

Given the fecundity of inventive and innovative ideas within an organization, the problem becomes that of rank-ordering proposals in terms of their impact on strategies and objectives and (as in R&D project selection) funding the highest ranking of these—which become known as tactical action plans (TAPs)—until the current funds allocated for TAPs have become exhausted. TI's long-term economic and innovative performance is clearly dependent upon the effectiveness of this screening and selection process.

Starting from the bottom and working upwards, individual proposals are screened and ranked at the tactical, strategic, and objective levels. Higher-level innovation management specifies the total funds available at various hierarchical levels so that a funding cutoff can be established. Proposals which look promising, but cannot be supported in competition with others because of insufficient funds, are placed in a creative backlog so that they may be supported should more funds become available. Furthermore, the selecting/screening process is repeated at the higher hierarchical levels. Strategy managers review the selections made at the tactical levels and Objective managers review the selection made at the strategic levels. In some cases, these higher-level managers may change the lower-level rankings, possibly in light of wider strategic and objective considerations. This repeated review of rankings is important since it ensures that a potentially radical, as opposed to incremental innovation, which may have a potentially broader impact upon strategies and objectives, receives adequate consideration at the appropriate hierarchical level. It also gives higher-level managers an opportunity to identify potential synergies between proposals coming from separated lower-level sources. Some earlier funded proposals (or TAPs) which have duration times which extend into the current cycle are reviewed as part of the selection process to see if they have fulfilled their earlier promise or are now less congruent with revised strategies and tactics. They may consequently be terminated and the unspent funding re-allocated to other proposals. TAPs thus face a continued competitive review process.

OST is a procedure for institutionalizing an organizational culture conducive to maintaining innovation and entrepreneurship as the organization grows. However, because it institutes an innovation management hierarchy which overlays the operations management hierarchy, it inevitably duplicates bureaucracy since OST itself is a bureaucratic process. Although OST has succeeded in institutionalizing an innovative spirit in TI, its bureaucracy inhibits some creative entrepreneurial individuals from championing their ideas competitively over the successive hurdles of review process. This will apply particularly to the younger man who may be inexperienced and impatient with company politics, and who may produce an idea for a radical innovation which, although offering commercial promise, is incongruent with the priorities of the current OST program. The company therefore also has an *IDEA Program* designed to accommodate such exceptions. It is introduced to its staff as follows:

The IDEA Program provides an opportunity for initial feasibility demonstration of concepts that do not fit within the immediate OST thrusts. This program will appeal to those of you who want to be entrepreneurs and innovators because it will provide an environment in which you and your ideas can flourish.[12]

Quoting the words of Patrick Haggerty:

Every effort is made to keep the IDEA System simple. There are no approval cycles, no delays, no reviews, and no reports. TI had named about 50 key persons around the company who have a proven record and shown that they know what a good idea is when they see one. Each of them is available to listen to new ideas and on their own, they can fund, without any further discussion or approvals, up to $25,000.

Two very successful products which were initially funded from the IDEA Program were the $19.95 digital watch launched in 1976 and the "Speak and Spell" learning aids based upon a TI engineer Gene A. Frantz's idea for electronically synthesizing the human voice using large-scale IC technology.

One impressive feature of the semiconductor industry, which is still little more than a generation in age, has been the rapidity with which unit prices (first with discrete devices and then ICs) have fallen as output yields have risen. In this situation, a company which can maintain continued increases in labor and resource productivities is more likely to maintain a competitive cost advantage. Furthermore, for a given market price level, manufacturing operations which are more efficient will incur lower costs, thus creating more profit to be re-invested in innovative ventures. Therefore, once OST had become established, TI expanded it to include a People and Asset Effectiveness Program (P&AE) which, the then President of TI, Mark Shepherd, Jr. described to stockholders in these words in 1973:

People and Assets Effectiveness is an improved ability to solve customer's problems through increased productivity of people and assets. It means improved planning and control systems. It means the application of automation to design, manufacturing and to management problems. And, most importantly, it means the fullest use of all the talents of their minds as well as their hands—to be involved in the planning and control of their work, not simply the doing . . .

The P&AE Program is, in effect, incorporated in the OST Program with goals set for increased productivity and growth. Proposals for improving productivity are evaluated through a similar selection process and, as the final sentence in the above quotation implies, it is designed to elicit contributions from blue— as well as white-collar workers. TI was one of the early companies to join the job enrichment movement, encouraging both manufacturing and clerical work teams to devise their own workplace organizations and procedures. The participative climate for stimulating innovation thus permeates all levels, an important factor since successful technological innovation often requires innovative thinking at the technician/craftsmen level as well as the scientific/engineering level.

OST generates an organization culture which has two important features relevant to our present discussion. First, since individuals know that their proposals will be evaluated in the context of corporate objectives, strategies, and tactics, they will be forced to conduct their own evaluations of their ideas, thus impelling them to seek a reconciliation of their personal goals with corporate goals. OST thus provides a framework for matching individual and organizational goals, and ensures that top management's perception of corporate goals permeates the organization, thereby increasing the likelihood of the effective implementation of a corporate technological strategy. Second, it forces individuals to wear two hats as both operations and innovation managers. This second feature will be discussed in a little more detail.

As was stated earlier, operations at TI are managed through PPCs. This management structure is decentralized, and local PPC managers have the autonomy to run their own businesses and are being held accountable for their own performances against criteria agreed upon between corporate management and themselves. Operating performance is measured using conventional reporting criteria of profit and loss statements and annual operating budgets. That is, when wearing their operations manager's hats, individual's report in a conventional manner. Successful proposals or TAPs are funded from a separate strategic budget and evaluated independently against agreed upon criteria so that operations and innovation programs are separated, then consolidated in an overall profit and loss statement (Figure 12.2). Typically, at any time, 75% of the managers have innovation as well as operations management responsibilities, and report simultaneously through innovation management and operations management hierarchies which are largely overlapping.

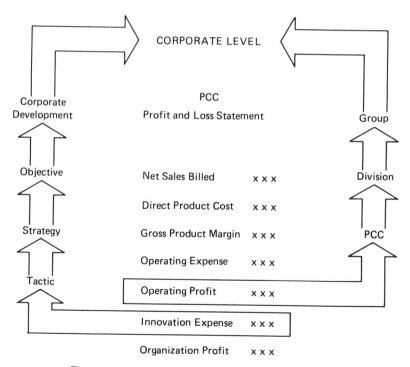

Figure 12.2. Operations and innovation P&L statement.

That is, an individual may report to a divisional manager wearing his operations management hat and a strategy manager wearing his innovations hat. In some cases, this individual may be a two-boss person (as in the conventional matrix structure) since divisional and strategy managers are different individuals. However, in many cases, an individual may report to only one boss since his superior will also be wearing two hats and enacting both the divisional and strategy managers' roles. Whichever situation exists, the dual-operations and innovation reporting structure permeates the entire management hierarchy. However, the enactment of an innovation as opposed to operations management roles imposed differing behavioral requirements.

Conventionally, it is generally recognized that a manager's spans of responsibility and authority should coincide if he is to perform effectively. However, virtually by definition, an innovative project or TAP will require the innovation manager to exercise a wider span of responsibility than his authority allows. The separation of responsibility and authority requires him to exercise entrepreneurial or project championing skills to secure through negotiation and cooperation, rather than authorization, the organizational resources that the project requires. The OST system creates a demanding but satisfying climate for technological entrepreneurship, and an individual may develop a successful management career based upon proven performance in innovation as well as operations management. This is summed up in the works of George H. Heilmeier, a VP of TI who formerly headed the Pentagon's Advanced Research Projects Agency.[15]

> TI seems to have discovered the style that allows a multi-billion-dollar corporation to grow at 15% a year. To match that, companies are going to have to share information with more people. They're going to have to maintain entrepreneurial spirit, and management is going to have to have good visibility in all the company.

Minnesota Mining & Manufacturing Company—3M

The company epitomizing an intrapreneurial culture is 3M, while the product epitomizing the intrapreneurial innovation process is its renowned Post-It Notes.[16] Norman Macrae, the former editor of *The Economist*, proposed the notion of the *future* successful organization as a *confederation of entrepreneurs*.[17] The 3M Co. could be viewed as the *present* successful organization as a *confederation of intrapreneurs*. 3M was known for its encouragement of internal new venture creation before entrepreneurship and intrapreneurship became "in" words. Hanan reported that, by 1969, 3M had launched over two dozen ventures (with six current operating divisions having grown out of the venture system).[18]

The 3M culture encourages intrapreneurship in several ways[19]:

1. Divisions must generate 25–30% of sales from products launched within the last five years.
2. 3M itself began with failure. The company was created to manufacture and market sandpaper made from corundum extracted from a mine its founders

purchased for that purpose. Unfortunately, the mine contained no corundum, so the founders had to seek other supply sources and products to survive this initial failure. This means that risk-taking and tolerance of failure have been embedded in its culture from its beginnings. Since some new product ideas fail, tolerance of failure breeds more new product attempts and more successes.

3. Individuals are encouraged to spend about 15% of their time on developing their own new product ideas.

4. If they believe that these ideas offer attractive new product or venture opportunities to the company, individuals can seek out sponsorship for them. If they work in R&D they may seek an R&D-sponsored venture. If they work in an operating division, they seek it there.[20] If their parent division refuses to support them, they may seek sponsorship from another division or from a New Business Ventures Division which is associated with Central Research. For ideas that do not fit into the above, the Genesis Program awards seed money up to $50,000 for the development of product prototypes. Consequently, if they possess a good idea plus the required personal entrepreneurial qualities, they are likely to find sponsorship somewhere within the corporation. As Park expresses it:

You get the sense that its **de rigeur** to jostle a bit for the limelight—especially if you want to become a project manager. Sometimes, at division level, the spot-light waivers—there are so many things to push along the managerial scrutiny must, perforce, skip lightly over some of the newer ideas. But even then, the man with an idea—especially if he's done his homework and can answer some hard-nosed questions—can bring his dream to an arena where the candlepower is a bit more intense. This is 3M's New Business Ventures Division . . .

In deciding whether to sponsor a new venture, 3M recognized the critical importance of the entrepreneurial qualities of the lead intrapreneur or product champion in the 3M vocabulary. Quoting the words of one former 3M technical manager:

The typical product champion at 3M turns out to be one-quarter technical man and three-quarters entrepreneur. He pushes a product idea by what amounts to conning the management—somewhat exaggerating its chances for success, hoping to get enough money to make the product fill the bill.[20]

Clearly, experienced top management knows how judiciously to discount an individuals exaggeration and when to support it. As one 3M president put it:

We're in the business of gambling on individuals. If a man has been a pretty good judge in the past of what he said he was going to do and has done it, then if he comes in and says I want to start importing moon dust, I guess I'm likely to let him try. In like manner, if a fellow who's been around here a great many years comes forward and says this year we've got a tremendous breakthrough that's going to cause all sorts of thing to happen, and he's had four or five years in a row of not delivering on what he said, well, I'm likely to give him an argument.

5. Once support is granted, intrapreneurs are encouraged to put together their venture team informally from their network of contacts within the company. Because of the reward system in place, a good news product idea is likely to attract a strong team. Rewards are linked to new product success. If a new product idea grows into a new venture, team members receive monetary and status rewards commensurate with its growth.

The Eastman Kodak Company

The company was founded by inventor–entrepreneur George Eastman in the last quarter of the nineteenth century. Performing empirical research in his mother's kitchen and funded by personal savings and loans, he sought to develop photography for the masses. One hundred years later, the single product Eastman Photographic Plate idea had grown into a diversified multinational corporation with 25,000 product lines in several business sectors and a billion dollar annual R&D budget. By now a mature corporation that had long outgrown its entrepreneurial beginnings, it exhibited some of the features of Schon's dynamic conservatism, so steps were taken to stimulate organizational renewal.

The first step taken was based upon a long-standing Research Proposal System (RPS) that allowed R&D staff to receive a prompt evaluation of new product ideas they submitted to top management.[21] In practice, some good ideas were quickly rejected because proposals were insufficiently refined to withstand top management scrutiny, so an Office of Innovation (OI) was created in 1979 to provide informal help to staff in preparing their proposals. It essentially consisted of innovation facilitators who provided staff with candid advice on improving their proposals and potential support sources within the company. This approach was very successful, with OI submissions increasing from 30 in 1979 to 599 in 1985, when six OIs had been set up in the United States, Europe, and Australia.

The OI approach worked well with innovations which could be readily integrated into mainstream operations, but not for those requiring offstream implementation. The next step, taken in 1983–1984 at the initiative of Robert Tuite (then assistant to the then director of research), was to create an internal corporate structure that could replicate the environment for the early stages of the venturing process described in Chapter 11. This structure, described as the *3-Phase Innovation Management Process,* is displayed in Figure 12.3.[22]

The OIs and Innovation Facilitators were incorporated into a new office of New Opportunity Development (NOD), with Robert Tuite as its Director. Support is divided into three stages:

1. The approximately 10% of promising proposals which were unsuited to mainstream operations can receive up to about $25,000 from NOD in support in the Idea Development Phase. Their originators now became venturers and could devote 20% of their time to the venture, working independently of their supervisors. In effect, the Idea Development Phase equates to the Preliminary Moonlighting Stage of Chapter 11.

3-Phase Management Process

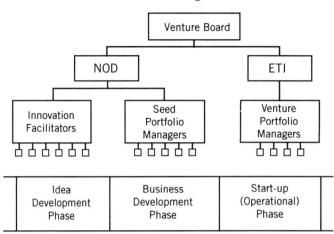

Figure 12.3. 3-Phase Innovation Management Process: The Eastman Kodak approach.

2. If still acceptable, the project next moves to the Business Development Phase. Now, the venturer leaves his original job and can receive up to $75,000 of funding. He will now be required to assemble a venture team, develop a business plan and prototype product, and identify target customers. He will receive advice and support from NOD staff through a Seed Portfolio Manager and an informal Venture Advisory Panel composed of suitable experienced people from throughout the company.

3. If still acceptable, the venture moves to the Start-up (Operational) Phase, with approval being granted by the Venture Board which can release up to $250,000 of funding. Further funding can be provided later for a venture, subject to quite demanding performance criteria. The venture now reports, through venture portfolio managers, to Eastman Technologies Inc. (ETI) which acts as both a holding company and venture capitalist sponsor for the start-up phase ventures. Although ETI is a fully-owned subsidiary of the parent company, the ventures themselves are formally severed from their parent. ETI acts as an incubator organization by providing financial and advisory support, as well as low-rental accommodation, but the ventures are formally distanced from the Company in the following ways:

a) NOD and ETI, together with the ventures, are located in a building in downtown Rochester which is physically separated from existing Kodak organizations.

b) Neither Eastman Kodak Company resources nor its name can be used to support or promote the ventures; nor must a venture failure adversely affect Company operations.

c) Team members must give up their former jobs in the Company, with no guarantees of re-employment should the venture fail.

ETI seeks at least a 25% return on its new ventures portfolio to maintain its own operations. Therefore, if successful after a few years of operations, ventures are expected to go public or find a purchaser so that ETI can achieve substantial capital gains. Only 10% of the original new product proposals receive initial support; of these, 90% are supported by mainstream operations as potential product or process innovations. Therefore, at most, only 1% of original new product ideas reach the Start-up Phase. By 1987, 14 new ventures had been launched and had experienced mixed fortunes. By 1989, two had been purchased back by Kodak and two were up for sale. Four failed, leaving a core portfolio of six operating ventures. The most successful was Sayett Technologies, launched to produce devices to project computer images onto a large screen. After only five quarters of operations, Sayett was valued at eleven times the cash invested in it. In fact, it was so successful that it briefly returned to the Kodak fold. It was purchased by Kodak Motion Picture and Audiovisual in 1987, but divested two years later because the market niche it occupied was too small to be of strategic interest to Kodak. Its two founders left it in 1987 to help run another ETI venture.

12.5   OTHER EXAMPLES

Having considered three examples in varying detail, we now examine some of the ways other mature high-technology organizations have stimulated intrapreneurship. As well as 3M, Hanan quotes Dow, Westinghouse, Monsanto, Celanese, Union Carbide, DuPont, and General Mills as corporations committed to the venture philosophy.[23] As was stated earlier, separate venture organizations may be established in one of several ways, the preferred choice being dependent upon the nature of the innovation in relation to the company's other product lines, patenting, and financing considerations and the strengths and weakness of the planned venture team, among other situational factors. Alternative models for establishing a venture are now discussed.

R&D Venture Organizations

When discussing technological strategies in Chapter 4, it was pointed out that many companies follow mixed strategies (that is, two or more technological strategies simultaneously) to achieve corporate goals. We demonstrated in that chapter that the enactment of an offensive/defensive strategy required all-around excellence throughout the technological base of the organization. On the other hand, an applications-engineering strategy can be enacted from a truncated technological base since it requires a D-intensive organization capable of swiftly exploiting specialist applications opportunities or new market needs perceived by field sales and servicing staff. Furthermore, changes in corporate, government, and social attitudes to science over the last decade or so have encouraged companies to perform market needs-oriented development rather than new knowledge-oriented research to achieve corporate growth goals. This attitudinal change was also reflected in the "consumer–contractor" principle for research prescribed by Lord Rothschild in Britain.[24]

The R-intensive central R&D facility of an offensive/defensive innovator is unlikely to employ staff with the orientation and speed of responsiveness to achieve the timely development results required to implement an applications-engineering strategy. For the same reasons, the facility may display a slow inertial response towards the changes in corporate and public research attitudes cited above. These considerations have led some companies to establish semiindependent satellites to perform contract R&D (with an emphasis on the D rather than the R) on a profitable venture basis.[25] Such a venture can operate as a semiautonomous entrepreneurial enterprise, performing commissioned development work for an operations facility in pursuit of an applications-engineering opportunity, contract R&D on the open market (that is, for any other customer who is willing to pay for their services), and possibly develop new products different from the company's current product/market lines. The British Oxygen Company went even further in 1971 when they replaced their central R&D facility with a new venture secretariat which selected R&D projects which were performed on operations divisional sales.[26] These projects were established as separate semiautonomous venture units with responsibilities for marketing their own products, the operation divisions merely acting as landlords.

Internal Individually or Jointly Sponsored Ventures

An internal venture, perhaps sponsored by one, two, or more operating divisions, is one approach which has proved to be useful in several large corporations. This format may be particularly appropriate when the innovation contains features which impinge upon the technologies, product lines, markets, or uses of two or more operating divisions. Hillier cites RCA's development of magnetic videotape, considered commercially unattractive by a division.[27] Their broadcasting affiliate of NBC, however, anticipated that an RCA competitor would introduce videotape equipment which NBC would then have to purchase to maintain its technological competitives with CBS and ABC, then the only other major U.S. commercial TV broadcasting networks. A sensible approach in this situation was for the equipment division and NBC jointly to sponsor a videotape equipment venture. It was developed internally in the equipment division, partially financed by NBC.

DuPont uses a similar approach to stimulate intrapreneurship.[28] DuPont's R&D activities are broadly categorized as exploratory research, improvement of established business, and new venture development. Exploratory research is long-term, and improvement of established business can be roughly equated with incremental innovations in existing markets—innovations which may be readily accommodated in the extant management structure of current operating divisions. When looking at more radical innovations in possibly new markets, other considerations apply. In many cases, however, promising new products or systems are visualized in R&D programs which do not fit well into the marketing and manufacturing capabilities of the department. It is then that a New Venture Development is considered.[28] At

DuPont, new venture groups are set within an extant department of the firm, but there is no attempt to standardize the approach. Again, in deciding to sponsor a new venture, the emphasis is placed on the venture's CEO, called the "New Venture Manager," who enacts the role of the lead entrepreneur or project champion. The emphasis on the entrepreneurial qualities of the venture leader is a recurrent theme in most of the articles on this subject. Hanan quotes the words of one company president:

> I have an unfailing test for identifying entrepreneurial types. I throw every candidate right in with the alligators. The establishment man complains he can't farm alligators in a swamp. The entrepreneur farms 60% of the alligators, markets another 30% for everything but their squeal, drains their part of the swamp, and leases the land for an amusement park overlooking "Alligatorland." The other 10% of the alligators? That's his delayed compensation.[18]

External Joint Ventures

Another approach is an external joint venture sponsored by the parent company and an independent second company. This format may be appropriate when the innovation contains features which infringe upon the other company's distinct markets or industries. Alternatively, our innovative company may buy a major or controlling share in the second company and use the latter's industry/market knowledge and image to spearhead its thrust into new markets. This is one typical strategy for interindustry invasions such as chemicals into textiles or electronics into watches. Such joint ventures or strategic alliances have become almost commonplace recently, and are discussed further in the next chapter.

External Supported Ventures or Severed Ventures

A further approach, which has been pioneered by GE in the United States, is to establish a virtually independent company outside the walls of its parent.[29] It is a spin-off venture which provides a compromise between either the internal or external joint ventures discussed above and the formation of a totally independent entity or spin-off company discussed in Chapter 11. The severed venture approach adopted by GE has the following features:

1. The severed team contributes their skills and some personal capital to the development of the innovation and holds an equity share in the venture.
2. The parent company holds a minority equity share based upon its contribution of equipment, inventories, and patents to the venture. GE does not contribute any capital to the venture, but (in principle anyway) a parent company could make a minority venture capital contribution, without attaining a majority equity position and effectively creating an internal venture.
3. Further equity capital is obtained from the venture capital market or other sources.

The problems faced by GE are broadly identical to those faced by other high-technology corporations and discussed earlier. Their rationale for setting up ventures is noted below.

An innovation may appear to be commercially promising, but offer too small a market and therefore an unattractive direct opportunity to the corporation for two reasons. First, a small market does not offer the commercial returns to justify the opportunity cost of investing further scarce corporate management and physical resources in the innovation. Larger returns can be obtained by investing these resources in other investment opportunities. Second, if such a large corporation were to enter such a small market, it would probably dominate it, with the possible consequent invocation of anti-trust legislation. On the other hand, if a promising innovation cannot be sold or licensed to a third party, its abandonment represents a potential loss to society at large (which does not enjoy the benefits of the innovation) and an actual loss to the corporation (since it has to write off the investment to date).

By formally spinning off the innovation into a severed venture launched by its former employees, GE still maintains the potential for social benefit of the innovation and the recovery of corporate-sunk investment from its minority equity share in the venture.

The approach was pioneered by David J. BenDaniel, who was foundation head of GE's Technical Ventures Operation which launches such ventures. To qualify for incorporation as a severed venture, an innovation must possess an identifiable market niche in which it offers a competitive advantage which can be exploited as an opening wedge for expansion into a larger market—with the consequential potential for equity growth for the corporation. It must offer an entry into a market which the corporation wishes to monitor without further investment or formal entry. It must be spun off by a severed venture team of former GE employees which includes committed project-champions or a viable entrepreneurial team—*"men who believe in the business enough to invest their savings and risk their futures in it."*[29] BenDaniel set up the Technical Ventures Operation in 1970, and by 1973, he estimates that GE had enjoyed a threefold growth in its equity shares in its severed ventures to $3 million–4 million.

12.6 PROBLEMS OF VENTURE DEVELOPMENT AND INTRAPRENEURSHIP

In the foregoing sections we have reviewed a number of mechanisms used by the larger high-technology firms to stimulate innovation and entrepreneurship, and it would be appealing to be able to report that these mechanisms have been uniformly successful. Unfortunately, this is not so, and we must now examine some of the problems of new venture development in high-technology companies. Kanter describes them as *newstream* as opposed to *mainstream* activities that face special hazards[30]:

It is the nature of newstreams to be uncertain, bumpy, boat-rocking, controversial, knowledge-intense, and independent—seeking their own course. These characteristics make them vulnerable to unique dangers. And in addition to the problems all start-ups share, newstream activities face additional vulnerabilities when they are carried out in the midst of a powerful mainstream.

Vesper and Holmdahl[31] studied the 100 companies in the Fortune 500 list which had the largest sales and had also won awards for introducing the most successful new technical products. They found that 65% of their respondents used venture management approaches, as compared with 36% for a sample drawn from the top 100 and 25% for a sample drawn from all the Fortune 500 firms. Thus, venture management approaches do appear to be positively associated with successful technological innovation. However, despite these sanguine observations, other studies have highlighted difficulties with such approaches.

Hlavacek[32] studied 21 product innovation failures which were nurtured using internal venture management approaches. These failures were selected from 12 corporations on the Fortune 500 list. He conducted in-depth interviews with the venture managers concerned and the corresponding top management (either at divisional or corporate level) to whom they reported. He found that the responses from the two groups were consistent, and the causes of failure might be summarized as follows:

Top management most commonly cited sunk costs as growing too large, whereas venture managers cited that top management was too impatient for results. Next, top management cited poor market evaluations and too small a market. This citation suggested that minimum ROI (the 3M criterion) rather than market size should be the criterion. Alternatively (although that author does not directly suggest it), the GE severed venture approach may be appropriate. Note also that both groups cited distribution difficulties, which again suggests inadequate attention to marketing issues. Third, top management often cited too narrow venture management experience as a cause of failure, which suggests injudicious entrepreneurial team selection procedures. Interestingly, venture managers cited that top managers (particularly if they were divisional as opposed to corporate managers) adopted insular viewpoints and saw new ventures as potential threats to existing operations. Again, this latter problem can be ameliorated by the 3M alternative sponsorship approaches.

Even with a well-conceived and delineated business plan and a strong venture team, internal ventures are vulnerable to failure from more subtle causes. Fast analyzed the life span of 18 new venture groups established and/or operating in the early 1970s, with a 50% survival rate in 1976.[33] He described the two main causes of venture failure as the *strategic reversal* and the *emergence trap.* We deal with each cause in turn.

The first of these reflects a conflict between the typical lead-time requirements for developing new ventures and dynamics of organizational economic and political climates. Fast postulates that four main driving forces dictate the launching of new internal ventures: a corporate strategy emphasizing diversification, a corporate money glut or positive cash flow which exceeds promising investment opportunities, an unfavorable outlook for the firm's main line of business, and entrepreneurially-oriented or risk-taking top management. When these four forces coincide, a venture group is probably established. When they disappear, that is, when the firm experiences a strategic reversal, the venture is probably disbanded. A new venture will typically be created in the fourth or fifth phases of Bright's innovation process—the development/pilot production stages. Thus, given the time scale of the process, it may well be 3 to 10 years before the venture becomes commercially viable. *It is very unlikely that either the internal or external environment of the firm remains unchanged over this time period.* Economic environments and business conditions typically experience cyclical changes over such time periods, and corporate strategies often change also, possibly associated with changes in top management. Then, the benign corporate climate may change for the worse and venture activities may experience a strategic reversal in their fortunes. A climate change of this kind may be severe enough to terminate a venture.

Fast quotes the example of a major chemical company which established a new venture division in 1968 when the first four driving forces pertained. By the early seventies, the climate changed. The demand for two of the company's major products increased dramatically, so it decided to concentrate on these. The company was also suffering from a shortage of cash so that capital had to be tightly rationed. The entrepreneurially-oriented president had been succeeded by a more conservative individual. Consequently, top management now decided to spin off or wind down its separate ventures and disband the new ventures operation. EMI's experience with the CAT scanner provides another example.

Thus, even if they are showing commercially promising results, internal ventures run the risk of premature termination from quite normal fluctuations in corporate climates. Fast found that there was a strong correlation between the percentage changes in the corporate profits over the period 1965–1976 and changes in the number of venture groups extant in the sample studied.

The second cause of venture failure is associated with what can be called envious and threatening success. If a venture group does not experience or survive a strategic reversal and launch several successful ventures, it faces other hazards. Its success may lead to top management favoritism. The venture group may be viewed as top management's "blue-eyed boys" or managerial corps d'élite and the leading contenders for top management succession. The expanding requirements of the new ventures may lead to territorial infringements. A growing venture may establish its own engineering, production, and marketing functions, both domestically and off-shore, which impinge upon the territories of established divisions. Thus, by being successful, it will be perceived as a threat by established power centers within the corporation. They will be resentful and jealous and perceive it as a dangerous new contender for scarce corporate resources and top management patronage. At best, these power centers will seek to subvert and undermine its status. They may well seek to destroy it—and be successful.

12.7 VENTURE MANAGEMENT AND MAINTAINING AN INTRAPRENEURIAL CLIMATE

Note that all the above problems might well have been avoided with a rigorous new venture planning process as discussed in Chapter 11, especially if top management defined and maintained a clear mission for corporate venture operations.

Hlavacek makes seven suggestions for more successful venture management which are comparable with the recommendations of other writers:

1. Both corporate and divisional managements should be made aware of the long-term growth benefits of venture operations.

2. Top management should develop a venture charter which specifies the functions, procedures, and boundaries of venture management.

3. Uniform formats for composing and reviewing venture business plans within agreed-upon time frames should be instituted.

4. A limited number of ventures, with independent budgets and at varying stages of development and maturity, should be sponsored.

5. Multiple sources of internal sponsorship for ventures should be maintained.

6. Top management should frequently review the organizational status of each venture in light of its performance and changing circumstances, and change this status if warranted. Thus, a venture might be upgraded to new divisional status, absorbed into an existing division, re-structured as a joint or severed venture, sold, or liquidated.

7. Product champions and high-caliber key team members should always be selected to lead and manage such ventures.

As Knight suggests, the first step in developing a venturing mission is to create a climate where everyone in the firm believes that they can be an intrapreneur through the promotion of support programs and publicization of successful examples and role models.[34] He also emphasizes the role of the sponsor or mentor (preferably a senior manager in the firm) who has the job of "blocking" for the intrapreneur in his efforts to "carry the ball for a touchdown." Knight's sponsoring or mentoring is similar to the coaching role in Chapter 7, and the intrapreneurial team also needs the critical functions skills of the project team of that chapter, with its new product proposal being evaluated using criteria discussed in earlier chapters.

Pinchot suggests ten *freedom factors* that stimulate an intrapreneurial climate, congruent with Hlavacek's seven points:

1. **Self-Selection.** Intrapreneurs appoint themselves and, given that their proposal is approved, should be allowed to champion their project.

2. **No Handoffs.** Ensure that intrapreneurs can remain with "their" project and are not required to hand it on to downstream functions.

3. **The Doer Decides.** Allow the team as much decision-making and resource-accessing freedom as possible. Minimize the bureaucratic clearance and permissions requirements from higher management levels.

4. **Corporate "Slack."** Allow intrapreneurs discretionary time and resources to explore their embryonic ideas.

5. **End the Home-Run Philosophy.** Rather than concentrating on a few larger well-studied and planned projects, support more smaller exploratory projects.

6. **Tolerance of Risk, Failure, and Mistakes.** Given the importance of error-elimination learning in technological innovation and entrepreneurship, blunders and false starts should be tolerated.

7. **Patient Money.** Don't look for fast bucks. Recall Chapter 2, and recognize that some innovations can take a long time. In backing a project, try to ensure that corporate support can be continued until its feasibility is tested.

8. **Freedom from Turfiness.** Because new ideas often span current organizational boundaries, encourage everyone to support promising new ideas rather than practice negative turf protection.

9. **Cross-functional Teams.** Since an intrapreneurial team's critical function requirements may usually require membership from several functional areas, try to ensure that turf protection does not limit team selection.

10. **Multiple Options.** Ensure that intrapreneurs have access to multiple resources. That is, that they are allowed to seek support and resources from outside their own division, or even the firm, if internal divisional support is lacking.

Even with venture management and intrapreneurial programs incorporating the above requirements in place. Smollen suggests that the risk–rewards structure in many corporations discourages successful venture development.[35] We saw, in Chapter 10, that entrepreneurs are moderate risk-takers, and they often feel that intrapreneurship, as opposed to entrepreneurship, is a riskier and less rewarding situation. This is because intrapreneurs have less control over their environments in the corporate situation. Despite their readier access to corporate financial or other resources, if a strategic reversal occurs, intrapreneurs lack maneuverability in seeking alternative financial funding sources. They are also likely to receive less financial reward if their venture is successful. A successful entrepreneur may enjoy significant to outstanding capital gains from their shares in the equity of their successful company. Although intrapreneurs may hold stock options in their parent corporation, even an outstanding performance of a venture which they have launched may have only a modest impact on corporate stock values, so their reward is correspondingly small. It is therefore apposite to examine corporate reward structures for intrapreneurs.

12.8 INTRAPRENEURIAL COMPENSATION

It is theoretically possible to offer the intrapreneur a special salary/bonus package linked to the performance on their own venture (particularly if it is a spin-off severed venture). Cook reports that some corporations have successfully employed such compensation plans,[36] but this is often difficult to implement without offending other managers in the parent corporation. We also pointed out earlier that intrapreneurs are likely to be hierarchically ambitious when compared with their independent counterparts. Ronald Oberlander, who as Executive Vice President of Abitibi-Price Inc. (one of the world's leading newsprint producers) was responsible for managing intrapreneurs, suggests that they may be distinguished from entrepreneurs by their attitudes towards monetary and ego risks.[37] Independent entrepreneurs are willing to risk investing their personal savings in a venture as well as to forgo the security and status of working for a major company. In contrast, intrapreneurs, while risking neither savings nor security, do take substantial career ego risks. Although successful venture management may offer a vehicle for advancement to top management, it does face the perils of the emergent trap. If a venture fails or is only moderately successful, intrapreneurs may find their corporate reputations sullied and corporate ambitions permanently constrained.

Because of such conflicting consideration, it is difficult to design an ideal compensation plan for intrapreneurs. In concluding a study of eight major corporate intrapreneurial venture plans, Sykes recommends that:

> The plan should be constructed so that there is congruence between the individual, venture, and corporate goals. The plan should be flexible enough to adapt to changes in corporate strategy. It should emphasize team versus individual rewards. And it should be perceived as fair by those outside as well as those in the plan.[38]

These considerations suggest that there are no simple ways of stimulating intrapreneurship, but that appealing risk–reward structures should be devised, and that venture managers should be allowed one or two failures (provided they are not too disastrous financially) without penalty to their long-term career prospects.

12.9 IN CONCLUSION: PINCHOT'S TEN COMMANDMENTS FOR INTRAPRENEURS

This chapter, like its predecessor, has shown that managing the intrapreneurial venture, like its entrepreneurial counterpart, is an always challenging, but often very rewarding, task. Before turning to a third type of venture, the strategic alliance, in the next chapter, we conclude this one by citing Pinchot's ten commandments for intrapreneurs[39]:

1. Be willing to risk being fired.
2. Circumvent orders aimed at stopping your dream.

3. Do any job needed to make your project work—forget your job description.

4. Find helpers.

5. In choosing helpers, rely on your intuitive people judgment and pick only the best.

6. Work underground as long as you can—publicity triggers the corporate immune system.

7. Never bet on a race in which you are not running.

8. Requests for forgiveness come easier than requests for permission.

9. Be true to your goals, but realistic about means.

10. Honor your sponsors.

FURTHER READING

R. A. Burgelman and L. R. Sayles, *Inside Corporate Innovation Strategy: Structure and Managerial Skills.* New York: Free Press, 1986.

G. Pinchot, III, *Intrapreneuring: Why You Don't Have to Leave the Corporation to Become an Entrepreneur.* New York: Harper & Row, 1985.

REFERENCES

1. D. A. Schon, *Technology and Change: The New Heraclitus.* New York: Delacorte Press, 1976, p. 320.

2. D. M. Collier, "Research Based Venture Companies—The Link Between Market and Technology." *Research Management* **17**(3), 16–30 (1974).

3. R. Biggadike, "The Risky Business of Diversification." *Harvard Business Review* **57**(3), 103–111 (1979).

4. G. Pinchot, III, *Intrapreneuring.* New York: Harper & Row, 1985, p. 4.

5. J. Chposky and T. Leonis, *Blue Magic: The People, Power & Politics Behind the IBM Personal Computer.* New York: Facts on File Publications, 1988.

6. R. C. Alexander and D. K. Smith, *Fumbling the Future: How Xerox Invented, Then Ignored, the First Personal Computer.* New York: Morrow, 1988.

7. H. B. Sykes, "Lessons from a New Ventures Program." *Harvard Business Review* **64**(3), 69–74 (1986).

8. R. A. Burgelman, "Designs for Corporate Entrepreneurship in Established Firms." *California Management Review* **26**(3), 154–166 (1984).

9. *The Economist,* November 30 (1991).

10. M. Jelinek and C. B. Schoonhoven, *The Innovation Marathon Lessons from High Technology Firms.* Cambridge, MA: Basil/Blackwell, 1990, p. 409 et seq.

11. R. P. Olsen and M. P. Diamond, "Innovation at Texas Instruments," Boston: Harvard University Press ICCH 9-672-036 (1971).

12. M. Jelinek, *Institutionalizing Innovation: A Study of Organizational Learning Systems.* New York: Praeger, 1979, Chapter 5.

13. P. E. Haggerty, "The Corporation and the Individual," Unpublished address in a series of lectures on *The American University: Community and Individual.* Dallas, TX: University of Dallas, Spring 1979.

14. J. Gardner, *Self Renewal: The Individual and the Innovative Society.* New York: Harper & Row, 1971.

15. "Texas Instruments Show U.S. Business How to Survive in the 1980s." *Business Week,* September 18, pp. 66–92 (1978).

16. A. Fry, "The Post-It Note: An Intrapreneurial Success." *SAM Advanced Management Journal* **52**(3), 4–9 (1987).

17. N. Macrae, "The Coming Entrepreneurial Revolution: A Survey." *The Economist* December 25, pp. 41–65 (1976).

18. M. Hanan, "Venturing Corporations—Think Small to Stay Strong." *Harvard Business Review* **54**(3), 139–148 (1976).

19. R. Mitchell, "Masters of Innovation." *Business Week* April 10, pp. 58–60 (1989).

20. F. Park, "Start More Little Businesses and More Little Businessmen." *Innovations* **5** (1969).

21. R. M. Kanter et al., "Engines of Progress: Designing and Running Entrepreneurial Companies; the New Venture Process at Eastman Kodak, 1983–1989." *Journal of Business Venturing* **6,** 63–82 (1989).

22. R. J. Tuite, "Strategies for Technology-Based Business Development" Unpublished paper. La Hulpe, Belgium, July 1989.

23. M. Hanan, "Corporate Growth Through Venture Management." *Harvard Business Review* **47**(1), 43–61 (1969).

24. Lord Rothschild, *A Framework for Government Research and Development.* London: H. M. Stationery Office Cmnd. 4814, 1971.

25. R. M. Hill and J. D. Hlavacek, "The Venture Team." A New Concept in Marketing Organization." *Journal of Marketing* **36**(3), 44 (1972).

26. B. C. Twiss, *Managing Technological Innovation,* 2nd ed. London: Longman Group, 1980, pp. 197–198.

27. J. Hillier, "Venture Activities in the Large Corporations." *IEEE Transactions on Engineering Management* **EM-15**(2), 65–70 (1968).

28. A. B. Cohen, "New Venture Development at DuPont." *Long Range Planning* **2**(4), 7–10 (1970).

29. S. Sabin, "A Nucleopore, They Don't Work for GE Anymore." *Fortune* December, 145. (1973).

30. R. M. Kanter, *When the Giants Learn to Dance.* New York: Simon & Schuster, 1989, Chapter 8.

31. K. A. Vesper and T. G. Holmdahl, "How Venture Management Fares in Innovative Companies." *Research Management* **16**(3), 30–32 (1973).

32. J. D. Hlavacek, "Towards More Successful Venture Management." *Journal of Marketing* **38**(4), 56–60 (1974).

33. N. Fast, "A Visit to the New Venture Graveyard." *Research Management* **22**(2), 18–22 (1979).

34. R. M. Knight, "Corporate Innovation and Entrepreneurship: A Canadian Study." *Journal of Product Innovation Management* **4**(4), 284–297 (1989).

35. L. E. Smollen, "Entrepreneurship in the Existing Enterprise," Unpublished paper. Belmont, MA: Institute for New Enterprise Development, 1975.

36. F. Cook, *A Piece of the Action—Venture Management Compensation.* New York: Frederick W. Cook.

37. R. Y. Oberlander, *Business Quarterly* Autumn, 7–16 (1988).

38. H. B. Sykes, "Incentive Compensation for Corporate Venture Personnel." *Journal of Business Venturing* **7**(4), 253–265 (1992).

39. G. Pinchot, III, *Intrapreneuring.* New York: Harper & Row, 1985, p. 22.

PART VI
THE STRATEGIC SETTING

Part VI returns to the corporate setting. Chapter 13 discusses the important role that external technology acquisition and partnering plays in technology strategies. Finally, Chapter 14 reviews the results of recent studies on successful technology—based firms and concludes with some prescriptions for the effective management of innovation and entrepreneurship suggested by these successes.

____13
INTERPRENEURSHIP: TECHNOLOGY ACQUISITION AND PARTNERING

Alliances are like marriages—they only work when both partners do.
KENICHI OHMAE, "THE GLOBAL LOGIC OF STRATEGIC
ALLIANCES" *HARVARD BUSINESS REVIEW*

Currently, the view on strategic alliances has shifted quite dramatically. Now, the issue is much more one of recognizing that complementary activities, leading to a win–win situation for both parties, can represent considerable savings in time and pooling of resources (particularly brain-driven), which otherwise might not occur.
PETER LORANGE AND JOHAN ROOS, *STRATEGIC ALLIANCES*

13.1   INTRODUCTION

In late 1992 and early 1993, reports in London indicated that America's Boeing Corporation and members of Europe's Airbus Industrie consortium, supported by Japanese manufacturers, were planning an alliance to develop a 600–800 seat "super-jumbo" airtransport.[1,2] By this time, Airbus had replaced McDonnell Douglas as number two behind Boeing in the global civilian air transportation market. Throughout the 1980s, Boeing and Airbus had been fiercely competing for sales of their respective planes to airlines throughout the world, amid allegations that Airbus received unfair subsidies from its European Community sponsors. Nevertheless, such a partnership made sense. On the one hand, major airlines needed a super-jumbo to satisfy future demands on their long-haul routes such as Los Angeles–Tokyo. On the other hand, the development costs of the super-jumbo dwarfed those of the original 747, reported in Chapter 8, and prohibited competing models being designed, manufactured, and marketed. Therefore, economic logic dictates that the two arch-rivals should enter into a strategic partnership to share the costs, risks, and rewards of its development. This alliance between rivals typifies many of the partnerships that have been created and executed over the past two decades. They may be divided into four broad categories:

1. *Technology Acquisition and In-licensing*. Faced with growing R&D costs, even large firms find it too expensive to generate all their new technology in-house and choose to access some of it from outside sources, that is, *buy* rather than *make* it.

2. *R&D Consortia*. In some circumstances, an alternative to the buy decision is to share the costs of new technology discovery and development with others through R&D consortia. Such consortia are playing increasing roles, especially in the performance of precompetitive research at the discovery end of the innovation process.

3. *Strategic Alliances between Rivals*. The above two examples focus on alternative forms of R&D alliances, that is, the sharing of resources to forge the initial *science* and possibly *engineering* links in the innovation chain of Chapter 2 *only*. Because of the rising costs of developing new technology in the aerospace, semiconductor, and other industries, the last decade has witnessed a burgeoning of collaborations between rivals, working jointly to forge one or more of the later links in the chain. That is, to share the costs and efforts of prototype, manufacturing, and sometimes the marketing of new technology-based products. Possibly the best known of such partnerships to be created in recent years is Sematech, the billion-dollar effort created by the U.S. microelectronics industry and government to develop the technologies needed to manufacture 16-megabit and later generation DRAMs that could compete with those produced by the Japanese semiconductor industry.[3] It is of interest to review briefly the background of its creation since it typifies the economic, global, and technical considerations that stimulate the creation of such alliances.

As we saw in Chapter 1, the development of the wireless telegraphy (and later) electronics industry was pioneered in both Western Europe and the United States. In contrast, following the invention of the transistor at Bell Laboratories, U.S. firms led the pioneering efforts in the semiconductor industry, with Western European firms such as Philips and Thomson CSF following essentially defensive strategies. Until the 1960s, Japanese firms played a relatively modest role in both industries, but by the early 1970s, they had become the world's leaders in consumer electronics, still largely based upon imported semiconductor devices. However, by the late 1970s, this situation had changed. Until then, Japan's domestic semiconductor manufacturers had followed an absorbant strategy, but they were now ready to become offensive innovators and challenge the U.S. leadership in the race to market the 64K RAM chip. Japan essentially won that race to claim global market leadership, with 40% of the worldwide sales of commodity chips, to force both Western European and U.S. semiconductor firms to recognize a new global "ballgame" in their industry. The development and "up-front" manufacturing costs of successive generations of RAMs and microprocessors was rising rapidly to outstrip the financial resources of any one firm. Therefore, strategic alliances (SAs) between rivals became an economic and technological imperative. In Western Europe this was reflected in the alliance between Philips and Siemens to fund 1M and 4M DRAM developments, as well as the agreement between the same partners and SGS-Thomson Microelectronics and other European firms in the $3.9 billion Joint European Silicon Submicron Initiative (JESSI) consortium. All three are mature, vertically-integrated firms, and the intent of these alliances was to enable the partners to meet the Japanese challenge and maintain an autonomous domestic European manufacturing capability for state-of-the-art commodity chips. The U.S. semiconductor industry faced a similar challenge, and created Sematech to meet it. However, as Tassey argues, this level of collaboration may be insufficient since the U.S. semiconductor industry is disadvantaged by being less vertically-integrated into the larger electronics industry than its Japanese counterpart.[4]

4. *Strategic Alliances Combining Complementary Assets.* As has been mentioned in earlier chapters, in contrast to the semiconductor industry, alliances that bring together the complementary assets needed to execute the innovation process have been a notable feature of the new biotechnology industry over recent years. They have also been used extensively in other industries, with the IBM-Microsoft collaboration being possibly the most notable example. Also, some firms, for example NEC and Corning, are using a *spider's web* or network of such alliances as a basis of corporate strategy,[5] while Boeing and Airbus have extensive experience of the last two types of alliances. 747s can include components that have been made or assembled by suppliers in up to 29 different countries, and Airbus is a consortium of several EC airplane companies that are competitors in other market segments of the industry.

We explore the rationales as well as some of problems of these collaborations in this chapter, beginning with those focusing essentially on R&D.

13.2 BUYING R&D

The planning process described in Chapter 4 (Section 4.4) should, implicitly or explicitly, include a comprehensive audit of the firm's technological capabilities. It will frequently identify gaps between these capabilities and the requirements specified by the technology plan. Such technology gaps must filled by either *making* it (that is, the pursuit of new technology development programs *in-house*) or *buying* it (that is, acquiring the needed technology from sources outside of the firm). Abetti[6] suggests that the choice between making or buying should be based upon the following considerations:

1. Making is preferable if the technical progress rate is slow and market growth is moderate with significant new entry barriers, since success could yield a profitable product monopoly. Abetti cites GE's exploitation of artificial industrial diamonds over many years to illustrate this option.
2. Buying is preferable if the technical progress rate is slow, but market growth is fast. If patent protection and other market entry barriers are high, exclusive in-licensing is preferable. If market entry barriers are low, the cheaper nonexclusive licensing optional is preferable.

3. If both the technical progress rate and market growth are fast, buying through the acquisition of an established company in the technology–market segment is preferable.

Cutler and Rubenstein review the alternative sources from which technology may be *bought,* which are now briefly discussed.[7,8]

University Research Laboratories and Centers

University research laboratories, particularly those supported by government or industry, are obvious and growing prime sources of new technology. As Martin and Othen argue, the "creeping" privatization of the publicly-funded universities mentioned in Chapter 3 is forcing many of them to rediscover their earlier missions.[9] They point out that the first precursor institution to the universities in Britain was founded in Llanwit Major in South Wales by Saint Illtyd in the sixth century. Moreover, the Welsh saint recognized that such institutions should combine both *sacred* and *mundane* missions. As well as founding a School of Divinity, he anticipated the agricultural extensions missions of the U.S. land-grant universities by some thirteen centuries in adapting a French plow for use in the farming conditions of South Wales. This was probably one of the earliest examples of technology transfer from "university" to "industry"! The above authors suggest that funding pressures may well force the leading public universities to shift their major emphases from a pure research mission towards a mission emphasizing the *deployment* as well as the discovery of new technology.

Cutler cites several examples of excellent university-affiliated applied research facilities, of which one is affiliated with a land-grant university which is placing increasing emphasis on technology deployment. This facility is the internationally renowned Ames National Laboratory, owned by the U.S. Department of Energy and managed by Iowa State University. Iowa State University also promotes technology transfer and its commercialization through its Center for Advanced Technology Development (CATD) and EDGE Technologies, Inc. The University is thus reasserting its original mission, as a land-grant institution, in deploying its internally generated technology to stimulate new technology-based job creation in the state. There are, of course, numerous other university-based centers supported by government and industry throughout the United States and elsewhere. Apart from supporting such centers, many firms have substantial programs to ensure close links with university researchers. Digital Equipment views universities collaborating in its External Research Program (ERP) as *virtual laboratories* for the corporation. Again, there are numerous other such programs, in both the United States and elsewhere. For example, in Canada, Bell Northern Research and its manufacturing affiliate Northern Telecom have built up an impressive program of university linkages.

Government Laboratories

At least since the Apollo Program of the 1960s, governments have been promoting the spin-off benefits of their R&D laboratories. Although some such laboratories have little success in generating commercially worthwhile new ideas, their more successful counterparts have become significant sources of new technology for industry. The Conservation Program of the U.S. Department of Energy (DOE) is one successful program. For example, its leadership of a project to develop low-emmissivity window coatings had, by 1990, saved U.S. consumers over $300 million in heating bills. To give another example ABCO, the small company cited in Chapter 6 (Section 6.4), employed a government agricultural research establishment as a *virtual laboratory* in its development of vegetable processing equipment.[10]

Contract and Supplier R&D Laboratories

Specialist R&D establishments that perform projects on a fee-for-service basis can also be employed as suppliers of customized new technology. Alternatively, suppliers often provide excellent sources. In developing its line of fiberglass-reinforced plastic product lines, ABCO implicitly drew upon the R&D capabilities of its much larger suppliers including Exxon and Ashland.

13.3 R&D COST SHARING

An intermediate category between making or buying R&D is cost-sharing through R&D consortia which usually perform precompetitive research. Prototype consortia have existed for many years in the United Kingdom in the form of industry-wide research associations. However, they have become a regular feature of the North American scene in the last decade or so, particularly consortia that involve the participation of universities and/or government laboratories as well as private firms. Fusfield and Haklisch[11] provide a useful overview of such consortia up to the mid-1980s, with Evan and Polk[12] providing a later update. Also, several texts have been published which have focused on their achievements and operating problems.[13,14]

One industry-based consortium is Microelectronics and Computer Technology Corporation (MCC), which was founded in 1983 by 16 computer and semiconductor companies and has since expanded to embrace 56 companies, such as Sematech located in Austin, Texas. Rhea[15] cites the names of 203 "horizontal" consortia registered in the United States between 1984 and the end of 1990. He also argues a case for the creation of "vertical" consortia based upon the concept of a consortium of noncompeting companies developing an invention through the total innovation process to commercialization. He cites as an example of the latter the proposal to create a consortium (American Superconductor Corp.) to perform the basic research and prototype development required to apply superconductivity to electric power distribution. He suggests that consortia can perform research to establish common standards, protocols, and interfaces for participant firms and their suppliers, and the participants can also conduct multifirm market trials to delineate and stimulate the growth of a new technology market.

Souder and Nassar[16] studied twelve effective and nine ineffective consortia to derive the following ten commandments for successful consortia:

1. Establish strong commitments to the consortium.
2. Implement strong decision controls.
3. Establish a strong consortium charter.
4. Set up systematic management processes.
5. Organize the consortium as a combination holding company and matrix structure.
6. Implement effective technology transfer practices.
7. Ensure the consortium generates private benefits.
8. Establish a governance philosophy.
9. Select projects relevant to the industry's needs.
10. Establish core members with complementary values.

These are broadly identical to the success requirements for strategic alliances which will be discussed shortly.

13.4 TECHNOLOGY TRANSFER PROBLEMS

The intrafirm technology transfer barriers discussed earlier in this book are potentially greater when interorganizational transfers are required. First, the NIH syndrome is much more likely to be present. Second, the technology supplier may be a university research facility or government laboratory staffed by researchers with radically different attitudes and value systems than industrial researchers, who are both geographically and *culturally distant* from the latter. In several separate studies, this author has reviewed well over 50 examples of technology transfers between either university or government researchers and private firms, and has found that geographical and/or cultural "distance" is often a major transfer barrier. This impression is supported by the observations of Smilor and Gibson[17] in their study of technology transfer issues in consortia. They suggest four factors which positively affect technology transfer outcomes from consortia (transmitters) to their company shareholders (receptors):

1. Good passive (paper) and, more importantly, active (personal) communication links between transmitters and receptors.
2. Close geographical and cultural proximity, as suggested above.
3. A low level of *equivocality* of the technology. The authors describe low equivocal technology as that which is easy to understand and integrate into the receptor company's products and processes.
4. High motivation of the technology receptors, based upon the positive recognition given to the technology transfer activities in the receptor firm and, by implication, the absence of the NIH syndrome.

The technology transfer of a project that exhibits a combination of high communication and motivation with low equivocality and close proximity will be successful or a *grand slam*. That of a project that combines low communication and motivation, together with high equivocality and distance between transmitters and receptors, will fail and be *dead in the water*. Projects exhibiting combinations between these two extremes could be expected to enjoy varying levels of success. Managers in both consortia and their shareholding firms should, as far as possible, foster transfer practices conducive to the creation of the first of the above combinations of factors. The most important point for the reader to recognize is that because the barriers are likely to be greater, interorganizational technology transfer will probably require greater attention than its intraorganizational counterpart if those barriers are to be surmounted.

13.5 INTERPRENEURSHIP THROUGH STRATEGIC ALLIANCES

As was stated earlier, apart possibly from vertical consortia, the above partnerships focus on the initial *science* and *engineering* links in the innovation chain, and we now examine the alliances that seeks to forge some or all of the remaining links in the chain. Larson[18] argues that smaller firms need to expand their entrepreneurial activities to embrace the management of a partnership network. This might be described as *interpreneurship,* or an alliance between entrepreneurial small firms and intrapreneurial larger ones. Saxenian describes the important role played by producer–supplier partnerships in pursing *reciprocal innovations* in the computer and semiconductor firms in Silicon Valley in the 1980s.[19] In these partnerships, the individual smaller firms retain their independent identities while contributing to the forging of the links of the chain by acting as innovative component suppliers to the large firms, such as Apple Computers, Sun Microsystems, and Tandem Computers. In contrast, a notable feature of the biotechnology industry in recent years has been for NBFs, and LMFs to form joint venture subsidiaries to forge the links of the innovation chain.[20,21] Forrest and Martin[22] explore the role of and life cycle of such alliances, using both the innovation chain and *temporary marriage* metaphors. As the term suggests, the life cycle can be divided into a number of successive stages.

Marital Purpose

A strategic alliance is the *child* of a marriage between two independent firms or *parents* which, like its biological counterpart, seeks to realize synergy from the new corporate (as opposed to genetic) ensemble created. As with human marriages, it is important that the trio involved (i.e., the two parents and their child) be viewed as a *family* since its success is dependent upon purposeful and sensitive interactions among all three family members. The achievement of synergy in an SA can happen, but only after a lot of hard work from all three of them. The ways and means of doing this can be viewed in the context of the technological innovation base of Chapter 3. In an SA between two rivals in (say) the semiconductor industry, most of the activities in the technological innovation bases of the parents may be replicated in their child, dependent upon the purpose and scope of the alliance. In contrast, in those between two partners providing complementary assets, it is likely that the child's technological innovation base will be provided by different components from each of its parents.

This is best illustrated in alliances between small new biotechnology firms (NBFs) and larger mature firms (LMFs) in the agricultural, brewing, biological, chemical, and pharmaceutical industries. The successful commercialization of new biotechnology, based upon recombinant DNA (rDNA), monoclonal antibodies (MAb), and biprocess engineering, requires the forging of links between the new scientific research knowledge, mainly embedded in university and other nonindustrial research institutions, and the knowhows located in other links of the chain, mainly embedded in LMFs in related industries. For example, two such NBFs, Genentech and Centocor, have established spider's web alliance networks to implement their technology strategies.[23] The founders of NBFs are typically talented scientists with entrepreneurial bents and academic or other nonindustrial research backgrounds. Thus, the NBF usually lacks the skills, knowhows, and other resources required to forge the remaining links in the chain. In contrast, the larger companies are typically lacking in the specialized biotechnology scientific research knowhow and innovative ability extant in the small firm, while possessing the remaining resources required. Therefore, the NBF will primarily contribute the R&D component, with the LMF contributing the remaining components (again, dependent upon the purpose and scope of the alliance) of the technological innovation base to their child. From the LMF's viewpoint, entering into an SA with an NBF should be based upon similar considerations to the make or buy decision discussed in Section 12.2 and the technology auditing and planning activities of Chapter 4. In general, an LMF seeks an SA to access important technological capabilities and develop new products required in its technology plan and strategy, but which it currently lacks. For a more detailed and exhaustive list of reasons, see Harrigan.[24] In contrast, the NBFs are typically seeking access to the clinical testing, regulatory requirements, manufacturing, and marketing skills they lack but need to commercialize their new technology inventions.

Dating Dancing Partners and Courtships

Once a firm has identified the complementary capabilities it is seeking in a partner, finding and successfully wooing that partner may not be easy. As in human affairs of the heart, the corporate equivalent of the perfect woman is often seeking its equivalent of the perfect man! Given their maturity, the choice of alliance partners in some industries may be limited, but this is not true for firms in the new biotechnology industry. There are now several hundred NBFs in North America and some tens of LMFs exploring or participating in NBF–LMF alliances, so the dating and courtship process can take some time. Firms must be prepared to court several (or even numerous) partners before finding a mate with complementary and compatible strategic objectives. This requires the patient maintenance of a clear-sighted vision of the strategic purpose of the potential alliance, coupled with the flexibility to be willing to modify it to accommodate the legitimate needs of a promising potential partner. Given the difficulty of evaluating emergent technologies and small, young companies, some experienced LMFs have special departments with the sole responsibility of seeking out and negotiating with potential partners. For their parts, the NBFs must become cognizant of the LMFs financial, development, and marketing resources and capabilities. As in human courtship and marriage, equitable give-and-take is often a prerequisite of success.

In assessing the benefits of a potential alliance between rivals, a firm must obviously assess its potential mate both as an ally and a competitor. One benefit of such an alliance is that the ally ceases to be a competitor in the particular technology–market it serves, at least throughout the duration of an alliance. However, this benefit must also be assessed against the longer-term cost of divulging firm-specific knowhow to the potential partner. Once the SA has been terminated, the partner may become a competitor once more, and be able to exploit the firm-specific knowhow it has acquired from the alliance. Such concerns are reflected in the discussion of the Ajimoto–Searle alliance to manufacture and market aspartame, as outlined in Chapter 1. They are also reflected in the *Trojan Horse syndrome*.[25] That is, during an alliance, one firm may acquire sufficient expertise in another's activities to surplant it in the marketplace or acquire it once the alliance is ended. Both to acquire diverse knowhow and access to resources, and to protect themselves from the Trojan Horse syndrome, prudent firms practice polygamy rather than monogamy in their SAs and form a *spider's web* of alliances, each satisfying a different need strategic intent.[24] We have already cited Corning in this context, and in an industry, such as new biotechnology which is arguably still in a turbulent fluid stage and where dominant designs have yet to be established and market structures are uncertain, the pursuit of multiple alliances of short duration offers the best strategy for success. In addition to Corning, other firms have built success upon longer-term alliances. For example, Xerox has forged a series of international alliances with Fuji Photo (Japan), The Rank Organization (United Kingdom), and Siemens (Germany), as well as with smaller companies over the past two decades to sustain its growth and diversification into complementary technologies and global markets.

Negotiating the Alliance and Defining the Marriage and Divorce Settlements

Once they have identified each other, partners must negotiate the alliance agreement. This typically takes up to a year or even longer to complete, depending upon the issues involved. As in human affairs, hasty marriages, or ones which are hurriedly negotiated by the top managers of both partners without the participation of other managers and R&D staff and without adequate attention given to how the alliance will be implemented, are unlikely to succeed. Rather, they may be compared with the often protracted negotiations which historically took place between royal houses contemplating dynastic marriages. In such negotiations, the objectives and obligations of both partners, together with the overall program of work for the alliance, should be clarified from the outset. They should define a mutually agreeable method for dissolving the alliance (exit terms or a divorce settlement), and if necessary, a sunset clause or the limitation of the agreement to a specific time period. Each partner's voting rights and equity shares in the joint venture need to be agreed upon, together with an appeal procedure for dealing with any future conflicts between them.

Careful thought must also be given to the employment contracts and remuneration schemes for the child's staff. Clearly, such individuals will be concerned for their long-term careers (especially if the agreement includes a sunset clause), and these concerns must be recognized if the parents are to elicit their commitments to the child's success. On the one hand, staff seconded to the child from both parent firms must have their long-term contracts with the latter maintained. On the other, it should be made clear that their performance evaluations will be based upon their contributions to the child's goals, rather than those of their respective parent firms.

Once mutually agreeable marriage and divorce settlements have been negotiated and signed, the partners can address the problems of implementing the alliance.

Consummating the Marriage

According to Killing,[26] *mutual trust* and *respect* is the golden rule for SA success, and with sufficient commitment from both parties, this should develop over time. The first point to note is that top management commitment and time involvement do not end with the signing of an agreement. Just as the nurturing and upbringing of a healthy human child requires the active involvement of both parents, so does that of the offspring of an SA. As the child grows, it will inevitably experience problems and require the help and guidance of its parents to solve the problems.[27] A key step to ensure continued top management commitment is the selection of the child's Board of Directors. It should include senior management representatives from both its parents, and these should be individuals who have the competencies, diplomatic skills, and time to participate in and guide its activities, as and when needed. Also, again as with its human counterpart, it will require more money as it grows, and the parents' representatives on this Board have the responsibility to their respective stockholders to ensure that this money is really needed and will be spent wisely.

The judicious selection of the child's operating staff is of equal importance. Staff are required who can work within the tripartite cultures involved. When seconding the parents' staff to the child, it is also vital to choose individuals with good track records and promising career potentials in their parent firms. Ideally, nomination of the child should be treated and recognized as a desirable career opportunity in the human resources development programs of both parents, so that the child attracts the quality staff it needs to succeed. In selecting seconded staff, their prior attitudes to the alliance should be considered. Obviously, volunteers are to be preferred to *pressed* men or women, so individuals should be given the opportunity to apply for positions in the child company. Also, as Alster points out, *alliance champions* and *deal or alliance killers* can often influence the chances of the child's success or failure.[28] The former, like *project champions* of Chapter 8, are determined individuals dedicated and committed to the success of the SA. In contrast, the latter are those who are perhaps equally determined to sabotage it. Therefore, the participation of the former individuals should be positively encouraged, while, if possible, the latter individuals should be excluded from any involvement in the child's activities.

Even if it enjoys full parental support and is staffed by able people, the child's success as a venture is obviously dependent upon a cohesive team effort by its members. Staff from both parents have to learn to work together to achieve common goals. Porter, in his study of alternative corporate diversification strategies, concluded that *transferring skills* and *sharing activities* are critically important.[29] Since the LMF is seeking technological diversification opportunities through the notional temporary acquisition of the NBF in the strategic alliance, his approach is directly relevant to the present discussion. A strategic alliance clearly provides a vehicle for transferring skills, or in this context, more often exchanging complementary skills between the two parents. This exchange will primarily occur in their technology development activities. However, some complementary skill transfers will be needed in the reverse direction, both to facilitate the technology transfer process between the two, and to enable the NBF's staff to acquire the development, production, and marketing knowhows they are seeking. Since a successful SA requires mutual trust and an attitude of give-and-take between its partners, it is most important that this climate be established and maintained equitably. One fear of firms participating in SAs, which is a manifestation of the Trojan Horse syndrome, is *technology bleedthrough*. That is, Partner A's acquisition of Partner B's technology which, after the alliance is terminated, A then uses as a competitive weapon against B. Doz *et al.*[25] use the metaphor of the *organizational membrane* to address this issue. They suggest that the differences in strategic intents of the partners provide the differences in pressure between the two sides of the membrane. They imply that, like a membrane that is designed selectively to allow the passage across it of some molecules and not others, the organizational membrane should be designed to allow the passage across it of only those skills and activities that have been contractually agreed between the partners.

Even if information exchanges stay within the spirit of the contract, the relative knowhow that each partner acquires will depend upon two other factors. First, as the above authors put it, this exchange is also dependent upon each organization's receptivity or *porosity*. Some people and organizations are more able to learn from others because they are more receptive or porous to new ideas. Since each partner is seeking skills from the other, each must have the porosity to absorb these skills. These will typically be a mix of explicit and tacit skills. For example, the manufacturing scale-up knowhow sought by an NBF in its alliance with an LMF embraces explicit equipment-embodied skills and tacit experienced-based knowledge of the problems of scaling-up chemical and biological processes. When implementing an SA, it is important that each parent recognizes the attributes of the knowledge or information and skills it is seeking from its partner, and that, insofar as possible, it tries to staff the child with people with the abilities to absorb them. Alster suggests another role requirement for the alliance to succeed.[28] This is enacted by *buffer* or liaison individuals between the child and its parents. Given the differences in cultures and decision-making dynamics that can often exist between the two parents themselves and also their child, this is a crucial role. Without its enactment, the SA may be ossified by the bureaucratic red tape that can develop between the participants. In common with other semiautonomous corporate ventures, the SA faces the twin threats of a *strategic reversal* and the *emergent trap* of Chapter 12.

Terminating or Extending the Alliance

Many SAs are designed to be temporary liaisons in pursuit of specific strategic objectives. Once these objectives have been achieved, they may be terminated under the conditions of the divorce settlement as defined in the original alliance agreement. If the SA has been successful, from both partners' viewpoints, the divorce settlement should be harmonious. If both partners found their marriage to be productive, they may wish to extend it or enter into a new one based upon evolving organizational, technological, and market needs and opportunities. They may even enter into an agreed merger or sellout deal. SAs offer the best option of survival to some firms. McKenna cites examples of computer firms that failed because they did not form alliances; Alster argues that well-designed SAs, of variable duration, often provide better mechanisms for technology commercialization to both parties.[28,30]

13.6 SUCCESSIVE FACTORS FOR STRATEGIC ALLIANCES

Despite such optimistic observations, some researchers take a pessimistic view of SAs, and claim that low chances of success can be expected from them.[26,31,32] Such conflicting observations are consistent with the results of a study by Forrest and Martin.[33] They studied 20 LMFs and 40 NBFs which, between them, had participated in almost 1500 alliances. The LMFs reported a success rate of 47.5% from 346 alliances. This is consistent with W. R. Grace's success rate of 40–50% with mainly R&D-based alliances.[34] In contrast, the NBFs reported a success rate of 83% from 1146 alliances. The lower perception of success for the LMFs appears to be derived, as will be discussed below, from their more rigorous standards of accountability. Respondents were asked to cite the reasons for success and failure in their alliances. The citings were similar for both DBCs and LMFs, and could be summarized as the three Cs of *continuing, compatibility,* and *commitment.* That is, there was agreement between the two parties on the *ends* to be sought, the compatibility to work well together, and the sustained commitment of the required resources to the alliance by both parties jointly to provide the *means* to achieve this end. It is of interest to note the interorganizational cultural compatibility did not figure largely as a success factor. What the results did suggest is that the ability of the two-partner companies to work together is dependent upon the presence of individuals who can bridge any gap between the two corporate cultures involved rather than upon an overall cultural compatibility. Many DBCs now have managers with large company experiences, who may be adept at *gap-bridging* or *boundary-spanning* roles. Such roles are analogous to impedance-matching circuits in electronic systems and synchromeshes in automobile gearboxes in facilitating effective linkages between two disparate subsystems. The absence of such individuals may expose an alliance to cultural disparities between its partners that destroys it. Burrill also reports successful experiences of SAs in the biotechnology industry and expresses similar ideas in stating:

> Something of a marriage between the best aspects of large companies and the best aspects of small companies is occurring in all technology industries, . . .[35]

Kanter expresses similar ideas in her *six Is* for partnership success.[36] The relationship is *important,* there is agreement on long-term *investment,* the marriage partners are *interdependent,* the organizations are *integrated,* each is kept *informed* of the plans and directions of the other, and the partnership is *institutionalized* by a framework of support mechanisms.

Such observations are echoed by Niederkofler in his discussion of the evolution of SAs.[37] He argues that an alliance will succeed if it constitutes a good strategic and operational fit to the goals of both parents. Therefore, the initial negotiations must identify the strategic fits for both parents, and the alliance agreement must lay a sound foundation for its operational implementation, as well as the creation and maintenance of goodwill and trust among all the parties involved. Since these goals may change as it evolves, the original contract should incorporate the flexibility to change both the strategic and operational fits of the alliance to its parents in order to accommodate to their changing needs. Such flexibility should be present if the climate of goodwill and trust has been built up between the parties involved. Obviously, a time may come when the alliance can no longer serve the needs of one or both of its parents. Should that situation arise, the appropriate exit terms of the alliance agreement should be followed, to dissolve the marriage by an "amicable divorce." In this situation, the child might continue to operate as an independent venture severed from its parents or be purchased by a third firm. Alternatively, it may be dissolved, and its component parts divided and shared between parents under mutually acceptable "community property" rules.

FURTHER READING

J. D. Lewis, *Partnerships for Profits: Structuring and Managing Strategic Alliances.*   New York: Free Press, 1990.
P. Lorange and J. Roos, *Strategic Alliances.*   Cambridge, MA: Blackwell, 1992.

REFERENCES

1. A. Lorenz, "Boeing and Airbus Join Up to Build 'Super' Jumbo Jet." *The Sunday Times* December 20 (1992).

2. "Now For the Really Big One."   *London: Economist* January 9, pp. 57–60 (1993).

3. B. Merrifield, "Strategic Alliances in the Global Marketplace." *Research Technology Management* **32**(1), 15–20 (1989).

4. G. Tassey, "Structural Change and Competitiveness: The U.S. Semiconductor Industry." *Technological Forecasting and Social Change* **37,** 85–93 (1990).

5. A. Nanda and C. Bartlett, "Corning Incorporated: A Network of Alliances." Boston: Harvard ICCH 1990.

6. P. A. Abetti, "Technology A Key Strategic Resource." *Management Review* February, pp. 37–41 (1989).

7. W. G. Cutler, "Acquiring Technology from Outside." *Research Technology Management* **34**(3), 11–18 (1991).

8. A. H. Rubenstein, *Managing Technology in the Decentralized Firm.* New York: Wiley, 1989, pp. 274 *et seq.*

9. M. J. C. Martin and D. A. Othen, "Developing University Technology: Some Observations and Comments." *Management of Technology. III. The Key To Global Competitiveness,* Proceedings of the Third International Conference on Management and Technology. Norcross, GA: Industrial Engineering and Management Press, 1992, pp. 48–56.

10. M. J. C. Martin and P. J. Rossen, "R&D Philosophy at ABCO." *Four Cases on the Management of Technological Innovation and Entrepreneurship,* Technological Innovation Studies Program. Ottawa, Ont.: Department of Industry, Trade and Commerce, 1984.

11. H. I. Fusfield and C. S. Haklisch, "Cooperative R&D for competitors." *Harvard Business Review* **63**(6), 60–76 (1985).

12. W. M. Evan and P. Olk, "R&D Consortia: A New U.S. Organizational Form." *Sloan Management Review* **24**(3), 37–46 (1990).

13. H. I. Fusfield, *The Technical Enterprise: Present and Future Patterns.* Cambridge, MA: Ballinger, 1986.

14. D. Dimanescu and J. W. Botkin, *The New Alliance: America's R&D Consortia.* Cambridge, MA: Ballinger, 1986.

15. J. Rhea, "New Directions for Industrial R&D Consortia." *Research Technology Management* **34**(5), 16–26 (1991).

16. W. E. Souder and S. Nassar, "Managing R&D Consortia for Success." *Research Technology Management* **33**(5), 44–50 (1990).

17. R. W. Smilor and D. V. Gibson, "Accelerating Technology Transfer in R&D Consortia." *Research Technology Management* **34**(1), 44–49 (1991).

18. A. Larson, "Partner Networks: Leveraging External Ties to Improve Entrepreneurial Performance." *Journal of Business Venturing* **6,** 173–188 (1991).

19. A. L. Saxenian, "The origin and dynamics of production networks in Silicon Valley." *Research Policy* **20,** 423–437 (1991).

20. Office of Technology Assessment, *Commercial Biotechnology: An International Analysis.* Washington, DC: U.S. Government Printing Office, 1984.

22. U.S. Department of Commerce, International Trade Administration, *High Technology Industries: Profiles and Outlooks—Biotechnology.* Washington, DC: U.S. Government Printing Office, 1984.

22. J. E. Forrest and M. J. C. Martin, "Strategic Alliances as Viable Growth Strategies for

High Technology Firms." *Proceedings of the 1987 IEEE Conference on Management and Technology.* Norcross, GA: IEEE, 1987.

23. J. Freeman and S. R. Barley, "Inter-organizational Relations in Biotechnology." In R. Loveridge and M. Pitt (Eds.), *The Strategic Management of Technological Innovation.* New York: Wiley, 1990, pp. 127–156.

24. K. R. Harrigan, *Strategies for Joint Ventures.* Lexington, MA: Heath, 1985.

25. Y. Doz *et al.,* "Strategic Partnerships: Success or Surrender?" Unpublished presentation. London: AIB-EIBA Joint Annual Meeting, November 20–23, 1986.

26. P. J. Killing, *Strategies for Joint Venture Success.* New York: Praegar, 1983.

27. K. R. Harrigan, *Strategies for Joint Ventures.* Lexington, MA: Heath, 1985, p. 375.

28. N. Alster, "Strategic Partners: Seeking the Right Chemistry" *Electronic Business* **15**(10) (1986).

29. M. E. Porter, "From Competitive Advantage to Corporate Strategy." *Harvard Business Review* **65**(3), 43–59 (1987).

30. R. McKenna, "Market Positioning in High Technology." *California Management Review* **27**(3), 82–108 (1985).

31. Y. Doz, "Technology Partnerships between Large and Small Firms: Issues and Pitfalls," Unpublished presentation. New Brunswick, NJ: Conference on Strategic Alliances, Rutgers University, October 24–26, 1986.

32. N. Alster, "Dealbusters: Why Partnerships Fail." *Electronic Business* **15**(7), 70–75 (1986).

33. J. E. Forrest and M. J. C. Martin, "Strategic Alliances Between Large and Small Research Intensive Organizations: Experiences in the Biotechnology Industry." *R&D Management* **22**(1), 41–53 (1992).

34. M. F. Wolff, "Forging Technology Alliances." *Research Technology Management* **32,** 9–11 (1989).

35. G. S. Burrill, *Biotech 89: Commercialization.* New York: Mary Ann Liebert, 1988, p. 44.

36. R. M. Kanter, *When Giants Learn to Dance.* New York: Simon & Schuster, 1989, Chapter 6.

37. M. Niederkofler, "The Evolution of Strategic Alliances: Opportunities for Managerial Influence." *Journal of Business Venturing* **6,** 237–257 (1991).

___14
EPILOGUE: TOWARDS SUSTAINED INNOVATION AND ENTREPRENEURSHIP IN TECHNOLOGY-BASED FIRMS

> *Innovative firms tend to have a style of management that is open to new ideas, ways of handling staff that encourage innovation, systems that are customer focussed and which reward innovation, skills at translating ideas into action and so forth . . . One has to really believe an organization cares in order to invest the energy needed to help it change. Such commitment derives from superordinate goals . . . this is probably the most underpublicized "secret weapon" of great companies.*
>
> RICHARD TANNER PASCALE AND ANTHONY G. ATHOS,
> *THE ART OF JAPANESE MANAGEMENT*

14.1 INTRODUCTION

In this final chapter we review some overall approaches to the effective management of the innovation and entrepreneurship process in technology-based firms. It would be utopian to expect to end this text by suggesting that there is some ubiquitous guaranteed system for managing the corporate technological innovation and entrepreneurship process. Like the proverbial free lunch, such a system cannot be expected to exist this side of paradise. Nevertheless, after acknowledging this reservation, it is possible to suggest some general guidelines for enhancing the process.

14.2 THE CTO AND THE INNOVATION MANAGEMENT FUNCTION

As was suggested in Chapter 4 (Section 4.2), it is useful to postulate an *Innovation Management Function (IMF)* typically led by the *Chief Technology Officer (CTO)* as the responsible vehicles for the formulation, coordination, and implementation of the innovation and entrepreneurial policies and strategies of the technology-based firm. Although the titles of IMF and IMCS may rarely exist, such roles are certainly enacted in many firms in practice, and the title of CTO is used in some of them.

Frohman, based upon a study of nine companies that have sought to exploit technology, found that successful companies selected projects which supported business goals, reinforced technological leadership, and solved customer's problems. The successful companies had planning/decision systems and organizational structures that reinforced technological strategies and senior management assumed responsibility for technological planning and decision making, which was integrated with business planning. He also suggested that the leadership of the overall technological development should be placed in the hands of such a CTO, who should report to the corporate CEO or someone immediately below that level.[1] Adler and Ferdowes[2] describe the roles and responsibilities of a sample of 25 senior managers holding CTO positions in U.S. Fortune 100 firms. Nineteen of them reported to the Chairman, President, Chief Executive Officer (CEO), or Chief Operating Officer (COO) in their firms, and the remaining six reported to the Executive Vice President or Vice Chairman. All emphasized that their responsibilities embraced product and process technology in both centralized laboratories and SBUs. These results are consistent with Bhalla's recommended CTO job description, which is:[3]

1. To be responsible for the development and execution of a business–technology pan for each SBU in the firm.
2. To tailor the technology organizations (such as corporate and divisional R&D laboratories) in the firm, including their human resources, to provide the most effective support for SBU and corporate objectives.
3. To review, develop, and administer the technology portfolio and budget of the firm, in collaboration with top management.
4. To maintain a technology overview role and forecasting function, in order to react to external developments. In particular, to identify potential technology threats and opportunities and exploit emerging synergies between SBUs.

Although Bhalla does not include them, two other requirements should be added to the above job description:

5. To identify external technology sources for the firm, opportunities for technology development cost sharing, and other potential synergistic alliances between SBUs and external organizations.

And, of increasing importance nowadays:

6. To ensure that future technological innovations are congruent with responsible environmental stewardship and sustainable development.

Needless to say, the preceding job description invests the CTO with considerable power and influence in a firm. This degree of power and influence may be strongly resisted by others there, particularly the CEOs and Vice Presidents of R&D, Engineering, and Technology of the individual SBUs in the decentralized firm. Therefore, although it is becoming increasingly recognized that the CTO role is needed in technology-based firms, even when such appointments are made, not all CTOs may be granted such broad job descriptions. Rubenstein discusses the role and responsibilities of CTOs in some detail, based upon his more than three decades of research and consulting activities in technology management.[4] He, too, is supportive of a strong CTO role in the high-technology firm, and makes the following observation:

> If top management really wants their corporate CTO to look after the entire technology management program of the company, to assure its technical vitality, to have the various labs complement each others' work, and to provide the company with ability to respond to technical threats and to capitalize on technical breakthroughs and opportunities, then it either will have to give him the formal power he needs to resolve major conflicts or must be prepared to pay the continuing role of arbitrator between the CTO and divisional management (Rubenstein,[4] p. 107)

Regardless of whether a CTO position and an IMF is *formally* established in a firm, such a role and function must, at least, be *informally* enacted. The success or failure of this role enactment will be influenced by the appositeness of the firm's organizational structure, and the remainder of the chapter discusses these two issues. Both issues can be usefully discussed within the evolutionary framework introduced in Chapter 2.

14.3 NATURAL AND ARTIFICIAL EVOLUTION

In the Chapter 7 Supplement we briefly discussed Margaret Boden's ideas on the role of that artificial intelligence research might play in improving our understanding of the creative process. Recent work, especially that in neural computing, has focused increasing attention upon the extent of the homology between natural and artificial intelligence. In Chapter 2 we suggested that technological innovation can be viewed in the context of Popper's evolutionary epistemology of science and that the highest status of any innovation (whether it be product, process, or service), like that of a scientific theory, is that of an as *yet unrefuted conjecture*. Popper's approach essentially invokes a parallel homology between biological (or natural) versus scientific and technological (or artificial) evolutions. Just as it is fruitful to compare and contrast natural and artificial intelligence, it is also fruitful to compare and contrast *natural* and *artificial* evolutions, to improve insights in both cases. Comparing natural (or biological) and artificial (or technological) evolutions, we can identify the following similarities and differences:

1. Natural mutations are spontaneous and random and survive or fail through a Darwinian learning process. In contrast, artificial mutations (recall Chapter 2, Section 2.10) succeed or fail through a Lamarckian learning process. Being

Lamarckian rather than Darwinian, artificial evolution is much *faster* than its natural counterpart.

2. In both cases, survival is the reward of the *fitting*, rather than the fittest (again recall Chapter 2, Section 2.10 and Chapter 6), often based upon symbiotic relationships with other organisms or artefacts (recall Chapter 3, Section 3.14).

3. In both cases, these processes encourage and reward creative activities which generate diversity and variety in alternative evolutionary paths to generate a natural ecosphere and artificial *technosphere* with rich variations in adaptations and scopes for nichemanship.

4. As well as being slow, natural terrestrial evolution is a largely incremental process with rare evolutionary discontinuities, perhaps precipitated by extraterrestrial interventions. For example, the revolutionary discontinuity in the ecosphere that led to the extinction of the dinosaurs more than 50 million years ago, may have been precipitated by the impact of an asteroid or comet on the Yucatan Peninsula of Mexico. In contrast, revolutionary discontinuities occur frequently in artificial evolution with major implications for the organizational structures required.

5. *Most important of all,* natural evolution and artificial evolution are *not mutually independent* processes. As indicated in Appendix 4.2, there is growing evidence that the artificial technosphere is degrading the natural ecosphere, with potentially disastrous impacts on the life on earth in future years. Therefore, increasing attention must be paid to the adoption of approaches to technological innovation which will ensure that this potentially catastrophic degradation is reversed and, ideally, eliminated. Approaches to the design of products and processes which seek to reflect and replicate the designs of natural ecosystems are being developed under the rubric of *industrial ecology* and *industrial metabolism*.[5,6] These approaches can be expected to play increasing roles in the designs of innovative products, processes and services in future years.

14.4 ALTERNATIVE ORGANIZATIONAL STRUCTURES

As in the natural ecosphere, which embraces a diverse variety of flora and fauna, there are differing artificial evolutionary radiations or industry groupings, which are shaped by the differences in technologies, applications, and markets. These groupings often also change with discontinuities in technological evolutions, and dictate the organizational requirements of the member firms in each of these groupings. They are primarily dependent upon two sets of factors:

First, three *time–cost* factors:

1. The *length* of Generational Life Cycle (GLC). In the civil airframe industry the overall GLC is long, though GLCs may be shorter in engines, avionics, etc. In contrast, GLCs are short in microelectronics.
2. The *start-up* cost of the GLC.
3. The *unit* cost of the product.

Second, three *diversity* factors:

1. The *diversity* of complementary assets and cotechnologies required to design the product.
2. The *diversity* of complementary assets and cotechnologies required to manufacture, distribute, and market the product.
3. The *diversity* of the product lines.

A *project organization or project design consortium* appears appropriate when generational life cycles are long, both upfront and unit costs are high and the diversity of complementary subtechnologies and assets is low. It is probably most applicable in the aircraft industry (recall Chapter 13, Section 13.1 and the cost of the superjumbo) and in engineering turnkey projects. Note that, although these situations typically require the assembly of complementary subtechnologies and assets, they are not diverse. Such industries typically have well-established supplier networks that can be mobilized to perform specific projects. Again as we saw in Chapter 13, the design consortium is becoming a feature of the semiconductor industry. When life cycles and costs are shorter and a portfolio of projects are perform, the *matrix structure,* as discussed in Chapters 8 and 9, becomes appropriate. As lifecycles shorten and costs fall, it may be appropriate implement innovations in a more traditional functional organization. However, as we saw in the operations of Texas Instruments (Chapter 12, Section 12.4), they are best realized in participative frameworks which have some resemblances to good university faculties. We might therefore label them *collegial functional organizations;* a term that will be discussed further later. As diversity increases, in product lines and the manufacturing, marketing, and distribution resources required, an *intrapreneurial organization* is sought, as exemplified by 3M. Finally, when innovation also requires a diversity of subtechnologies and complementary assets, an *interpreneurial organization,* able to cooperate in a network or spider's web of alliances, as exemplified by Corning, the biotechnology industry, and to a lesser extent more recently, the computer industry.

14.5 SOME OBSERVATIONS FROM RECENT STUDIES

The conceptual generalizations of Sections 14.3 and 14.4 are of little practical value unless can be supported by specific observations on current practices in successful technology-based firms. Therefore we now link them to some recent studies of successful firms, to conclude the chapter with suggested prescriptions for the *successful* management of innovation and entrepreneurship in technology-based firms.

From their study of the management practices of a sample of Japanese and U.S. firms, Pascale and Athos[7] drew conclusions broadly congruent with those of Frohman cited earlier. They expressed these conclusions in their 7-S framework of management. They suggested that successful Japanese and U.S. firms (including TI and 3M) have the style, staff, skills, systems, structure, and strategy to stimulate and nurture innovation. These six Ss are embedded in the seventh—the superordinate goals or the values enshrined in the organizational culture which elicit a commitment to innovation from the firm as a whole and each of its individual members. There are alternative approaches to enshrining this culture, as reflected in the contrasting approaches of TI and 3M, but they suggest that this is (no longer) the *secret weapon* of the successfully innovative Japanese and Western companies. Another analysis of corporate practices for stimulating innovation and entrepreneurship, based upon the 7-S framework, was *In Search of Excellence*.[8] Received with critical acclaim when first published, it lost some favor after the appearance of an article reporting the declining performances of its *excellent* companies.[9] Nevertheless, many of its observations remain valid and are reflected in those of later studies. Excellent firms, like excellent baseball bats, can be expected to experience losses of form and experience less than excellent seasons. It is their hitting lifelong RBIs that ensures their ultimate entries into baseball's "Hall of Fame." High technology could also claim to have its "Hall of Fame" firms that sometimes stumble in their pursuits of excellence. Two later studies are of particular interest since, when considered together with the above and other writings, they suggest a useful set of guidelines for managing innovation and entrepreneurship in technology-based firms. First, Jelinek and Schoonhoven[10] conducted in-depth studies of five leading U.S. based high-technology firms in the electronics industry, Hewlett-Packard (HP), Intel, Motorola, National Semiconductor, and Texas Instruments (TI). Second, Majone also conducted an in-depth study of three U.S. based firms, two in the electronics industry, GE Medical Systems (GEMS) and Motorola Communications (COMM), together with the diversified Corning Incorporated.[11]

Learning-by-Striving to be Best in Class

Several writers have echoed Popper's error-elimination epistemology in emphasizing the role of organizational learning.[12,13] For example, Majone, in his in-depth study of three U.S. technology-based firms (GE Medical Systems, Motorola Communications and Corning), emphasizes *learning-by-trying* as the key to success.[11] Some of these writers mainly concentrate on *continuous improvement* learning, specifically associated with learning or experience curves. However, Botkin *et al.*[14] describe two levels of learning:

1. *Maintenance* learning associated continuous improvement and an advance down the experience curve, which can be viewed as an exercise in the Japanese *kaizen*.

2. *Innovative* learning which is required to deal with the frequent episodes of discontinuous and sometimes turbulent changes that characterize typical contemporary technological evolutions. It can be identified with generational innovations and also with more radical or revolutionary changes. Innovative learning is characterized by the capacity to anticipate and participate in change. This requires imaginative insight and foresight into technology and market coevolutions, to identify both the nature and *timing* of discontinuous

change. Witness the debates over the last decade or so concerning the convergencies of computer, telecommunications, and media technologies and the emergence of the $3.5 trillion information superhighway-based industry. It will also be required in pursuing the, by no means, utopian ideals of industrial ecology and metabolism for responsible environmental management.

Firms that Majone describe as *Best in Class* practice both maintenance and innovative learning. The first through incremental improvements in their established lines; the second by anticipating and participating in technological discontinuities, through pioneering the introduction of new products and product lines. That is, rather than being satisfied with developing products, organizational structures, and strategies that fit the evolving technosphere, they actively and creatively *seek and strive* to lead and shape this coevolution by conserving, stretching, and levering their technology and management competencies. Hamel and Prahalad[15,16] express similar ideas in their concepts of *expeditionary marketing* and strategy as *stretch and leverage*. It is also manifested physically in Rothwell and Gardiner's *robust design* concept in, for example, the Boeing 747 Series that have evolved over a 35 year life span.[17]

Successful firms possess the capacity and will to create continuously radical, generational, and incremental innovations, typically following an offensive first-to-marketing strategy. All five firms in the Jelinek-Schoonhoven and Majone studies can claim to have pioneered radical and generational innovations. Although GEMS entered the emergent CAT scanner market as a defensive innovator, following its introduction of the CT 8800 and purchase of the EMI-Thorn operation, it moved to a first-to-market strategy. After a late entry, it also followed an offensive strategy in the magnetic resonance (MR) medical electronics market segment. Intel's pioneering efforts in designing, manufacturing, and marketing successive generations of microprocessors are well known, but most readers are probably less familiar with similar achievements by Corning in optical fibers. Majone points out that Corning not only pioneered radical optical fiber innovations, but introduced seven generational innovations in processing technology over 15 years! *Kaizen* or continuous improvements through incremental innovations is particularly associated with Japanese firms, but was again manifested in these sample firms. TI and National Semiconductor are especially identified with learning or experience curve cost/price reduction strategies in the semiconductor industry, but the other firms showed that they were able to protect their leaderships in pioneering radical or generational innovations with successions of incremental improvements in products and processes. The tight interplay between product and process design innovations and improvements is a notable feature of the semiconductor industry where, in Jelinek's and Schoonhoven's words, success depends upon winning the *end game of manufacturing*. Also, as Corning's above-cited achievements attest, the skilled management of manufacturing plays an important role in all firms. The sustained continuous creation of innovation must be based upon the management of the interfaces among all three functions of R&D, manufacturing, and marketing that form an eternal triangle or *ménage à trois* (Figure 14.1). Note that firms do not necessarily

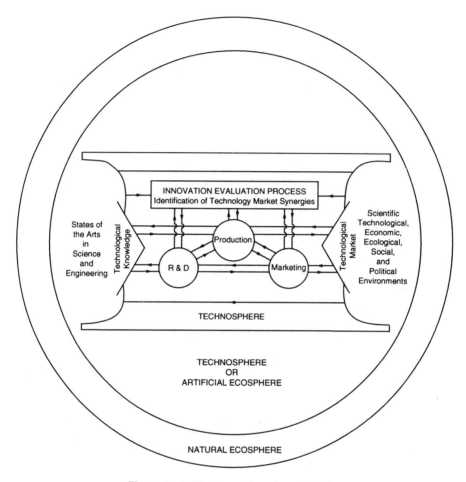

Figure 14.1. The innovation eternal triangle.

have to grow large to strive in this way. A small company pursuing a technological nichemanship strategy can strive to the *Best in Niche* firm.

Continuous Creation Based Upon Local Knowledge

Both maintenance and innovative learning must be based upon creative thinking, and Wagner[18] uses a metaphor from cosmology to illustrate this requirement. The theory of continuous creation of matter was conjectured in the 1950s to explain observations in our expanding universe. Most cosmologists claim that subsequent observations have refuted it in favor of the conjecture that our universe began with the explosion of an immensely dense concentration of matter in a big bang, more than 10 billion years ago, and that none has been created since then. However, some, including Sir Fred Hoyle, still support the continuous creation theory, arguing that matter may be created in certain localities in the universe rather than diffusely throughout space–time. Clearly, the diversity of outputs and approaches, needed to ensure sustained success for technology-based firms, must be based upon the *continuous creation* of innovation, analogous to continuous creation in cosmology. This too is typically based upon local pockets of knowledge within the organization.

In their study, Jelinek and Schoonhoven argue that, given the rapid rate of technological change, most senior managers are unlikely to have an up-to-date grasp of the state-of-the-art, but must rely on *local knowledge*. That is, engineers, scientists, and others who may be quite junior in the organizational hierarchy are likely to be most aware of evolving technological possibilities, manufacturing methods and market needs. This knowledge is typically context specific, so it exists in local pockets of expertise. Moreover, since individuals participating throughout the innovation process should be encouraged to make creative contributions, it must be both broadly and deeply rooted with successful contributions, from whatever sources, given commensurate recognition and rewards. It also implies that new employees should be quickly integrated into the corporate culture so that they feel able to make real contributions while their prior experiences are still up to date.

Jelinek and Schoonhoven also argue that such firms must induce a strong culture that strikes a careful balance between challenge and support, encouragement and demanding standards. This can be stressful for all concerned, and can sometimes lead to domestic disruptions and "hi-tech burnout." Since learning-by-striving generates *failures* as well as successes, the culture must also practice tolerance of failure. From his study of 150 new products, Gulliver[19] concluded that failure is the ultimate teacher; Thomas Edison, one of history's leading inventor-entrepreneurs, claimed that he failed his way to success. Although Jelinek and Schoonhoven emphasize the importance of a strong culture, it should be noted that not all firms entirely share this view. Cannon and Honda, for example, encourage the creation of *multiple* cultures by hiring significant numbers of midcareer managers from other firms to provide countercultures, to avoid the dangers of excessive inbreeding that is a risk of a single dominant organizational culture.[20]

Garvin suggests the following as important organizational learning attributes:

> A learning organization is an organization skilled at creating, acquiring, and transferring knowledge, and at modifying its behavior to reflect new knowledge and insights. . . . Organizations that do past the definitional test . . . become adept at translating new knowledge into new ways of behaving. These companies actively manage the learning process to ensure that it occurs by design rather than by chance.[21]

That is, as well as creating and acquiring knowledge, the organization must be skilled at transferring and disseminating knowledge, particularly if it is locally based. Jelinek and Schoonhoven also emphasize this point. In their sample of firms, the formal organizational structures are supported by numerous quasiformal and informal structures to facilitate this dissemination. The former are temporary teams or task forces, usually with dotted-line linkages to the formal organizational structure, formed to undertake specific missions in support of overall operations. They are essentially informal *collegial* groupings that come together to attack specific problems that lie outside normal operating procedures. Since participations in these groupings are likely to based upon the members' problem-solving abilities, they can often prove the foundation of local knowledge needed for sustained innovation.

Focused Leadership and Stratocracy: Domes and Jazz Orchestras

Although continuous creation may be fostered by the organizational culture which stretches individuals and accesses local knowledge, that alone is insufficient to ensure sustained innovation and corporate success. Innovations unquestionably must be deeply rooted in the organization if they are to thrive, but the adage to *think locally and act globally,* must also be honored. Innovations must also simultaneously be effectively filtered by higher-level managers to ensure that such proposals are congruent with a corporate focus, and often a global perspective coupled with responsible environmental stewardship. Establishing and maintaining this focus, as well as the organizational culture, is a leadership imperative for senior management. A consistent persistent strategic focus is a characteristic of successful firms. Nevertheless, the requirements of local knowledge-based innovation dictate a departure from the leadership and decision-making structures of the past. The joint requirements of strategic focus and local knowledge dictate that senior managers should communicate with individuals well down the corporate hierarchy if both parties are to play effective roles. It is also implies that senior managers should be open and willing to describe the rationales of their decisions, to encourage further new ideas, and to some extent, to give others genuine power to help steer and influence the organization. Bahrami and Evans[22] describe this hierarchical spanning requirement as *stratocracy.* Tomasko[23] suggests that this spanning requirement is best achieved by integrating organizations in a *dome-like* rather than *pyramid* organizational structures. Clearly the collegial linkages cited above are likely to be more effective in a dome-like structure, with strong lateral as well as vertical struts, rather than a pyramid-like organizational structure. Matrix structures, interpreneurial and networking alliance approaches can also be viewed as methods for structuring the organization as a dome rather than a pyramid.

The need for this organizational flexibility or fleet-footedness is also exemplified in Kanter's *When Giants Learn to Dance,*[24] and Drucker uses a symphony orchestra metaphor for organizational structure to convey a similar idea.[25] Rather than work through a rigid hierarchy, like orchestra conductors coordinating individual instrumentalists, senior managers must interact with individual pockets of local knowledge within the framework of an overall score. Although this metaphor is compelling, it is also inexact. Given that innovation is based upon local knowledge combined with strategic direction, the score of a classical music symphony, which imposes quite rigid requirements on both instrumentalists and conductor, does not apply. Rather, the metaphor of a large orchestra playing jazz is more apposite. The conductor may choose an established orchestral work, say *Porgy and Bess* or *Rhapsody in Blue,* but allow individual jazz instrumentalists to extemporize around the score in light of their preferences, playing skills, and current musical fashions. In other words, perhaps successful corporate innovation is closer to Basie than Beethoven!

Frequent Organizational Mutations

Although their conductors interact directly with individual orchestra members, whether they are playing classical or jazz music, one should recall that orchestras have hierarchical structures based upon the conductor, orchestra leader, first violin, and so on. Similarly, despite the importance of local knowledge and stratocracy, a formal hierarchical organizational structure is required both to implement new innovations and to maintain ongoing present operations in a strategic framework. Given the rapid rate and unpredictable, sometimes turbulent, nature of technological changes, this formal structure is itself subject to frequent changes. Such changes do not reflect dissatisfaction with the past performance of the current structure, but the recognition that changes in the firm's technological, social, and market environments require a changed organizational fit. They can be viewed as deliberately engendered *organizational mutations* designed to maintain their adaptiveness to their technosphere. The technology-based firm in some industries is perhaps homologous to a *potent virus* in that it often has to undergo frequent *organizational mutations* to maintain and consolidate its competitive advantage in its technosphere. This adaptiveness is particularly important given the often unpredictable *chaotic features* of both the innovation process and technological change that favor strategic incrementalism over rigid forecasting and planning; Peters goes so far as to recommend that firms should be in a permanently revolutionary state.[20,26,27] Reflecting such considerations, most of the Jelinek–Schoonhoven and Majone firms in the electronics industry indulge in frequent internal structural reorganizations for this reason, while Corning is more likely to seek such changes through changes in its spider's web alliance network.

Based upon these observations and studies, we can conclude with some comments on the effective management of innovation and entrepreneurship in the technology-based firm.

14.6 THE EFFECTIVE MANAGEMENT OF INNOVATION AND ENTREPRENEURSHIP IN TECHNOLOGY-BASED FIRMS

Whether or not a technology-based firm has a CTO formally appointed and an IMF formally instituted, all must have a cadre of top managers who implicitly enact these roles. The function of this cadre is to oversee and, where appropriate, participate in the activities considered throughout this book. It should ensure that the firm practices the following *seven vital virtues* to stimulate and encourage innovation and entrepreneurship.

1. **Ensure** that continuous innovation creation, locally conceived but universally nurtured, is respected and accepted by all as a necessary condition for sus-

tained corporate success. Based upon the innovation chain and a critical functions venture team, it will require entrepreneurial, intrapreneurial, or interpreneurial linkages to ensure success. Therefore, the CTO and IMF must ensure that these linkages are present to be available for use as and when needed.

2. **Ensure** that the corporate culture or cultures be consciously engendered and sustained by top management to encourage this continuous creation. It should simultaneously provide individuals with support and encouragement, together with challenge and demanding standards to stretch their capabilities.

3. **Recognize** that a "Best in Class" performance requires more than fast and sustained maintenance learning. It is achieved by striving to stretch the technological and managerial capabilities of the organization above and beyond their expectations, by making maximum use of its competences and resources, and by shaping the changing requirements of technology–market coevolutions more than matching them.

4. **Ensure** that top management provides a focus and strategic direction for the above, not through a rigid hierarchical pyramid, but rather through a dome-like organizational structure that encourages multiple lateral formal and informal connections. The structure should itself be flexible, so allowing for the possibly frequent structural changes needed to match the changing requirements of technology–market coevolutions.

5. **Recognize** that successful participation in technology–market coevolution is dependent upon fast and sustained organizational learning through technological and managerial learning. This is based upon persistent learning-by-trying with failures as well as successes. Learning-by-failing should be tolerated in the context of the overall learning-by-trying process.

6. **Recognize** that fast organizational learning is also based upon fast innovation. The organizational structure and culture should engender fast innovation.

7. Given that underlying chaotic mechanisms may be inherent in technology–market coevolutionary processes, it is imprudent to place total confidence in the even best laid future visions and strategic plans.[20] As the Prussian General von Block reputedly put it: *"No plan survives first contact with the enemy."* Therefore, *ensure* that the signals of change are continuously monitored to detect technological discontinuities and paradigm shifts early enough for the firm to participate in and shape them profitably. Above all, *ensure* that the firm and its members are open minded towards the future, and are *prepared to be prepared* for the unexpected. In a world of rapid technological change, the unexpected is the only certainty!

14.7 AFTERWORD

The Preface of this book began by commenting that public and business concern for technological innovation had been aroused by the problems of declining growth and increasing unemployment. Many writers, like Freeman and Soete,[28] take the view that the stimulation and implementation of technological change offers the best means of ameliorating, if not eradicating, these problems. Writing this book will have been worthwhile if it helps present and future technological innovators and entrepreneurs create wealth for themselves, work for others, and worth for society. To any such readers, whether they are planning or have already launched their own innovative ventures as entrepreneurs, intrapreneurs, or interpreneurs, go this author's good wishes for success in their endeavors.

REFERENCES

1. A. L. Frohman, "Technology as a Competitive Weapon." *Harvard Business Review* **60**(1), 97–104 (1982).
2. P. S. Adler and K. Ferdowes, The Chief Technology Officer. *California Management Review* **32**(3), 55–62 (1990).
3. Bhalla, *The Effective Management of Technology.* Reading, MA: Addison-Wesley, 1987.
4. A. H. Rubenstein, *Managing Technology in the Decentralized Firm.* New York: Wiley, 1989, Chapter 3.
5. H. Tibbs, *Industrial Ecology: An Environmental Agenda for Industry.* Cambridge, MA: Arthur D. Little, 1991.
6. R. U. Ayres and U. Simonis, *Industrial Metabolism.* New York: U.N. University Press, 1992.
7. R. T. Pascale and A. G. Athos, *The Art of Japanese Management.* New York: Simon & Schuster, 1981.
8. T. J. Peters and R. H. Waterman, *In Search of Excellence.* New York: Harper & Row, 1982.
9. "Who's Excellent Now?" *Business Week* November 5, pp. 76–78 (1984).
10. M. Jelinek and C. B. Schoonhoven, *The Innovation Marathon Lessons from High Technology Firms.* Cambridge, MA: Basil/Blackwell, 1990.
11. J. G. Morone, *Winning in High-Tech Markets: The Role of General Management.* Boston: Harvard Business School Press, 1993.
12. M. Jelinek, *Institutionalizing Innovation: A Study of Organizational Learning Systems.* New York: Prager, 1979, Chapter 5.
13. P. M. Senge, *The Fifth Discipline.* New York: Doubleday, 1990.
14. J. W. Botkin, M. Elmandjra, and M. Malitza, *No Limits to Learning: Bridging the Human Gap.* Elmsford, NY: Pergamon, 1979.
15. G. Hamel and C. K. Prahalad, "Corporate Imagination and Expeditionary Marketing." *Harvard Business Review* **69**(4), 81–92 (1991).
16. G. Hamel and C. K. Prahalad, "Strategy as Stretch and Leverage." *Harvard Business Review* **71**(2), 75–84 (1993).
17. R. Rothwell and P. Gardiner, "Re-Innovation and Robust Designs: Producer and User Benefits." *Journal of Marketing Management* **3**(3), 372–386 (1988).
18. H. E. Wagner, "The Open Corporation." *California Management Review* **33**(4), 46–60 (1991).

19. F. R. Gulliver, "Post-Project Appraisals Pay." *Harvard Business Review* **65**(2) (1987).

20. R. Stacey, "Strategy as Order Emerging from Chaos." *Long Range Planning* **26**(1), 10–17 (1993).

21. D. A. Garvin, "Building a Learning Organization." *Harvard Business Review* **71**(4), 78–91 (1993).

22. H. Bahrami and S. Evans, "Stratocracy in High-Technology Firms." *California Management Review* **30**(1), 51–66 (1987).

23. R. Tomasko, *Rethinking the Corporation.*   New York: AMACOM Books, 1993.

24. R. M. Kanter, *When Giants Learn to Dance.*   New York: Simon & Schuster, 1989.

25. P. F. Drucker, "The Coming of the New Organization." *Harvard Business Review* **66**(1), 45–53 (1988).

26. J. B. Quinn, "Innovation and Corporate Strategy: Managed Chaos" *Technology in Society* **7**(3), 167–183 (1986).

27. T. Peters, "Crazy Ways for Crazy Days." *BBC Television Broadcast* December 6 (1993).

28. C. Freeman and L. Soete (Eds.), *Technical Change and Full Employment.* Oxford: Basil-Blackwell, 1987.

INDEX

DATE DUE
